TECHNOLOGY
IN YOUR WORLD

TECHNOLOGY
IN YOUR WORLD
SECOND EDITION

Michael Hacker
New York State Education Department

Robert A. Barden
Electronics Engineer and former Member,
Technology Education Advisory Council

DELMAR PUBLISHERS INC.®

NOTICE TO THE READER

Cover Photo Credits
Fiber Optics photo courtesy of AT&T Bell Labs
Compact Disk photo courtesy of General Electric Company
Thermographic image of hands courtesy of NASA
Sunraycer Automoblie photo courtesy of General Motors Corporation
Space Shuttle photo courtesy of NASA
Hero Robot photo courtesy of Heath Company

Delmar Staff

Associate Editor: Christine Worden   Art Supervisor: John Lent
Production Editor: Christopher Chien   Art Coordinator: Mike Nelson
Production Supervisor: Karen Seebald   Design Supervisor: Susan C. Mathews

For information address Delmar Publishers Inc.
2 Computer Drive West, Box 15-015
Albany, New York 12212

COPYRIGHT ©1992
BY DELMAR PUBLISHERS INC.

Printed in the United States of America
Published simultaneously in Canada
by Nelson Canada,
a division of The Thomson Corporation

10   9   8   7   6   5   4

Library of Congress Cataloging-in-Publication Data

Hacker, Michael.
    Technology in your world/Michael Hacker, Robert Barden.—2nd ed.
      p.   cm.
    Includes index.
    ISBN 0-8273-4425-2 (textbook)
    1. Technology.   I. Barden, Robert A.   II. Title
T45.H33   1991
600—dc20   90-46607
                                                                              CIP

CONTENTS

(Courtesy of NASA)

(Courtesy of International Public Affairs Division, Toyota Motor Corporation)

(Courtesy of Simpson Industries, Inc.)

(Photo by Ruby S. Gold)

CHAPTER 13   CONTROLLING THE SYSTEM   362

(Courtesy of Grumman Aerospace)

CHAPTER 14   IMPACTS AND OUTLOOKS   392

(Courtesy of NASA)

INTRODUCTION

The world in which we live is in a state of constant change. For better or for worse, people have been modifying their environment since prehistoric times. At first, the rate of change was very slow. During the lifetime of our grandparents there have been more changes than in all of the rest of the time that humans have lived on the earth. Each succeeding year brings change at a faster pace.

Because technology is developing so rapidly, it is virtually impossible to predict exactly what kinds of technological skills will be needed in the future. As soon as we have learned about a form of technology, something new takes its place. Therefore, people must develop skills that are useful despite the changes that will no doubt occur. Such skills include learning how to understand new technologies as they evolve.

TECHNOLOGY EDUCATION

Although technology can improve the quality of our lives, it can also create undesirable consequences. Many of today's major social issues involve the impacts of technology.

School is the ideal place to begin understanding technology and the role it plays in our culture. Students should learn that technology is a human endeavor. Whether it is used to benefit or to destroy our society is our decision. Students can learn that people can and must control the development and application of technology.

Modern curriculum specialists have seen the need to develop new technology programs. The National Science Board's Commission on Precollege Mathematics, Science and Technology Education has recommended that all students study science and technology for one year in grades seven and eight. Ernest Boyer's Carnegie report also calls for a study of technology by all students. The Jackson's Mill symposium resulted in the development of a comprehensive curriculum base for teaching about industry and technology, and many states have now introduced Technology Education as a part of their total school program. As the International Technology Education Association's Mission Statement indicates, "Technology Education may be viewed as a national concern, as a mission for education, and as a stimulus for a new curriculum with new goals directed toward technological literacy." The development of a technologically literate population has indeed become a major priority for education. The second Jackson's Mill Symposium, which was completed by a task force of ITEA in May 1990, suggested organizing technology content for study in a way presented by this book. The authors were among the contributors to this task force.

Technology Education provides an opportunity to apply mathematics and science concepts in a laboratory setting. This type of "mission-oriented" activity leads to technological innovation. A study of technology without an applications phase does not present clear interpretation of technology's dimensions. In order to make the study of technology exciting and relevant, content should be presented through a laboratory-based, activity-oriented approach. This textbook has been written to provide the conceptual base necessary to support such a program.

Technology Education, as distinct from "technical education" or "vocational education," is the means through which general knowledge about technology is provided. The goals of other technically or occupationally oriented programs are to provide students with job-specific skills. A significant distinction must be drawn between programs designed to provide vocational training and those designed to provide technological literacy for all students.

What can Technology Education accomplish? Educated people need to recognize that technology is not magic, nor is it beyond their control. Technology can be understood and therefore designed, modified, and influenced by intelligent, voting citizens. Technological decisions, and public policies stemming from them, must not be left to a technological elite.

If the United States is to retain its leadership in the world community, new applications of existing technology and new technological products and services must be promoted and encouraged.

SCOPE OF THE BOOK

Technology in Your World fully discusses the technological processes central to communication, manufacturing, construction, biotechnical, and transportation systems. All technologies are presented as systems, with common recurring elements. These elements (inputs, resources, processes, outputs, feedback, and control) provide the basis for an organized study of technology.

This book is unique in the way it relates technologies to the system concept. The system concept provides students with a conceptual tool with which all technologies can be analyzed.

Problem solving is treated as a central theme in *Technology in Your World*. A good deal of the conceptual problem-solving methodology is derived from work done within the Technology and Design movement in the United Kingdom. This text harmonizes the British approach to problem solving with the systems approach popular in the United States. A method is presented that provides an opportunity for students to work within a formal problem-solving system.

Technology in Your World is divided into fourteen chapters. The first three chapters introduce the student to the study of technology and focus on problem solving. Chapter 1 introduces the study of technology through an historical overview. Students are shown how technology has been used as a means through which people satisfy their needs and wants.

Chapter 2 presents a problem-solving model that includes seven steps: Define the problem, specify results in a design brief, gather information, develop alternative ideas, select the best solution, implement the solution, and evaluate the solution and make changes.

Chapter 3 focuses on communicating solutions to design problems. Graphic methods are detailed including sketching, drawing, rendering, illustrating, and using color. Modeling techniques are covered as well.

Chapter 4 introduces students to the seven basic types of technological resources: people, information, materials, tools and machines, energy, capital, and time. How people make choices among resources is also discussed. Such considerations as appropriateness, cost, and availability of resources are covered. The trade-offs people must make in choosing technological resources are explained.

Chapter 5 details the processing of resources. Materials, energy, and information processing are discussed, and the use of tools and techniques to process resources is covered in detail.

Chapters 6-12 are devoted to a study of the various technological systems. In Chapter 6, systems concepts are introduced. Technological systems familiar to students are illustrated by means of the basic system diagram. Students are thus provided with a tool that helps them analyze unfamiliar systems. The problem-solving model is shown to be analogous to the basic system model.

Chapter 7 looks at electronics and the computer. The tremendous impact of integrated circuits and microelectronics is described. Computer systems, from microprocessors and personal computers to supercomputers, are explained in simple terms.

Chapter 8 covers communication and information systems. Topics range from the most primitive forms of writing to modern electronic telecommunications. Examples show how all communication processes include a transmitter, a channel, and a receiver.

Chapter 9 deals with the manufacturing systems that produce goods in a workshop or factory. The craft approach is contrasted with mass production and the assembly line. Impacts of the factory system on society are discussed. Automated methods of manufacturing such as CAD/CAM, and Computer-Integrated Manufacturing (CIM) are detailed.

Chapter 10 explains construction systems used to produce a structure on a site. The three construction subsystems—designing, managing, and building—are presented. Structures described include bridges, buildings, dams, harbors, roads, towers, and tunnels.

Chapter 11 involves a study of the transportation systems used to move people and goods from one location to another. Land, sea, air, space, and nonvehicular means of transportation are described.

Chapter 12 discusses biotechnical systems, including biotechnology agriculture, food production, and medical technology. Ultramodern techniques such as genetic engineering, laser surgery, and prosthetics are clearly explained.

Chapter 13 gives us a look at the devices and methods used to control technological processes. The feedback loop of the technological system is presented as having three components: sensors, decision makers, and controllers. Examples of each component are provided, as are illustrations of common systems employing various kinds of control.

Chapter 14 reminds us that technology has both positive and negative effects on people and on the environment. Desired, undesired, expected, and unexpected results of technology are described. This chapter takes an exciting look at the future of communications, manufacturing, construction, transportation, and biotechnical systems. The chapter closes with examples of techniques used to forecast the future.

SPECIAL FEATURES

Technology in Your World includes many special features to assist the reader in becoming technologically capable.

- The use of four-color printing in the photographs and illustrations throughout the book helps clarify the explanation of technological concepts.
- The unique systems approach shown in numerous diagrams and examples, provides a "tool kit" students can carry mentally to help understand new technologies.

- The introduction of problem solving and the communication of design solutions to problems provide a cutting-edge treatment of these important topics.
- Each chapter begins with a set of major concepts. These objectives prepare the reader to be aware of key ideas while reading.
- Each major concept is repeated in the margin adjacent to its related discussion in the text.
- Key terms are highlighted and carefully explained.
- Interdisciplinary connections to mathematics, science, and social studies are presented throughout the text.
- Complete summaries, included at the end of each chapter, review the chapter's major concepts.
- For each chapter, a specially designed "crosstech" puzzle provides a motivating review of key terms and again reinforces the major concepts.
- Numerous boxed articles containing features of special interest relate technological concepts to real-life applications.
- Technology learning activities are included for each chapter. These problem-solving activities relate directly to major concepts and serve to provide the hands-on application so necessary to effective learning.
- A series of design briefs accompany each chapter and motivate the student to apply the chapter material through additional hands-on experience.
- A glossary of terms with accurate definitions of words is provided.
- A Technological Time Line is included for historical reference.
- Opportunities for student youth leadership and participation in special projects and events are discussed in the Technology Student Association section.

ACKNOWLEDGEMENTS

The authors dedicate this book to the technology teachers who have forged a vision for change into a viable and exciting Technology Education program. There are colleagues from England who deserve our most sincere thanks. Anthony Gordon, Education Inspector from the Staffordshire LEA, assisted us in conceptualizing our problem-solving approach and provided us with samples of materials generated by students in Staffordshire. One of these students, Mark Townsend from Cannock Chase High School, permitted us to redraw his design folder and use it to illustrate the problem-solving chapter. We also owe thanks to Mr. P.H.M. Williams, County Inspector in Staffordshire, for providing some of the design briefs used in each of the chapters.

Grateful thanks are given Henry Harms and Neal Swernofsky, both friends and exceptionally talented technology teachers and teacher trainers who contributed the activities that so significantly enhanced our work. Sincere gratitude is also extended to Rose Ambrosino, Charles Chew, Jane Cappiello and William Fetsko, who contributed interdisciplinary connections in mathematics, science, and social studies. For their assistance in reviewing the manuscript for this edition, grateful acknowledgement is given the following people:

Tom Barrowman
Queensbury Middle School
Queensbury, New York

Robert Bauer
Springhill Junior High School
Akron, Ohio

M. James Benson
Dunwoody Institute
Minneapolis, Minnesota

Anthony T. Gordon
Education Inspector for Design
   and Technology
Staffordshire, United Kingdom

Rodney Gould
Unadilla, New York

Donald P. Lauda
California State University
   at Long Beach
Long Beach, California

David Magnone
Matoaca Middle School
Matoaca, Virginia

David Pullias
Richardson Independent School
   District
Richardson, Texas

John Ritz
Old Dominion University
Virginia Beach, Virginia

Margaret Rutherford
Howell Intermediate School
Victoria, Texas

George Trombetta
West Hollow Junior High School
Melville, New York

Peter Tucker
Highland High School
Highland, Illinois

Robert Warren
Hidden Valley Junior High School
Roanoke, Virginia

We also wish to express our deepest appreciation to Christine Worden, Associate Editor at Delmar Publishers, who gave us constant support during the revision process and with her wit and good humor, kept us on task. Thanks also are due Christopher Chien, our Project Editor, who helped to make this work approach his own personal high standards of excellence, and to Cindy Haller, who helped conceptualize the review process and guided our initial steps.

To all our friends and colleagues in Technology Education, we offer this work in the hope that it will contribute to the technological capability of our students and to the growth of our discipline.

ABOUT THE AUTHORS

Michael Hacker is a twenty-year veteran of secondary school teaching in Long Island, New York. His commitment to Technology Education has shaped his career. As early as 1969, his technology-based junior high school program had received national attention. As a teacher, Mr. Hacker was a team leader and curriculum writer for the New York State Education Department's "Introduction to Technology" syllabus. He has authored a dozen articles in national journals and has presented at numerous state and national conferences. He is past president of the New York State Technology Education Association and has served on and chaired various International Technology Education Association (ITEA) committees on behalf of the Technology Education profession. In 1985, he was named an Outstanding Young Technology Educator by the ITEA. In his present capacity as Associate, Division of Occupational Education, New York State Education Department, his responsibilities include the development and implementation of Technology Education curricula.

Robert A. Barden is an electronics engineer specializing in the design and development of high capacity data, voice, and video communication systems. His work includes the integration of fiber-optic, microwave, and satellite technologies. Mr. Barden was the chairman of the New York State Futuring Committee, which recommended the establishment of Technology Education as a mandated discipline in New York. As a curriculum team leader for the New York Technology Education curriculum project, he was instrumental in the conception and development of the state syllabus. On the national level, Mr. Barden has served as a member of the National Advisory Council of the ITEA and has published articles and presented seminars on systems and technology for teachers and teacher trainers. He is a senior member of IEEE and has served on its Committee on Precollege Mathematics, Science and Technology Education.

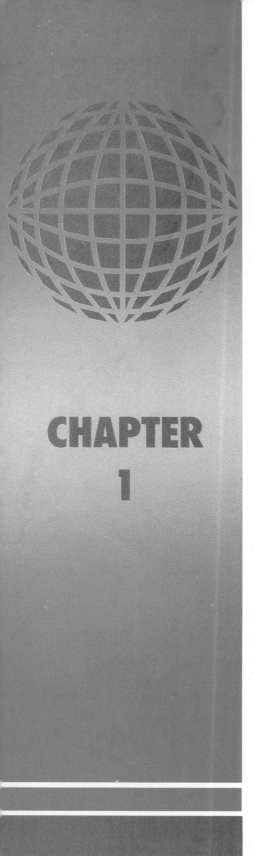

CHAPTER 1

CREATING OUR OWN WORLD

MAJOR CONCEPTS

- Technology affects our routines.
- Science is the study of why natural things happen the way they do.
- Technology is the use of knowledge to turn resources into goods and services that society needs.
- Science and technology affect all people.
- People create technological devices and systems to satisfy basic needs and wants.
- Technology is responsible for a great deal of the progress of the human race.
- Technology can create both positive and negative social outcomes.
- Combining simple technologies can create newer and more powerful technologies.
- Technology has existed since the beginning of the human race, but it is growing at a faster rate today than ever before.

INTRODUCTION

For most of us, life is made up of **routines**. At a certain time every morning, we wake up, wash up, and eat breakfast. Then we go to school or work. In the afternoon or evening, we come back home for dinner, work or relax or both, and go to sleep. Bedtime for most of us is the same every night. The cycle of routines is finished in one day and begins again on the following day.

WAKE UP AT 7:00 A.M.

GO TO SLEEP AT 11:00 P.M.

WASH UP

HAVE AN AFTER-DINNER ACTIVITY

EAT BREAKFAST

RETURN HOME FOR DINNER

GO TO SCHOOL OR WORK

Diagram of a twenty-four-hour cycle: a common daily routine.

Why do we follow the routines that we do? For most of us, almost every part of our routine depends on, or is affected by, our **technology**. For example, when we wake up and when and how we wash up depends on technology. Some cultures have a primitive technology. People awake when the sun rises. They wash up when they can in nearby lakes or streams. In our culture, artificial lighting makes it possible to be up and busy before dawn. We shower in warm water simply by turning on a faucet, anytime we like.

People in some parts of the world base their routines upon the solar cycle. (Photo by Michael Hacker)

This book is about technology and how it affects people's routines and lifestyles. Our routines and the way we live are greatly affected by the devices, products, and services provided by technology.

WHAT IS THE DIFFERENCE BETWEEN SCIENCE AND TECHNOLOGY?

Science is the study of why natural things happen the way they do.

Technology is the use of knowledge to turn resources into goods and services that society needs.

Science and technology are different. However, they are related in the following ways:

- Scientists learn how rocks and minerals in the earth were formed. Technologists use these materials to make useful objects.
- Scientists study space to learn basic facts about the sun, the planets, and the stars. Technologists build and launch space shuttles and satellites.
- Scientists discover how the human body works. Technologists make artificial hearts and limbs.

Scientists and technologists work together. What scientists discover, technologists put to use. Biologists, chemists, and physicists are scientists. Architects, engineers, product designers, and technicians are technologists.

Science and technology affect all people.

Technology is interpreted differently by different people.
(Courtesy of Feedback, Inc.)

Scientists and Technologists

SCIENTISTS	TECHNOLOGISTS
DO RESEARCH	DEVELOP SYSTEMS TO COMMUNICATE INFORMATION AND IDEAS
ASK QUESTIONS ABOUT EVENTS THAT OCCUR IN THE NATURAL WORLD	CONSTRUCT BUILDINGS, BRIDGES, AND TUNNELS
PROPOSE NEW THEORIES	MANUFACTURE PRODUCTS
EXPERIMENT UNDER CAREFULLY CONTROLLED CONDITIONS	TRANSPORT PEOPLE AND GOODS
CONFIRM OR DENY THEORIES	BREED NEW FORMS OF PLANT AND ANIMAL LIFE

Technology Is . . .

- The sum of human knowledge, used to change resources to meet people's needs.
- Making things work better.
- A way of helping our species survive.
- The means by which people control or change their environment.
- "The application of knowledge and the knowledge of application."—Melvin Kranzberg
- "That great growling engine of change."—Alvin Toffler
- A process that uses many kinds of resources to meet people's needs.
- Our way of adjusting to our environment.

WHY STUDY TECHNOLOGY?

People create technological devices and systems to satisfy basic needs and wants.

Technology affects the lives of all of us. The telephone, television, automobile, refrigeration, new medicines, polyester clothing—the products of technology are all around us. Technology fills our needs and makes our lives more comfortable, healthy, and productive.

To use technology wisely, we must understand what it can and cannot do for us. It can give us tools to increase our powers. We can reach higher with ladders. We can swim under water with scuba tanks. We can travel quickly on the ground and in the air and in space. We can cure disease and heal injury. We have been able to shape our environment to fit our needs and wants. With technology, we create a different world than that we receive from nature.

Technology is responsible for a great deal of the progress of the human race.

But technology causes problems, too, such as pollution and crowded highways. Solving one problem with technology very often creates another problem.

WHEN DID TECHNOLOGY BEGIN?

People often think of our times as technological times and other eras as before technology. In fact, people have always used technology, as you can see from the technological time line on pages 422-423.

For at least a million years, people used tools and invented ways to use them on various materials. The kinds of tools that have been important in different times have given us a way to classify historical periods.

The **Stone Age** lasted from about 1,000,000 B.C. to about 3000 B.C. During this period stone was used for many tools. Other materials used for tools were bones and wood. The earliest known dugout canoe was made about 6500 B.C.

Morning Routines

MORNING ROUTINES NOW	MORNING ROUTINES THEN

(Courtesy of Kohler Co.)	
Hot water for a bath or shower comes from a modern plumbing system. The system sends warm water through pipes to various places in the house.	In the middle and late 1800s, a charcoal-burning heater was placed in a tub of cold water. The heater was removed when the water was warm.

(Courtesy of Kohler Co.)	(Courtesy of Science Museum Library, London)
Electrical appliances and new counter-top materials make cooking and clean-up much easier.	Meals were cooked on wood or coal stoves. Cooking a meal required starting a fire and was a job that took hours.

Technology Satisfies Our Needs

TECHNOLOGY SATISFIES OUR NEED TO PRODUCE FOOD

Agricultural Technology Now

(Courtesy of Sperry Corp.)

Farm machines can harvest enough wheat in nine seconds to make seventy loaves of bread.

Agricultural Technology Then

Years ago, farmers had to use human or animal muscle power to plow their fields.

TECHNOLOGY SATISFIES OUR MEDICAL NEEDS

Medical Technology Now

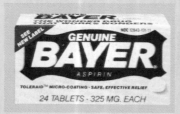

(Courtesy of Glenbrook Laboratories of Sterling Drug Inc.)

Many medicines are available for headache relief.

Medical Technology Then

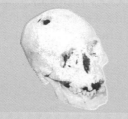

(Courtesy of Science Museum Library, London)

Early people thought headaches were caused by evil spirits in the head. To get rid of the spirits, holes were drilled in the skull with drills made of shark's tooth or flint. This is known as trepanation. Surprisingly, some patients survived more than one of these drillings.

TECHNOLOGY SATISFIES OUR NEED FOR MANUFACTURED ITEMS

Manufacturing Now

Manufacturing Then

(Courtesy of Renault, Inc.)

In a modern automobile plant, cars are produced at the rate of one every minute.

(Courtesy of Ford Motor Co.)

In 1885 in Germany, Karl Benz built the first gasoline-driven automobile. Henry Ford built his first car in 1896 and founded the Ford Motor Company in 1903.

TECHNOLOGY SATISFIES OUR NEED FOR ENERGY SOURCES

Energy Sources Now

Energy Sources Then

(Courtesy of Bob Klein)

Modern technology produces energy using oil, gas, coal, and nuclear fuels.

In the Middle Ages (about A.D. 1100), water wheels provided most of the power for production. They were used to grind grain into flour.

TECHNOLOGY SATISFIES OUR NEED TO COMMUNICATE IDEAS

Communications Now

Communications Then

(Courtesy of NASA)

Today, satellites bring television signals from other countries into our homes. The signals are sent from ground stations to the satellites, which rebroadcast them throughout the world.

In 1774, in Geneva, Switzerland, George Lesage set up a telegraph using one wire for each letter of the alphabet. The telegraph sent a message along wires to another room.

TECHNOLOGY SATISFIES OUR TRANSPORTATION NEEDS

Transportation Now

Transportation Then

(Courtesy of NASA)

(Courtesy of Smithsonian Institution—National Air and Space Museum)

Today the space shuttle and rockets make it possible to move objects and people between the earth and space.

In 1903 at Kitty Hawk, North Carolina, Wilbur and Orville Wright's biplane flew 852 feet and stayed up for 59 seconds.

CONNECTIONS: Technology and Social Studies

- The Industrial Revolution caused cities to grow in number and in size. People had to move to where the factories were located and little towns became cities.
- Abuses of employees led to the uniting of workers in order to protect their rights and advance their interests. Labor unions were formed to deal with safety issues, working conditions, and wages through collective bargaining.
- Technological change today is happening so quickly that people continually need to readjust to new products and ideas. Often, as people begin to feel comfortable with a new technology, it becomes outdated!
- With so many new products entering the marketplace each year (day!) we have become a "throw-away" society. Things are rarely repaired, just replaced. This increase in waste disposal has created a major problem for this country.

During the **Bronze Age** (starting about 3000 B.C.), people made tools and weapons from bronze. Bronze is a mixture of copper and tin. Before the Bronze Age, copper was used for decorative purposes. Bronze Age people learned that copper could be heated with charcoal to yield pure copper. By melting other ores with the copper, stronger metals like bronze could be produced.

The **Iron Age** was the period when iron came into common use. It began around 1200 B.C. in the Middle East and about 450 B.C. in Britain. The process of making iron from ore is called **smelting**.

People have been using technology for over a million years.

The early iron-smelting furnace was a clay-lined hole in the ground. Iron ore and charcoal were placed in the hole. Air was pumped in with a bellows. The air made the charcoal burn hot enough to turn the ore into a spongy mass of iron. This spongy mass was hammered into shape while it was red hot.

Prehistoric people used fire for cooking meat and protection from wild animals. They used the bow for hunting. They made needles from splinters of bone to turn animal skins into clothing.

During the Stone Age, there were few villages. Most people lived nomadic lives, wandering from place to place. When they used up food sources, they would move to a new location. People hunted animals and gathered plants, fruit, seeds, and roots for food.

The plow, developed in Egypt, let people grow their own food. People began to settle into towns. They were no longer nomads. They became farmers who grew flax, which provided fibers to make linen cloth. They used oil lamps for light at night. They invented the wheel and vehicles with wheels, and they built roads and stone houses. They had systems for irrigation to increase plant production. They had sewage systems to safeguard health.

One of the most significant of the early inventions was the water wheel. Human muscle power was replaced by water power. Before the water wheel, people ground grain by hand, using two large stones. This was such hard work that people were sometimes made to do it as punishment.

The water wheel started the machine age. It provided power not only for grinding grain to make flour but for other purposes as well. It was used also in pumping water, making cloth, and smelting iron.

The use of machinery instead of hand tools was perhaps the most important technological development of the Middle Ages. In Britain, there were 5,624 water wheels—about one for every fifty families in the country at the time.

THE INDUSTRIAL REVOLUTION

Starting about 1750, many new devices and processes were invented. New ideas spread through the Western world. This marked the beginning of the Industrial Revolution and the factory system of work.

The Industrial Revolution got its name from the great changes that took place in the way products were made. Before this time, craftspeople used their own tools and their workshops to make things. With machines, however, factories could be set up. Products could be made faster and cheaper. Craftspeople could not

This water wheel operates a set of bellows for iron smelting.

CONNECTIONS: Technology and Science

- The development of science proceeded at a different rate than technological developments. Science, as we know it today, developed in the 1500s and 1600s. During this time period people began to change the way they studied their surroundings. Scientists began to ask the types of questions that could be checked by experimentation.

  Measurements and carefully developed experiments became important. Instead of relying on simple observations, people began to devise carefully controlled experiments. As a result of this new scientific method, rapid progress occurred in all scientific fields. The Scientific Revolution had begun.

- Technology and science are closely related. Many technological devices such as the microscope, vacuum pump, and telescope were important in the development of various sciences. Scientific discoveries such as X-rays and radioactivity influenced new technologies.

compete. Many of them went to work in factories. They worked for wealthy people who had the money to buy machinery and set up factories. Workers used their employers' tools and machines instead of their own and received wages for their work.

Henry Ford is often called the "father of mass production" because he was the first American to use the assembly line on a large scale. But the assembly line did not start with Ford. In England, more than a hundred years before Ford's assembly line, Richard Arkwright developed the first manufacturing system. He linked together machines of his own design. They changed raw cotton into thread in a sequence of steps. Each machine carried out one job and prepared the material to be processed by the next machine.

Arkwright was the inventor of the **factory system**. He built many mills, turning what had been hand work into work done by machines operated by people.

Arkwright's factory system produced cloth cheaper and faster. However, it also caused some harm. Men, women, and children worked fourteen hours a day under crowded and un-

Technology can create both positive and negative social outcomes.

This machine helped convert raw cotton into thread.
(Photo by Michael Hacker)

safe conditions. Workers and others pushed for changes in the way workers and their families were treated.

The Industrial Revolution brought about social changes as well as technological change. Trade unions were organized. Laws were passed about working conditions, hours, and wages. Child labor laws were passed, forbidding factory owners to hire children below a certain age. Parents were often gone from the home all day, so schools were built to educate and care for children during the workday.

Before the Industrial Revolution, change had taken place slowly. After it started, things happened quickly. People began to expect change.

Patent laws were developed to protect inventors. People who invented a new device or improved on an old one could patent their ideas so no one could use them without paying for them. The profit motive began to push people to invent and improve, and change became planned.

Combining simple technologies can create newer and more powerful technologies.

The widespread use of the printing press also encouraged technological change. Books and manuals provided people with the ideas of others, ideas they could use, build on, and combine. Transportation systems helped by spreading printed materials all over the world. The sharing of knowledge allowed people to combine ideas. A few simple ideas could be used as building blocks for newer and more powerful technologies.

EXPONENTIAL CHANGE

During prehistoric times, technological change came very slowly. It took many thousands of years to change from stone tools to metal tools. As time passed, people had new ideas, and invented new tools and devices. These ideas could be combined to come up with more new tools and devices. The more ideas there were, the faster new tools and devices were invented.

The rate of change in technology became faster and faster. Today it is faster than it has ever been before. Some people say our knowledge is doubling every four years. We say there is an **exponential rate of change** because changes are happening at a faster and faster rate.

What does "exponential rate of change" mean? Let's look at an example. Maria found a store that was selling cassette tapes at a very good price. Her mother agreed to let her buy ten cassettes a month. Her brother, Steven, wanted some, too. He got his mother to agree to this plan: he would buy two cassettes the first month, and double the number he bought each month thereafter. Who do you think would have more cassettes after six months?

Combining Technologies

Shipbuilding techniques + the steam engine = the steamship

Optics + chemistry = photography

Telegraph + printing technology = the newspaper

Internal combustion engine + wagon and carriage construction technology = the automobile

Synthetic materials + medical technology = the artificial heart

Kite and glider technology + light-weight gasoline engine = the airplane

New materials (titanium) + jet engine technology = spy aircraft

Photographic technology + satellite technology = exploration of the earth

Robert Fulton's steamboat, *Clermont*, sailed from New York City to Albany in the early 1800s.
(Courtesy of The New York Public Library—Astor, Lenox and Tilden Foundations)

Cameras carried by satellites can determine the best locations to prospect for oil or natural gas.
(Courtesy of NASA)

At the end of the first month, Maria owned ten cassettes. Each month, she bought the same number of tapes as she did the month before. After six months she had sixty cassette tapes. Her tape supply increased by the same rate (10 tapes) each month. Therefore, the change in the number of tapes was a **linear rate of change** (like a straight line).

Steven started out by buying two tapes. He doubled his purchases each month thereafter. In the second month, he bought four and in the third, eight. Steven's tape supply increased faster each month, while Maria's increased at the same rate every month. His supply increased exponentially. Their mother

MONTH NUMBER	NUMBER OF TAPES PURCHASED EACH MONTH	TOTAL NUMBER OF TAPES OWNED
1	10	10
2	10	20
3	10	30
4	10	40
5	10	50
6	10	60

Maria's tape purchases:
a linear rate of change.

MONTH NUMBER	NUMBER OF TAPES PURCHASED EACH MONTH	TOTAL NUMBER OF TAPES OWNED
1	2	2
2	4	6
3	8	14
4	16	30
5	32	62
6	64	126

Steven's tape purchases: an
exponential rate of change.

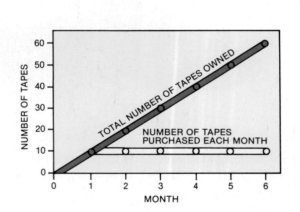

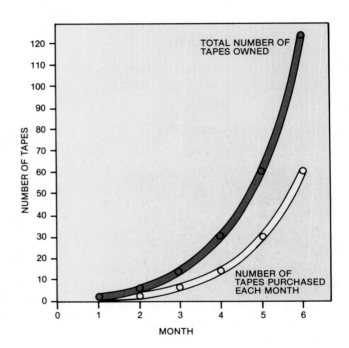

Technology has existed since
the beginning of the human
race, but it is growing at a
faster rate today than ever
before.

soon realized she would be spending all her money on tapes for Steven, and put an end to the arrangement.

When something changes at an increasing rate, we say that it changes exponentially. This is true of technological knowledge. It took about a million years to go from using stone tools to using tools from bronze, but only about 5,000 years to go from using bronze tools to using machines. The growth of technological change is still getting faster and faster.

You can get an idea of how fast technology has changed by looking at a time line of technological change. A time line is a type of graph. It lists events that have occurred over a given period of time. Over the last hundred years, technological change has come even faster than before.

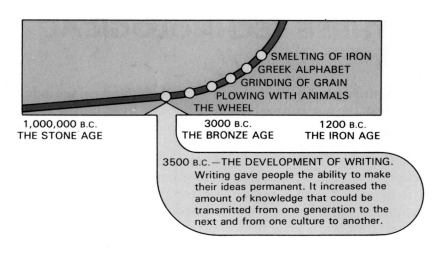

1,000,000 B.C.
THE STONE AGE

3000 B.C.
THE BRONZE AGE

1200 B.C.
THE IRON AGE

SMELTING OF IRON
GREEK ALPHABET
GRINDING OF GRAIN
PLOWING WITH ANIMALS
THE WHEEL

3500 B.C.—THE DEVELOPMENT OF WRITING.
Writing gave people the ability to make their ideas permanent. It increased the amount of knowledge that could be transmitted from one generation to the next and from one culture to another.

Time line: From Stone Age to Iron Age. During the prehistoric period (1,000,000 B.C. to 3,500 B.C.), technology changed exponentially, but at a very slow—almost linear—rate. After 3,500 B.C., the rate of technological change began to increase. It is still increasing today.

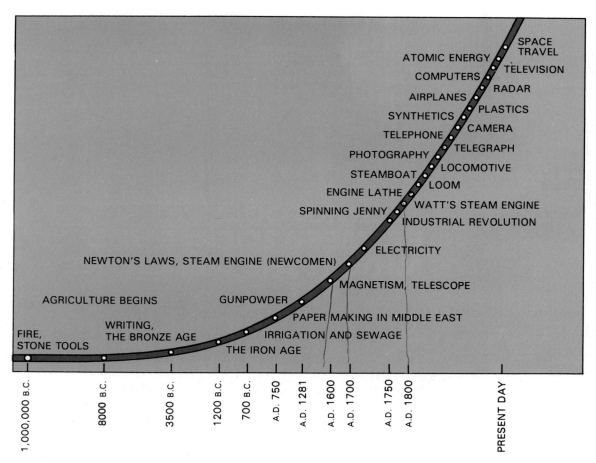

Technological time line: 1,000,000 B.C. to present day.

THE THREE TECHNOLOGICAL ERAS

Human history can be divided into three technological periods. The first was the **agricultural era**. Most people lived off the land. Many tools and discoveries had to do with growing and harvesting crops. The **industrial era** began with the Industrial Revolution in the late 1700s. During this era, many new machines were invented. Today, we are in an **information age**. Many of today's inventions are based on electronics and the computer.

During the agricultural era, most people were farmers. They grew their own food. They used their own muscle power or that of animals to do jobs like pumping water and plowing fields.

During the industrial era, many people were employed in factories. Machines replaced human and animal muscle power. Steam and electricity were used to run motors, which ran machinery.

We have now entered the information age. Many jobs depend on workers being well-educated and staying informed about changes in technology.

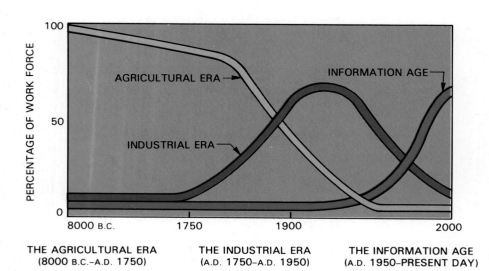

Shift from agricultural era to industrial era to information age.

TECHNOLOGICAL LITERACY

Because technological change has increased so quickly, technology is an important force in our society. To make sense of our world, we must understand technology and how it affects us. We must become **technologically literate**.

A person who is literate in English can read and write. A person who is literate in technology understands technology. Technologically literate people know that technology is not magic. They understand how processes work, and how products are made. They know that technology is created by people to fill human needs. They are able to make informed decisions about technology.

What kinds of decisions are these? One is choosing from among all the products and services that are available. All of us are consumers. We buy the products of technology.

Another kind of decision has to do with our role as citizens in a democracy. We can affect how technology is used in this country. We can write letters, vote, and speak out about what is happening. We can encourage the use of helpful technologies. We can oppose uses of technology that might harm us or the environment.

A technologically literate person is a good consumer of technology. (Courtesy of Tandy Corp.)

CONNECTIONS: Technology and Mathematics

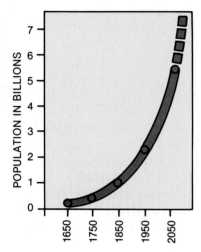

- Population, like technology, is growing "exponentially." In ancient times population doubled about every 1000 years. Recent predictions estimate that population will double every 37 years. The doubling sequence: 2 4 8 16 32 ... represents the powers of 2: 2^1 2^2 2^3 2^4 2^5 ... The exponents written to the upper right of the number cause the sequence to grow at an increasing rate. ($2^3 = 2 \times 2 \times 2$)
- In the early days of the industrial revolution, people worked in factories 14 hours a day. 14/24 is about 58% of the day. Today's workers spend approximately 8/24 or 33-1/3 of the day on the job.
- From 1750 to 1900 the percentage of the workforce in industrial jobs varied directly as the rate of change of technology. This means that as one increased, the other increased. Today a similar thing is happening with the percent of people working in information processing. The percentage of the workforce in this field increases as technology increases.

SUMMARY

Technology is as old as the human race. Since prehistoric times, technology has filled people's needs. Technology has filled our needs for food and medical care, our needs for shelter, clothing, and manufactured products, and our need to communicate information.

We can classify historical periods by the kind of technology that was available at different times. During the Stone Age, people used tools made of stone. During the Bronze Age and the Iron Age, tools and other items were made of metal. The Industrial Revolution began with the development of machines.

Before the Industrial Revolution, technological change took place slowly. As time passed, the profit motive encouraged invention of new devices and improvement of old ones. Technological change became very rapid.

Technological change is increasing at an exponential rate. Each new technological development can be the start of new ideas, resulting in more invention, more changes.

Human history can be divided into three technological periods. In the agricultural era, most people were farmers. In the industrial era, most people worked in factories. In the present era, called the information age, most people use computers and communications in their places of work. This book will help you become more technologically literate, so you can live successfully in our technological world.

1. Explain the difference between science and technology.
2. Describe how technology affected your routine this morning.
3. List five needs that people have and explain how technology helps to satisfy those needs.
4. Explain two ways in which technology satisfies our need to communicate ideas or process information.
5. Give two examples of technologies that have developed from more simple technologies.
6. Draw a technological time line that illustrates how technology is growing at an exponential rate.
7. Describe one major example of how technology has made life easier for people.
8. Define agricultural era, industrial era, and information age.

(Photo by Michael Hacker)

(Courtesy of GMF Robotics Corporation)

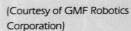

(Courtesy of Gregory Heisler/Pacific Gas & Electric)

CHAPTER KEY TERMS

Across

1. Era when most people lived off the land.
7. Period when iron came into common use.
8. The ability to do work.
11. The era when people discovered that metals could combine to form an alloy.
12. Regular patterns of behavior.
13. A technology used to send information.
14. The earliest technological era.
16. Using knowledge to transform resources to meet human needs.
17. A system used to produce products.

Down

2. Many machines were invented during this period, which began about 1750.
3. The study of why natural things happen the way they do.
4. Technological _____ happens at an exponential rate.
5. Technology of building structures.
6. A combination of things that act together.
9. Things that are turned into goods and services by technology.
10. A very rapid rate of change.
15. Extra-Terrestrial (abbreviation).

Agricultural era
Bronze Age
Change
Communications
Construction
Energy
ET
Exponential rate of change
Factory system
Industrial era
Information Age
Iron Age
Resources
Routines
Science
Stone Age
System
Technology
Technologically literate

SEE YOUR TEACHER FOR THE CROSSTECH SHEET

ACTIVITIES

TECHNOLOGY TIMELINE

Problem Situation

Technology is not new; in fact, it has developed over more than a million years. Early humans developed tools, such as axes, during the Stone Age, that look like tools we use today. Ceramic materials first used to make pottery around 6000 B.C. remain important. In addition to dinnerware, ceramics are now used to make dozens of building products, insulating tiles for the space shuttle, and nuclear fuel rods. Several automobile manufacturers are also testing ceramic engines.

The pace of technological development continues to increase. Look at the time line on pages 422-423 and compare the number of technological developments listed before and after 1900. How many new developments do you think we will be able to add during the next ten years?

Design Brief

Choose an event or invention from the Technological Time line found on pages 422-423. Design and construct a hanging mobile that describes the technological development you selected.

Suggested Resources

Fishing line
1/8" and 1/4" dowel
Markers
Cardstock in a variety of colors
Computer
Printer
Rubber cement or white glue
Word processing and graphics programs
Scissors

Procedure

1. List three alternative topics from the time line.
2. Select the theme for your mobile after reviewing reference materials that are available in the technology laboratory and your school library. Have the topic approved by your teacher.
3. Gather information and take notes.
4. Use the computer and printer to produce a sign that includes the name of the event or invention, and the date on which it occurred.
5. Make an illustration describing the technological development. Add color using markers.
6. Make a list of important facts about the technological development. Print this information using a computer and a word processing program. Use large type so that the list will be easy for others to read.
7. Print a list of the impacts of the event or invention. Be sure to include positive and negative impacts. Try to identify how later technological developments were influenced.

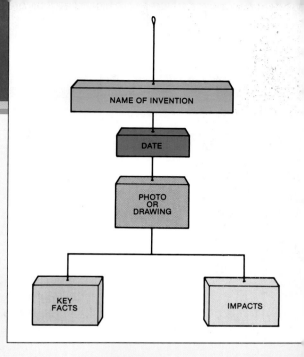

8. Use scissors to cut out the items. Fasten each to a different color of card stock using white glue or rubber cement.
9. Assemble the mobile using dowels and fishing line. Adjust so that each part of the mobile hangs properly.
10. Hang the mobile where directed by your teacher.

Classroom Connections

1. Technology is responsible for a great deal of the progress of the human race.
2. Complex technological systems develop from more simple technologies.
3. Technology is growing at a faster rate than ever before.
4. The Stone Age, Bronze Age, and Iron Age were periods of history named to reflect the most important technology of the day.
5. The Industrial Revolution brought about social changes as well as changes in production techniques. People moved from farms into the cities. Long hours and unsafe working conditions were common. Public school systems were needed to educate and supervise children.

Technology in the Real World

1. The current era is often referred to as the Information Age.
2. New developments in electronics continue to bring about rapid changes in communication and other technologies.
3. Technological changes have positive and negative impacts on everyone's life. All people need to become technologically literate to understand these impacts and make informed decisions.

Summing Up

1. Name several important impacts made by the event or invention you chose for your mobile.
2. Describe the importance of two other inventions or events that were described by mobiles made by other students.
3. List five new technological developments you expect to be using ten years from now.

ACTIVITIES

SPREADING THE NEWS

Problem Situation

People have always had a need to communicate ideas, feelings and thoughts. The invention of the printing press during the 1500s helped spread knowledge throughout Europe and other areas of the world. Through papers, books, and manuals people were able to read about other people's experiences and technological achievements. New ideas could then be built upon old discoveries, enabling technology to grow at a faster pace. As books became more available, more people became aware of new technologies.

Design Brief

Select any flight from the Mercury, Gemini, Apollo, or space shuttle programs and compose a front page headline and story about that mission for your paper.

Be sure your front page contains an interesting headline, sub-headline, picture, and well-researched story about the flight.

Using computer software, generate the headlines and text you will need for your front page. Using these materials, create a paste-up that can be used to duplicate 100 copies of your front page. Each student's front page will be collated to form a book of front-page stories about the history of American space flight.

Suggested Resources

Computer
Monitor
Printer
Software — Multiscribe, Mouse Paint, Print Shop, word processing program
Some type of duplication system — offset, mimeo, screen process, photocopy, Gestetner process
Binding materials
Paste-up materials
Texts for research

Procedure

1. Select a mission from the Mercury, Gemini, Apollo, or space shuttle programs.
2. As a reporter covering the launch and mission, you will be required to write an article about the flight that will appear on the front page of your newspaper. Use books or old newspapers and magazines as references to find out what actually took place during the mission.
   Remember, a good news article always answers the questions:

   - **who** was involved
   - **when** did it happen
   - **what** took place
   - **where** did it happen
   - **why** did it happen

3. Using software and a computer, word process your story so it may become part of your front page.
4. Use the computer to make headlines and subheadlines in different styles or fonts and sizes that will make the reader want to read your front-page story.

5. Prepare your original copy by pasting up your headline, text, subheadline, and pictures. Duplicate your original 100 times.

6. Collate all the front pages you and your classmates have produced into one book. You may want to draw an attractive cover for your book. The title might be "The History of American Space Flight."

Classroom Connections

1. Technology satisfies our need to communicate ideas.
2. People create technological devices and systems to help meet our needs and wants.
3. Technology changed rapidly because inventors, engineers, and scientists were able to combine ideas that had already been developed with new ideas and thus create new technologies.
4. In English class you learn to communicate using words, sentences, and paragraphs. A news reporter must compose paragraphs that are short, accurate, and easy for the paper's readers to understand. In order for people to understand your article, you must pay strict attention to grammar, spelling, and sentence structure.
5. President John Kennedy committed the United States to the goal of landing a man on the moon. His goal was accomplished in 1969 with the landing of Apollo 11. The Mercury, Gemini, and Apollo missions all helped in making this dream a reality.

Technology in the Real World

1. The use of satellites and telecommunications allows newscasters to send their stories to the newspaper office for publication instantly from any place in the world.
2. Video disk technology allows us to store thousands of graphic images on a single disk. Using computers, we can retrieve any of these images in seconds.

Summing Up

1. How many different type fonts can you count on the front page of your local newspaper?
2. What method was used to duplicate your original copy?
3. When writing about an event, what are the five basic questions you must answer in your story?

THE LUNAR LANDING NEWS

January 27, 1967

ASTRONAUTS KILLED IN FIRE

Fire
On Pad #34

It was a quiet and somber day on Launch Pad #34, Cape Canaveral, Florida. Clean up crews and experts slowly sorted through the wreckage of the burnt out Apollo I capsule searching for the cause of the deadly fire. What was supposed to be a routine test for astronauts Virgil Grissom, Edward White, and Roger Chaffee ended in the death of the three men as their spacecraft turned into a fiery inferno. NASA investigators are looking into the possibility of a short circuited electrical wire that ignited the pure oxygen atmosphere as the cause of the fire.

Cape Canaveral

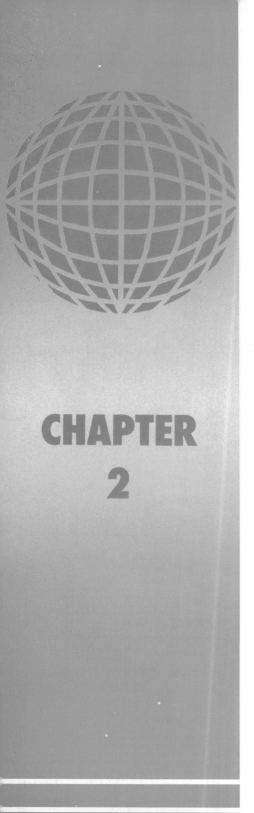

CHAPTER 2

PROBLEM SOLVING

MAJOR CONCEPTS

- People must solve problems that involve the environment, society, and the individual.
- The best solutions are those that work well, are economical, and cause the least harm to people or the environment.
- A carefully thought-out, multi-step procedure must be used to solve problems.
- As you go through the technological method of problem solving, you can keep a record of all your ideas and drawings in a design folder.
- A good solution often requires making trade-offs.
- Values affect technological decisions.
- Technological decisions must take both human needs and the protection of the environment into consideration.

PROBLEM SOLVING

Humans have always been faced with problems to solve. They have always needed food, clothing, shelter, and health care. These needs are met through technology. People solve their problems using available resources and knowledge.

Headaches, for example, have been a problem for humans for a long time. Do you remember trepanning, discussed in Chapter 1? People used tools to drill holes in a person's skull. This was supposed to let evil spirits escape. People did not know enough about the cause of headaches to solve this problem.

Early humans solved problems of food and shelter by using the materials around them. Caves were used as homes. People gathered roots, fruit, and seeds of plants to eat, and made tools to hunt animals for meat.

Today's problems are also those of human needs and wants, but there are many more problems and they are more complex than ever before. Some of these problems involve society and the environment, such as:

- How can we dispose of wastes without harming the environment?
- How can we produce enough energy to meet our increasing needs?
- How can we assure a continuing supply of clean, safe water?

Other problems have to do with the individual. For example:

- How can we improve the sound quality of recorded music?
- How can we help vision-impaired people to see better?
- How can automobile drivers communicate with each other?

Waste disposal has become a major problem. (Photo courtesy Stephanie Zarpas/NYSDEC)

Will driver-to-driver communication of this sort ever be possible?

Extended-wear contact lenses are now available that can stay in the eye for a period of weeks. (Courtesy of Schering-Plough Corporation)

Problem Situation 1:   Young children enjoy playing with toys that involve balance and dexterity.

Design Brief 1:   Design and construct a toy that will either balance on, or move along, a taut piece of string. The finished object should be attractive and safe for a small child to use.

Problem Situation 2:   Waste disposal has become a societal problem. Some communities are running out of space to bury their garbage and trash.

Design Brief 2:   Design and construct a model of a bin that can separate garbage from trash so that the trash (paper, aluminum, plastic, glass) can be salvaged and recycled.

Problem Situation 3:   Falling leaves in autumn create problems for gutters and drains.

Design Brief 3:   Design a device to remove leaves from drains and/or gutters.

GOOD DESIGN IN PROBLEM SOLVING

The best solutions are those that work well, are economical, and cause the least harm to people or the environment.

Solving a problem is rarely quick and easy. It takes time and thought. We must understand exactly what is needed to solve the problem. We must make our solution as cheap and easy to use as possible. We must figure out how long the solution will last. Very often, there are a number of different solutions to a problem. These solutions must be compared so that the best one is chosen. The best ideas are further refined and improved. Good solutions to problems are those that work well, are inexpensive, and cause little or no harm to the environment and people. When a technological solution causes harm, people must decide whether the solution's benefits outweigh the problems.

EXAMPLES OF WELL-DESIGNED TECHNOLOGICAL SOLUTIONS

Engineers, designers, and scientists solve a huge number of different technological problems. These can range from redesigning a stereo to creating new life forms like a plant that repels insects. Here are some well-designed technological solutions:

The Dame Point Freeway Bridge in Jacksonville, Florida. (Courtesy of Brendrup Corporation)

Hard-skinned tomatoes
(Photo courtesy Heinz USA)

Genetically engineered drugs
(Photo courtesy Cetus Corporation)

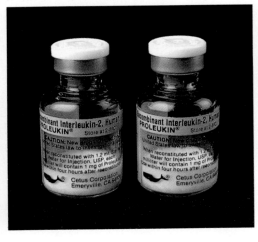

Personal stereos (Courtesy Sony Corporation of America)

Telephones (Courtesy of Radio Shack, a division of Tandy Corporation)

Problems That Still Need Solving

Engineers and scientists are trying to solve problems in many fields. Just one example of these involves **superconductors**.

Materials that conduct electricity, like copper wires, all have electrical resistance. Electrical resistance opposes the flow of electricity. Resistance produces heat, so energy is lost as electricity flows. Resistance is caused by the movement of atoms in a conducting material. The moving atoms oppose the flow of electricity.

Superconductors are materials with no resistance. That means they are perfect conductors of electricity. When an electric current starts flowing in a superconducting wire, no energy is lost.

Using superconductors, very strong electrical currents can be made to flow. Powerful electromagnets can be made using these strong currents.

For many years, it was thought that superconductors could only be made at temperatures around −432° Fahrenheit and below. To cool materials this far required liquid helium and was too costly to be practical.

In 1986 and 1987, researchers around the world developed materials that became superconductors at much higher temperatures. Some worked at temperatures as high as −234° Fahrenheit. To reach these temperatures, liquid nitrogen can be used. It is cheap enough to make such a superconductor practical. However, these new superconducting materials are brittle. They are hard to make into wire. Also, scientists are not sure whether they can carry large electrical currents. What is needed is a material that can superconduct at room temperature. This is a problem that scientists and engineers are hoping to solve someday.

This small piece of superconducting material has been cooled to −284° Fahrenheit. At that temperature, it becomes superconducting and floats in the air above a magnet (the Meissner Effect). (Courtesy Bellcore & NYNEX; photo © Bellcore 1987)

This high-speed magnetic levitation (maglev) train works because electromagnets in the trains and tracks repel each other and provide a nearly frictionless ride. With superconducting magnets, the trains could be made to travel at speeds over 300 miles per hour. (Courtesy Magnetic Transit of America, Inc.)

CONNECTIONS: Technology and Science

■ Heating power and light power can be produced using electricity. The equation is

$$P = I^2R$$
where P = power
I = electric current
R = resistance

Thomas Edison was a problem solver. He made a light bulb that took advantage of high resistance. Edison could have searched for a filament in a light bulb that would have produced light by having large amounts of electric current flowing through it ($P = I^2R$). Reasoning that too much current in a house could be dangerous, Edison decided to make his light bulb using a filament that had a high resistance to the flow of electricity.

■ Discoveries are sometimes made by accident. X rays were discovered accidentally by Wilhelm Roentgen. He was studying ultraviolet light coming from a discharge tube. Ultraviolet light is the kind of radiation that gives us sunburns. Roentgen was using certain kinds of crystals that fluoresce (give off light) when they are hit by ultraviolet light. He noted one day that crystals fluoresced even when they were far from the tube and the tube was wrapped in black paper. Roentgen studied these rays, which he called X rays.

HOW PEOPLE SOLVE TECHNOLOGICAL PROBLEMS

Problem solving is faster and easier, and results are better, if people follow a procedure. The procedure given below is only one of many. You may find that when you solve a problem, you go back and forth among the listed steps, and do not follow them in order. But the list covers the activities and thought processes that are always part of the problem-solving process.

THE TECHNOLOGICAL METHOD OF PROBLEM SOLVING

There are seven steps in problem solving:

1. Describe the problem as clearly and fully as you can.

2. Describe the results you want in a design brief.
3. Gather information.
4. Think of alternative solutions.
5. Choose the best solution.
6. Implement the solution you have chosen.
7. Evaluate the solution and make necessary changes.

A carefully thought-out, multi-step procedure must be used to solve problems.

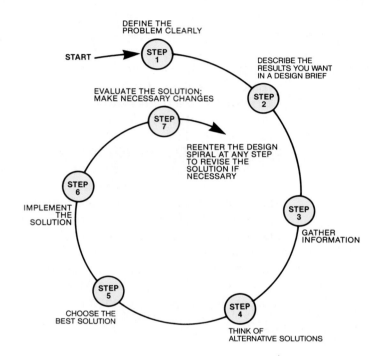

THE PROBLEM-SOLVING SPIRAL

CONNECTIONS:　Technology and Social Studies

■ The continuous need to engage in research and development has led to the creation of a major industry. Large sums of money and millions of people are devoted to seeking advances in such areas as manufacturing, medicine, and education.

■ Conflicting values can make decision making difficult. The person who wants to keep a piece of land "forever wild" is going to have a hard time coming to an agreement with the person who sees the same region as a source of oil.

■ Introducing a new product into the marketplace may be risky. Whether the product is a life-saving drug or a technological innovation, careful testing must be done to ensure the safety and well-being of the public. Government agencies and private watchdog groups try to protect society from products that may be harmful.

KEEPING A RECORD OF PROBLEM-SOLVING STEPS

As you work on solving your problem, keep a record of all your ideas and drawings and the information you collect in a **design folder**. This folder can be used to show the work and thought that went into your solution. It can be used for your own reference in case you want to change your project, or make another similar project later. Your folder can also be shown to parents, teachers, and friends. Later, you can use it when you apply to college or look for a job. The design folder should have a separate section for each of the steps you follow.

Step 1: Describe the Problem Clearly and Fully

As you go through the technological method of problem solving, you can keep a record of all your ideas and drawings in a design folder.

To solve a problem, we must first understand it. What has caused the problem? What situation caused the problem and requires us to think about finding a solution? A statement describing the problem gives you a way of thinking about the problem in a very clear way. Here are some examples of clearly stated problems:

- People with arthritis in their fingers have a hard time gripping small objects. They need an easy way to carry out such tasks as unlocking a door.
- Colored drawing pencils roll off a desk easily and break, and hunting for the right color of pencil in a box is a nuisance.

SITUATION

I NEED A MULTI-PURPOSE CONTAINER TO CARRY ITEMS RANGING FROM PAINTS TO FISHING EQUIPMENT.

Mark Townsend, a technology student in Staffordshire, England, likes to fish and to paint. This is the situation that Mark finds himself in.

Once we know what the problem is, we can decide what to do about it. We might decide that the problem is just too big for us. It will have to wait until a group of engineers with enough money and time get interested in solving it. Or we might decide that the problem is uninteresting or unimportant. We might decide we don't want to work on it.

Most of the time, we need to solve problems that are presented to us. Sometimes we want to solve them because they are important to us. Sometimes we have to solve them because parents, teachers, or friends ask us to. Sometimes we want to solve problems because they are challenging and we think it will be fun to come up with a solution.

Step 2: Describe the Results You Want in a Design Brief

Once you understand the problem you can decide what to do about it. You can specify the results you're looking for. One way to do this is to write a short **design brief**. A design brief is a simple statement that states in very general terms the solution you think might solve the problem. The design brief should include the design criteria and the constraints. The design criteria are the specifications the designer must follow. Constraints are limits designers must work within.

Here are some design briefs for the problems listed previously:

Problem Situation 1: People with arthritis in their fingers have a hard time gripping small objects. They need an easy way to carry out such tasks as unlocking a door.

Design Brief 1: Design and make a device that will help in using a house key when opening a door. The device must be lightweight and small enough to fit into a purse or pocket, and must cost less than fifty cents to produce.

Problem Situation 2: Colored drawing pencils roll off a desk easily and break, and hunting for the right color in a box is a nuisance.

Design Brief 2: Design and make a holder to stand on a desk or table. It must hold up to ten colored pencils and make selection easy. The holder may be made from any material, but the selling price must be under two dollars.

BRIEF

TO DESIGN A CONTAINER TO CARRY THE ITEMS LISTED IN THE SITUATION.

WHAT IS REQUIRED OF THE CONTAINER?

1. The material of the container should be plastic, metal, or fabric

2. The container should have a number of compartments

3. The compartments should have easy access

4. It should have a handle so it can be carried

5. It should be lightweight

6. The container should have a lock

7. It should be unbreakable – as much as possible

8. It should be secure so that the paints etc. will not move around

9. There should be enough room for a range of equipment

Mark Townsend's design brief and specifications.

Step 3: Gather Information

An important part of solving any problem is collecting information about it. This information gathering is called **research**. By knowing how other people have approached similar problems, you will learn about good and bad solutions. If a product is to be used by people, you may want to collect data from the library on people's heights, weights, length of reach, or other design factors that will make the product easier to use.

Some companies and government agencies constantly perform **basic research** into the nature of different materials and processes. They don't expect to produce any products immediately from what they learn, but they save the results and hope that the new knowledge will be useful at a later time.

Companies often do **market research** to determine if customers will like a new product. Companies may ask potential buyers to fill out a questionnaire to find out what they like or don't

The library is a good place to go to do research.

like. For example, if a company wants to develop a toothpaste for teenagers, the questionnaire might ask teens what they like or don't like about the toothpaste they're using now. Do they like the taste? Do they like the way it feels in their mouths? What kind of dispenser do they prefer to use, a tube or a pump? What colors do they prefer for the toothpaste, tube, and box? The company will use the results of this research to design the product so it appeals to the greatest number of people.

Step 4: Think of Alternative Solutions

The research done in Step 3 may give you one or more possible solutions to the problem. There is almost always more than one solution to every problem. We can suggest several ideas, each one of which might do the job. These different ideas for solutions are called **alternatives**.

Developing new alternatives is one of the most important parts of the problem-solving process. How do we come up with them?

One way to develop alternative solutions is to use our **past experience**. When we do research as described in Step 3, our information comes from the past experiences of others; by using our own experience and thinking of how we might have solved similar problems in the past, we may find a new way to solve the problem.

Another way of coming up with alternatives is called **brainstorming**. During brainstorming, each person in a group can

CONNECTIONS: Technology and Mathematics

■ Problem solving in mathematics is very similar to problem solving in technology.

Problem: How many rectangles can you find in the figure at the right?

Step 1 *Understand the problem.* What is given? What are the conditions? (Assumptions: Squares are rectangles. Rectangles can overlap.)

A	B	C
D	E	F

1 PART: A, B, C, D, E, F
2 PARTS: AB, BC, DE, EF, AD, BE, CF

3 PARTS: ABC, DEF
4 PARTS: ABED, BCFE
6 PARTS: ABCDEF

Step 2 *Devise a plan.* Find an organized way to count. Label the parts.

Step 3 *Carry out the plan.* There are 6 rectangles having one part; 7 rectangles consisting of two parts; 2 rectangles having three parts; 2 rectangles having four parts; and 1 rectangle with six parts. There are 18 rectangles.

Step 4 *Looking back.* Is the answer reasonable? Have I counted things twice? Did I miss any rectangles?

suggest ideas. One person writes all the ideas down; no one is allowed to laugh at or criticize an idea, no matter how foolish or unusual it might seem.

The brainstorming process is used to help people think more creatively. People feel free to share any wild ideas they may have. Sometimes one person's wild ideas will open up someone else's mind to a totally new approach. After many ideas have been proposed, the group reviews them all. The best ideas are then developed further.

A third way to develop alternatives is by **trial and error**. This is the way most people do jigsaw puzzles. When putting together a puzzle, we are really solving a problem by trying out different ways of placing the pieces. Eventually, all the pieces are put in the correct places and the problem is solved. When solving real problems by trial and error, the end result may not fit together as perfectly as a completed jigsaw puzzle, but the process used to solve it is similar.

A fourth way to develop alternatives is to use what psychologists call **insight**. Have you ever had an idea just pop into your head? These sudden ideas are usually followed by the "Aha!" response ("Aha! I've got it!"). Insight comes from being

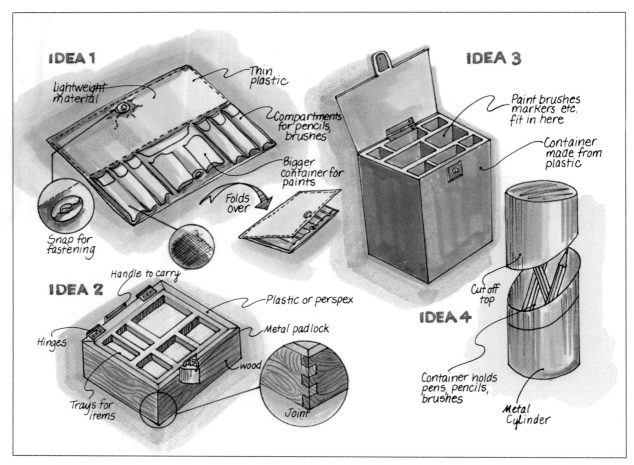

Here are a variety of designs Mark Townsend thought he might use for his container.

thorough in researching the problem, and from being creative in thinking about the problem from many different angles. Even when you are not consciously thinking about the problem, your brain may still be working on it.

Still another way alternative solutions are discovered is by **accident**. Some of the most important discoveries, like penicillin, occur when the inventor goes as far as possible and still doesn't solve the problem. A chance happening then provides the answer. In other cases, the solution to a problem is discovered by someone who is looking for the solution to another problem. It takes a person with insight to recognize when a solution has been discovered by accident.

Step 5: Choose the Best Solution

Once you have developed a list of alternative solutions to the problem, you need to select the best possible solution. Each alternative must be examined to see if it meets the design

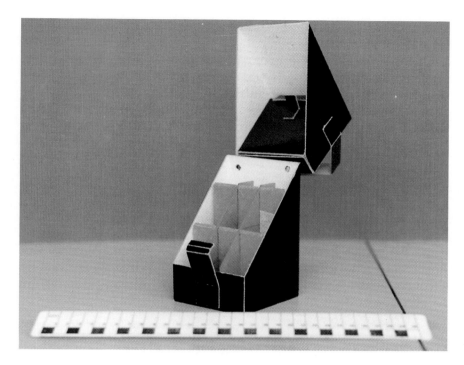

The finished product

SOLVING REAL-WORLD PROBLEMS

The problems a student might solve in a school technology class require the same methods used in the outside world. Solving such problems helps prepare students to solve other problems.

Social and Environmental Concerns

Engineers and designers must keep in mind what effect their solutions will have on society and the environment. For example, suppose a new airport is being planned. It cannot be built too near a residential area. The noise would disturb and upset people.

In the real world, the needs of society or the community must be considered. A good technological solution meets the needs of people and preserves the environment.

Politics

Often, solutions to real-world problems are affected by politics. Groups of people have special interests. For example, some people oppose the building of nuclear power plants. They say that radioactive material could poison the environment in case of an accident. They point out that there is no safe way to get rid of radioactive waste.

Other groups favor nuclear power plants. They believe that nuclear power can provide the energy our country needs. They think it will make us self-sufficient so that we don't have to depend on other countries for oil to fuel our power plants. They think the risks of an accident are small. They are willing to trade these risks for the benefits of a reliable source of energy.

Risk/Benefit Trade-Offs

A common trade-off made in solving large problems is a risk/benefit trade-off. To obtain wanted benefits, we accept some risk. We try to keep the risk as low as possible. We may not implement a solution if risks are too high. When you travel by car, you accept a very low risk of being hurt in an accident. You receive the benefit of traveling quickly and comfortably.

Need for Continued Monitoring

Often, solutions to real-world problems must be monitored over many years to make sure there are no unwanted outcomes. For example, between 1958 and 1961, many pregnant women took a drug called thalidomide to help them relax. After several years doctors began to see that the drug had harmful side effects. Some of the women who took the drug had babies with birth defects. The birth defects were later traced to the drug.

When thalidomide was first given, it seemed to be a fine solution to the problem of helping pregnant women relax. Only after years had passed did people realize that it was not the best solution. All the effects of a technological solution may not be known until long after the solution is implemented. That's why it's necessary to keep on monitoring and studying results.

Values

Values affect technological decisions.

Our values influence our decisions. The way we feel about something makes us decide in favor of or against it. If we think of automobiles only as transportation, we might decide to buy a basic car that gets good gas mileage. If we feel that automobiles are neat and driving is fun, we might decide to buy a sports car or a luxury car.

Most problems that can be solved in a school technology class do not involve politics, the environment, or cultural traditions. Class problems are not as complex as the technological problems of the real world. Real-world problems mean more limitations—limitations that are related to social, environmental, and political factors.

SUMMARY

People have always been problem solvers. Human needs have caused people to use technology to make life easier. There are still problems. They involve society, the environment, and the individual.

Good solutions must be carefully thought out and designed. There is usually more than one good solution. Alternatives must be compared to choose the one that works best, is most economical, and causes least harm to people and the environment.

Problem solving is most effective if a step-by-step procedure is used. One such procedure has seven steps: 1. Describe the problem clearly and fully. 2. Describe the results that are wanted in a design brief. 3. Gather information. 4. Think of as many alternative solutions as possible. 5. Choose the best solution. 6. Implement the solution you have chosen. 7. Evaluate the results of trying your solution and make changes, if necessary.

As you carry out the problem-solving steps, you can keep a record of your ideas and designs in a design folder. A separate part of the design folder should be made for each of the steps you follow.

Modeling is a problem-solving technique. Models are used to test ideas without risking a great deal of time and capital or endangering the public.

The technological problems presented in school are less complex than real-world problems. Real-world problems involve political and environmental issues, as well as those relating to values.

1. Give one example each of problems involving society, the environment, and the individual.
2. Propose a workable and economical solution to a problem involving a personal issue.
3. What are the seven problem-solving steps listed in this chapter?
4. What are five ways of coming up with alternative solutions?
5. What is a design folder? How should it be organized?
6. Your city has run out of land for landfill (refuse disposal). City government has chosen to build a very expensive incinerator to handle the garbage problem. What are some trade-offs that were made in reaching this decision?
7. Give an example of how a person's values might affect his or her decision about the kind of car to buy.

Our values affect our choices.

CHAPTER KEY TERMS

Across

1. This is done to determine if people will buy a product.
3. A method of finding alternatives by trying one thing after another.
5. A place to keep a record of all your ideas and drawings.
9. A model which is made before the product is mass produced.
10. The problem-solving step when the solution is tried out under actual conditions.
11. A method of finding alternatives where a group of people generates ideas.
13. A method of finding alternatives where an idea pops into your head.

Down

2. Possible solutions.
4. A formal process consisting of seven steps.
5. A simple statement which spells out the problem.
6. Making an alternative work as well as it can.
7. One way of gathering information.
8. A compromise you make when solving problems.
12. Not the actual solution, but something that represents it.

Alternatives
Basic research
Brainstorming
Design brief
Design folder
Implementation
Insight
Market research
Model
Optimization
Problem solving
Prototype
Research
Trade-off
Trial and error

SEE YOUR TEACHER FOR THE CROSSTECH SHEET

ACTIVITIES

ROUGH TERRAIN TRANSPORT

Problem Situation

Throughout history people have developed and modified transport vehicles to accommodate a variety of terrains. During the early 1800s, Gravity Trains delivered coal from Pennsylvania to New York. They were pulled up the mountainside by mule and then fell free on a track from the mountain's peak to the next lift station.

In July of 1971, *Apollo 15* touched down on the moon's surface, bringing with it the Lunar Roving Vehicle (LRV). The LRV was designed to transport the astronauts and their gear while they explored the lunar surface.

In the development of these transport vehicles, designers and builders had to solve endless technical problems. What source of energy would power the vehicle? What type of surface would it have to travel on? How would they control the vehicle's direction? Solving any technical problem requires us to make many design decisions.

Suggested Resources

Illustration board
Foam board
Styrofoam
Wood scraps
Dowels
Hot glue
Rubber bands
Balloons
Straws
Solar voltaic cells
1.5-volt DC motor

Design Brief

Design and build a prototype for a rough-terrain transport vehicle. The vehicle must be able to cross a sixteen-square-foot area made up of one-inch-high corrugations. The corrugations are spaced one inch apart from peak to peak. The vehicle must also stay within the following specifications:

1. It must fit into an ordinary shoe box, with the lid on.
2. It may be powered only by a rubber band, a balloon, or a solar cell connected to a 1.5-volt DC motor.

Procedure

1. Through brainstorming, you should come up with a variety of methods to power and steer the vehicle. Select the one solution you feel will stay within the constraints of the activity.
2. Make some freehand sketches showing what the body of the vehicle might look like, how the power supply will be connected to the chassis, and how the wheels and axles will be attached to the vehicle.
3. Select modeling materials that will be used to build your prototype.

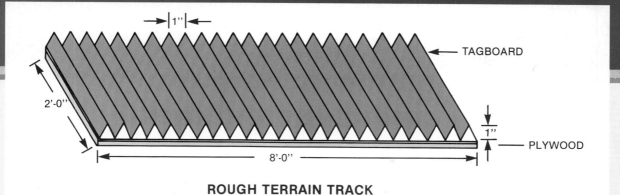

1" ← →

2'-0"

TAGBOARD

1"

PLYWOOD

8'-0"

ROUGH TERRAIN TRACK

4. Using safety procedures described by your teacher, construct your vehicle and test it on the corrugated surface. Record your observations.

Classroom Connections

1. Solutions to technical problems require us to transform the resources of technology into a product or service that satisfies a need or a want.
2. Engineers, architects, and designers all must have excellent writing skills so they can convey their ideas to others.
3. Newton's Third Law of Motion states that for every action there is an equal and opposite reaction. It explains how air escaping from a balloon (the action) can be used to propel a vehicle (the reaction).
4. Gear trains and pulleys are used to increase force, increase speed, or to change the direction of force.
5. From the Viking longboats to the Wright brothers' historic first flight, developing new ways to travel over land, sea, and air has been a goal of people.

Technology in the Real World

There are many different kinds of special purpose vehicles, including:

- Superfast trains using a magnetic field to levitate them just above the track are being developed to transport people and goods.
- Automobiles containing four-wheel steering are available today.
- Robotic transport vehicles controlled by computers and floor sensors move goods through crowded warehouses.

Summing Up

1. What is the advantage of large-diameter wheels over smaller-diameter wheels when traveling over rough terrain?
2. Describe how weight can be a factor when traveling over rough surfaces.

ACTIVITIES

ON THE CUTTING EDGE

Problem Situation

In Technology Education you will have the opportunity to learn about problem solving and design by doing activities. Sometimes you will be given a problem and be asked to design a solution. Usually the problem will have more than one possible solution.

In this activity you will design a new cutting tool and will be able to apply design and problem-solving concepts. The problem-solving system described in this chapter will help you approach problem solving in an organized way. As you work on this activity, follow these steps:

1. Define the problem situation clearly.
2. Describe the results you want in a design brief.
3. Gather information.
4. Think of alternative solutions.
5. Choose the best solution.
6. Implement the solution.
7. Evaluate the solution; make necessary changes.

Design Brief

A local company has donated 1,000 hacksaw blades that were accidentally cut to 6" lengths during the production process. Design and construct a useful tool that utilizes the 6" hacksaw blade. The tool should be comfortable to hold and safe to use. The blade should be replaceable. As you work on the problem, record your ideas and drawings in a design folder.

Suggested Resources

Design folder
6" piece of hacksaw blade
Assorted modeling materials

Procedure

1. Organize a design folder so that you will be able to record your ideas in drawings and notes.
   NOTE: Your teacher will provide guidelines for recording your progress as you complete each step of the procedure.
2. Make sure you understand the problem situation. Ask your teacher to explain anything you are not sure of.
3. Write your design brief for this activity. Use your own words to clearly state the desired results.
4. Gather information that will help you design the new tool. Review catalogs and books, and interview several people who use tools at home or at work. Visit a hardware store. Summarize the information on a page in your design folder.

5. Prepare a series of sketches to record your design ideas. Write a brief paragraph to explain each drawing.
6. Select your best design and make detailed drawings for it.
7. Use available modeling materials to make a prototype of your tool. Follow the safety procedures taught by your teacher. Use the tool and then ask several other students to test it. Instruct them in the safe use of your tool first.
8. Ask the students who tested your tool to provide feedback. How does your design work? Is it safe to use? How can it be improved? Make changes and test again.
9. Complete your design folder so that a person reading it will understand how you used the design and problem-solving process to design the tool.

Classroom Connections

1. Formal methods are used to solve technological problems.
2. Modeling techniques are useful problem-solving tools.
3. A good solution to a problem often requires us to make trade-offs. This means that some solutions have to be discarded for a variety of reasons. Compromises may have to be made.

Technology in the Real World

1. The problem-solving method can be used to solve problems in school and in the outside world.
2. The best solutions are those that are useful, economical, and do not have harmful effects on people and the environment.

Summing Up

1. Is the problem-solving process you used to design this tool useful for solving other types of problems?
2. Explain how you used feedback from other students to improve the tool you designed.
3. What materials did you use to produce your prototype? Why did you select these materials?
4. Give the tool you designed a name and sketch three possible packages for it. Use markers to add color to your best design.

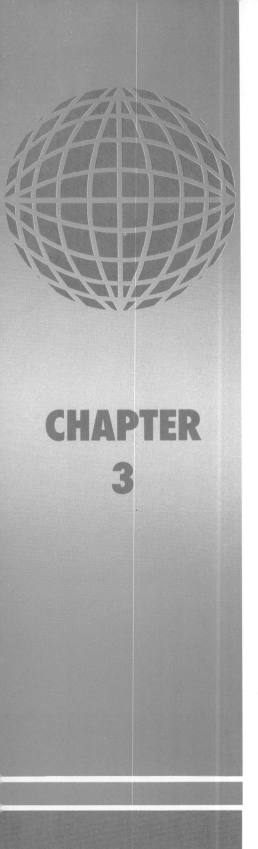

CHAPTER 3

COMMUNICATING DESIGN SOLUTIONS

MAJOR CONCEPTS

- Designing is the process of planning and producing a desired result in the form of a process, product, or system.
- Designers must consider functionality, quality, safety, ergonomics, and aesthetics.
- Eight communication elements are shape, line, texture, color, proportion, balance, unity, and rhythm.
- Technical drawing is done with special instruments and tools.
- Graphic techniques such as sketching and freehand drawing, technical drawing, and computer-aided drafting are used to communicate design ideas.
- Modeling techniques are useful problem-solving tools.
- Computer-aided drafting (CAD) uses computers to produce designs and drawings quickly and accurately.

Running shoe designs
(Courtesy of Nike, Inc.)

WHAT IS DESIGNING?

Designing is a process people use to plan and produce a desired result. The result may be a product, process, or system that meets a specified human need or solves a particular problem. We find examples of design in works of art, advertising, fashion, and product design. We also use design principles in architecture and engineering.

Sometimes, designs come from original ideas and require creativity. Designers must use their knowledge and experience in new and different ways. An invention like contact lenses is an example. Another is a new advertisement for a product.

At other times, designs are based upon existing knowledge and ideas. For example, civil engineers use the experience they have gained over many years to design a highway system. They know that when they use particular types and amounts of concrete and steel to build roadways, the road will support certain loads safely.

Designing is the process of planning and producing a desired result in the form of a process, product, or system.

DESIGN CRITERIA

Design criteria are the requirements that must be met by a solution. When people design, they must consider functionality, quality, safety, ergonomics (ease of use), and aesthetics (appearance).

Designers also work under certain **constraints**. Constraints are limits. They may include cost, effect on the environment, and availability of workers, materials, and time.

Functionality

Functionality refers to the ability of the product, system, or process designed to fulfill its intended purpose over its desired life span. For example, a light bulb manufactured to give 1,000 hours of service is expected to supply light for this period of time.

Problem Situation 1:   Most Technology teachers attempt to point out to students what the dangers are in Technology labs. The result is that there is a wide range of warning notices posted on the walls of the lab.

Design Brief 1:   Design and make a poster listing the main dangers to be found in school technology labs in a visual way. Use color and rendering techniques.

Problem Situation 2:   Birds annoy gardeners when they persist in eating newly planted seeds.

Design Brief 2:   Design and sketch a wind-driven device that could be erected in a garden that would drive the birds away.

Problem Situation 3:   Travelers occasionally have the need for minor first aid when riding in a car.

Design Brief 3:   Design a small, portable first-aid kit to be used in a motor car. Communicate the layout of the first-aid items inside the kit with sketches. Keep a record of your ideas and the alternative designs in a design folder.

Quality

The product, system, or process must be designed to meet certain minimum standards of **quality**. For example, medicines must have the same formulation for every batch produced. The quality must not vary.

Safety

The product, system, or process must be designed to be safely used by consumers. **Safety** codes and regulations must be met. For example, cooking utensils that are made of heat-conducting materials should have handles made of materials that do not conduct heat well.

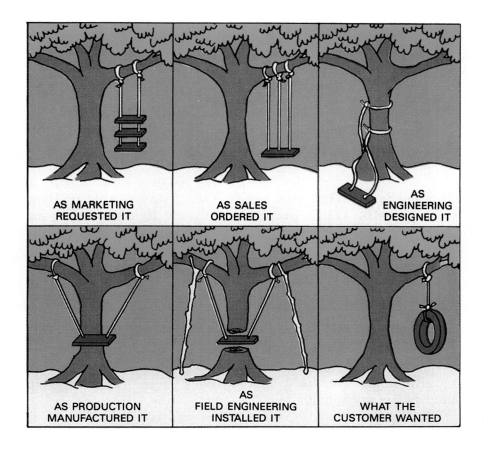

AS MARKETING
REQUESTED IT

AS SALES
ORDERED IT

AS
ENGINEERING
DESIGNED IT

AS PRODUCTION
MANUFACTURED IT

AS
FIELD ENGINEERING
INSTALLED IT

WHAT THE
CUSTOMER WANTED

This cartoon illustrates the need to state the design criteria exactly.

Ergonomics

Ergonomics, also called **human factors engineering**, deals with designing for ease of use by people. Ergonomics combines knowledge of the human body with the techniques of design and engineering. A good design fits the user's size and capabilities. For example, a desk chair must be designed for human comfort. An automobile dashboard must show information in such a way that a driver can read it at a glance. A computer keyboard must be suited to the size and functioning of the human hands and fingers.

When designing for people, the designer must keep in mind that people come in many sizes and shapes. There really is no average person. Most often, designers work in the middle of the range of body sizes. They neglect the bottom 5 percent and the top 5 percent of dimensions. For this reason, a pocket calculator may have buttons too small for people with very large fingers.

Designers depend upon the field of **anthropometry** for information about the size and shape of people's bodies. Anthropometry is the science of measuring people.

This automobile dashboard was designed by human factors engineers and is well matched to human capabilities. All the gauges and instruments are easy to read and in plain sight. (Photo courtesy General Motors Corporation)

These dummies are used to determine how best to prevent injury to humans in the event of an automobile crash. (Courtesy U.S. National Highway Transportation Safety Administration. "Vince & Larry" copyright 1985 by U.S. Department of Transportation. Characters used by permission.)

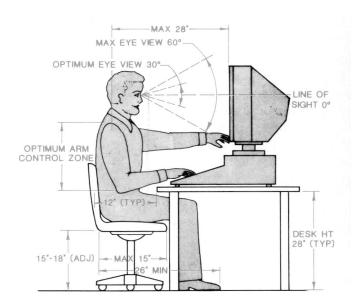

Computer workstations must be designed to be the right size so people are comfortable when working at them. (Reprinted from DRAFTING IN A COMPUTER AGE by Wallach and Chowenhill, © 1989 by Delmar Publishers Inc.)

The wire frame model represents the space within which a person with short arms can comfortably reach. (Courtesy of Eastman Kodak Company)

Aesthetics

Aesthetics refers to the way something looks and how that affects people's feelings. Think about how you are affected by the appearance of a certain car or particular kind of clothing. Designers know that to sell a product, they must make sure people like the way it looks.

The **form** of an object is often determined by its **function**. Tables hold plates of food, so they are flat. Because drinking glasses are usually held in one hand and contain liquids, they are higher than they are wide. It would make no sense to design tables with slanting tops or drinking glasses too wide to grasp. Can you think of other examples of objects in which **form follows function**?

Designers must consider functionality, quality, safety, ergonomics, and aesthetics.

COMMUNICATING A DESIGN

Eight elements help to communicate designs effectively. These elements are shape, line, texture, color, proportion, balance, unity, and rhythm.

The **shape** of an object is defined by its outline. Shapes can be those of regular geometric figures like squares, cubes, rec-

CONNECTIONS: Technology and Mathematics

- In order to stay in business, corporation managers need to decide how much of each product to make so that they can earn the greatest profit. They need to keep in mind their limited resources of personnel, materials, and time. Linear programming is an organized method of mathematical modeling and decision making. It uses equations and inequalities to help managers predict the results of alternative solutions so they can find the best solution.

- Sketching diagrams often help us to see relationships in geometry.

  Example 1 Find the area of the largest square that can be *inscribed* (drawn inside) in a 2-inch-diameter circle.

  Example 2 How many square inches of waste metal will there be when the largest circle is cut from a 12-inch square?

  Example 3 The length and width of a patio is doubled. How many times as large is the new patio compared to the old one?

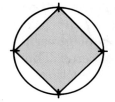
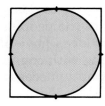

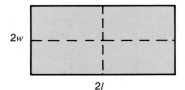

tangles, circles, triangles, pyramids, and polygons. Designs can be irregular shapes such as those found in nature, or combinations of regular and irregular shapes.

Line is often used to define shape. Lines put boundaries around space and form objects. Lines can lead the eye from one place to another, creating a sense of movement within the design. Lines may be of different thicknesses and lengths, and can be straight or curved.

Texture refers to the way the surface of an object looks and feels. Different materials have different textures. Some are rough, some are smooth. Texture is often used as a surface finish to improve the way something works. For example, a rough leather finish on an automobile steering wheel provides a good grip for the driver and makes driving safer. A smooth finish on a countertop makes for easier cleaning.

Color makes designs exciting, and more interesting. Color also produces some psychological effects. For example, blue colors make an object appear colder; red colors make an object seem warmer. Colors can relax us or upset us. Colors can make objects appear larger or smaller, closer or farther away, and lighter or heavier.

Proportion involves the sizes within the design. Certain proportions are pleasing to the human eye. The **Golden Rectangle**, discovered by the Greeks, has a pleasing shape. It is 3 units by 5 units.

Balance refers to the way the various parts of the design relate to one another. Some designs have **symmetry**. That is, all the parts of the design on one side of the center of the design have corresponding parts (a mirror image) on the other side of the center of the design.

Unity refers to the way in which all parts of a design look like they belong together and produce a single general effect.

Rhythm, as in music, has to do with movement. In the case of a design, it relates to the way the eye of the viewer moves around the entire work.

Eight communication elements are shape, line, texture, color, proportion, balance, unity, and rhythm.

CONNECTIONS: Technology and Social Studies

- The way a product is packaged can help gain the attention of potential purchasers. Packaging engineers try to design the best way to present an item so that it will attract the people's interest.
- Historically, differences in political ideas have often kept governments from cooperating when dealing with common concerns. This has led to a duplication of efforts and unnecessary competition in the attempt to solve society's problems.

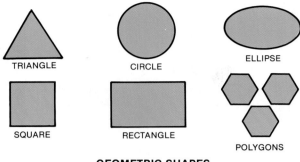

GEOMETRIC SHAPES

Yellow, magenta, and cyan are three process colors used in printing; process colors can be combined in different ways to make all other colors. (Reprinted from PRINTING TECHNOLOGY, 3rd Edition, by Adams, Faux, and Rieber, © 1988 by Delmar Publishers Inc.)

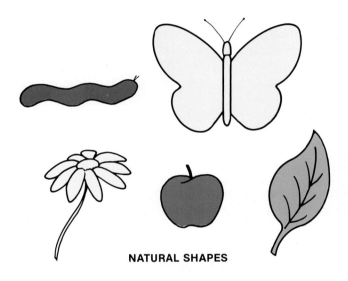

NATURAL SHAPES

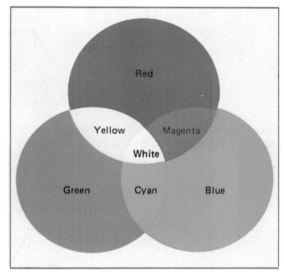

Various geometric and natural shapes.

GRAPHIC COMMUNICATION METHODS

Before a design becomes a product, it must be communicated to the people who will produce it. **Graphic** techniques are most often used for this kind of communication. These techniques include **sketching, freehand drawing, technical drawing**, and **computer-aided drafting**. Pictures are better than words when it comes to turning a design into the real thing.

Sketching and Freehand Drawing

Sketching is a way of beginning to draw designs. Sketches are simple drawings, done quickly. They give only the basic outline and a few details. Sketching is usually done **freehand**, without drawing tools or instruments.

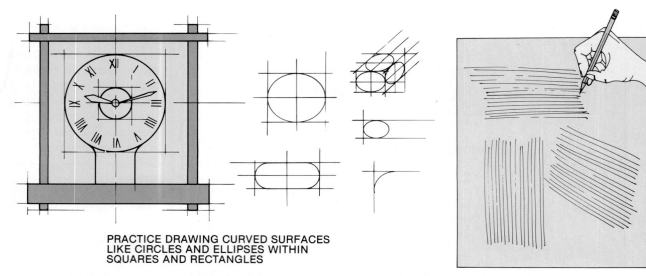

PRACTICE DRAWING CURVED SURFACES
LIKE CIRCLES AND ELLIPSES WITHIN
SQUARES AND RECTANGLES

Most objects can be sketched using geometric shapes as the basis. Notice
how curved surfaces are drawn within squares or rectangles.

Practice drawing horizontal,
vertical, and diagonal lines.

Try doing some exercises that will help you learn to sketch. Draw about 20 parallel horizontal lines on a piece of paper. Then draw some vertical lines and some diagonal lines. Try to move your whole arm when you draw. Now draw some geometric shapes like squares, rectangles, and triangles. After you draw a square, draw a circle inside it. Draw a rectangle with an ellipse inside it. Most objects you might try to draw can be made by combining these shapes. The technique of working from box-like geometric shapes is called **crating**.

Technical Drawing

Drawings are done with more care than sketches. They are usually more detailed. An artist can do fairly accurate freehand drawings. To make more precise drawings, **technical drawing** is used. This kind of drawing is carried out using special instruments and tools. Artists, engineers, architects, designers, and drafters do technical drawing.

Technical drawing can be done manually, or with computers. To do technical drawing manually, a drawing board, a T square, plastic triangles, paper, and a pencil are used.

Technical drawing is done with special instruments and tools.

The T square is held tightly against the edge of the drawing board. It is moved up or down so horizontal lines can be drawn that are exactly parallel to each other. The plastic triangle is placed against the long edge of the T square and used as a guide to draw vertical lines or lines at an angle. These instruments can be used to draw objects very realistically.

When we draw flat shapes, we draw in two dimensions. These dimensions are usually length and width. Two-dimensional drawings let us look at the front, side, top, or bottom of an object. They are used when important details in one view of an object must be shown. However, 2-D drawings do not give a realistic picture of what the object looks like.

Three-dimensional drawings are more realistic than two-dimensional drawings. They are called **pictorial drawings** because they give a clear picture of the object. The three dimensions in 3-D drawings are length, width, and height.

There are three kinds of pictorial drawings. These are **isometric**, **oblique**, and **perspective** drawings.

An **isometric** drawing is drawn within a framework of three lines, called an isometric axis. These three lines are like the edges of a cube. The two base lines of the axis are drawn at an

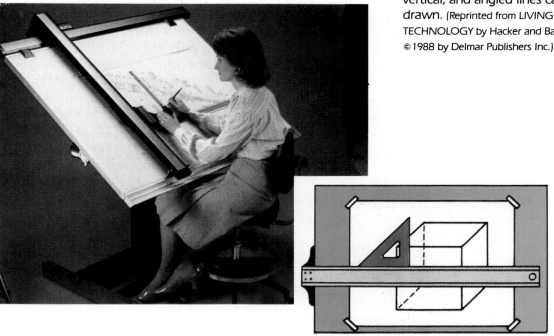

a) In technical drawing, a relatively small number of tools are used to produce complex drawings. (Courtesy of Keuffel & Esser Company, Parsippany, NJ 07054)

b) By moving the T-square and triangle on the drawing board, horizontal, vertical, and angled lines can be drawn. (Reprinted from LIVING WITH TECHNOLOGY by Hacker and Barden, © 1988 by Delmar Publishers Inc.)

An example of one-point perspective. (Courtesy of Association of American Railroads)

angle of 30 degrees to the horizontal. An **oblique** drawing shows one surface of an object with a straight-on view. The other two surfaces are shown at an angle. A **perspective** drawing is the most realistic. In a perspective drawing, parts of the object that are farther away appear smaller.

Rendering is a way of making drawings look more realistic. Rendering may be done by shading and using texture. **Shading** is using lighter and darker tones to show how light falls on an object. The light and dark parts of the drawing make it appear more like the real object.

Surface **textures** can be shown by drawing crossed lines (crosshatching), dots, small circles, and narrow and widely spaced lines. Dots can be used to show a rough surface such as sandpaper. Small circles can be used to show leather.

Color can be used to enhance drawings and designs. To make a design stand out, a designer might use a splash of color as a

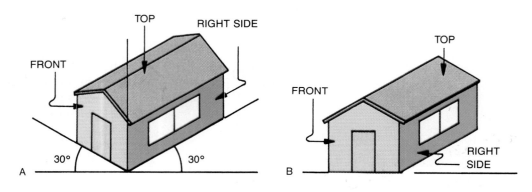

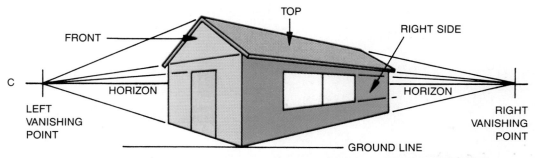

a) A three-dimensional view (an isometric drawing) of a house.
b) In an oblique drawing, the object is drawn with a straight-on view of one surface.
c) A perspective drawing makes an object look natural. If the horizontal lines on the drawing are extended, they will meet at two points called vanishing points. This is an example of two-point perspective. (Reprinted from LIVING WITH TECHNOLOGY by Hacker and Barden, ©1988 by Delmar Publishers Inc.)

background. To make a design look more realistic, colors can be used to highlight the object. Colors can be used in different shades, giving the impression of light and dark.

Color can be applied with felt-tipped pen, magic markers, colored pencils, or watercolor paints. Special kinds of paper must be used with some paints, such as watercolors.

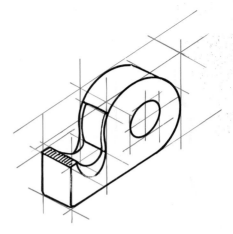

Try sketching some common objects in 3D.

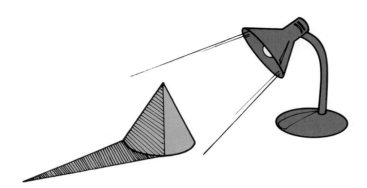

Rendering by shading objects and adding shadows.

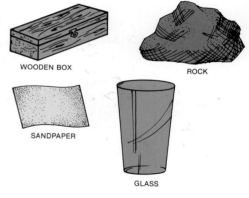

WOODEN BOX

ROCK

SANDPAPER

GLASS

Rendering a drawing by adding texture.

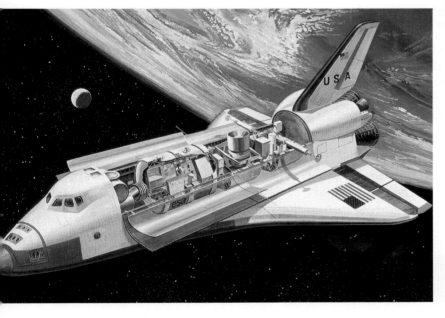

Color is used to make the drawing more visually appealing and realistic.
(Courtesy of NASA)

COMPUTER-AIDED DRAFTING

The computer has made drafting much easier. Computer-aided drafting is a new and important tool for engineers, drafters, and designers. **CAD** stands for computer-aided drafting or computer-aided design. **CADD** stands for computer-aided drafting and design. Such systems are used by drafters, CAD tech-

Graphic techniques such as sketching and freehand drawing, technical drawing, and computer-aided drafting are used to communicate design ideas.

Computer-aided drafting (CAD) uses computers to produce designs and drawings quickly and accurately.

nicians, or engineers, depending on the job to be done.

A CAD workstation has many parts. It includes the computer, keyboard, display screen, drawing tablet, and plotter. A person can draw on the screen by using the keyboard or drawing on the tablet. CAD computer programs make it possible to produce clear drawings of almost anything, such as part of a machine or a whole building. The drawings can be changed using keyboard commands or the drawing tablet. They can be "turned" on the screen to see how they look from different angles. A person can "zoom in" for a closer look at one part of a drawing.

When the design is ready, the plotter produces the actual drawing (hard copy) on paper. At the same time, CAD systems can print out a list of the parts that are needed and their cost. A CAD system can also tell how well a design will work and point out problems.

Using a CAD workstation, a designer can create very accurate drawings that can be stored in a computer's memory. CAD drawings can show mechanical parts, architectural designs, electronic circuits, and many other things.

A CAD system has many advantages over a hand-drawn design.

1. It saves time by combining design and drafting. It is two to eight times as fast in producing designs as hand drawing.
2. Information stored in the computer can help prevent errors in the design.
3. It saves time in changing and improving designs.
4. Designs are more accurate. They are also more consistent from one design to the next when changes must be made.

CAD systems free designers, drafters, and engineers to spend more time thinking about design. They can spend more time creating and less doing actual drawings.

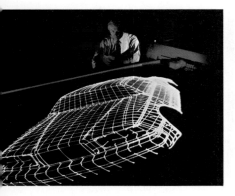

A CAD operator using a CAD system to model an automobile. *(Courtesy of Ford Motor Company)*

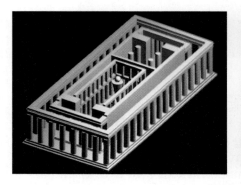

College students developed computerized solid models of the Parthenon in Athens, Greece. (Photos courtesy of Temple University and Structural Dynamics Research Corporation)

I-DEAS

I-DEAS is a computer program that was written by Structural Dynamics Resources Corporation (SDRC) in Milford, Ohio. The program lets people draw 3-D solid models of a design idea on the computer screen.

A solid model helps designers study a design to see how it may best be manufactured. It lets designers ask questions about the way the product will work and what it will be like. They can make any needed changes on the computer screen. They will avoid spending money on costly models only to change them later.

Using the program, designers can carry out an **analysis** to see how the product will perform.

Computer-aided **testing** is also part of I-DEAS. The program can be used to simulate conditions under which the product will be used, and testing will be carried out to see how it performs.

Using this information, engineers can study how a machine will work without building a model.

After design criteria are met, I-DEAS provides a computer-aided **drafting** program, so hard copies can be made of the computer design.

The model and drawings are then used to learn how best to **manufacture** the product.

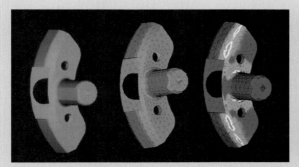

Analysis

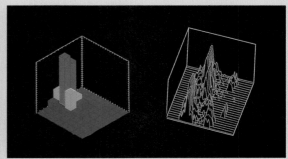

Testing

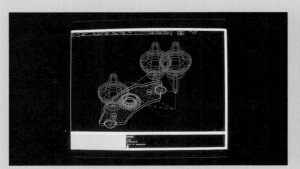

Drafting

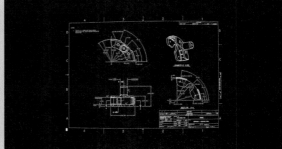

Manufacturing

(All photos courtesy Structural Dynamics Research Corporation)

OTHER WAYS OF COMMUNICATING DESIGN IDEAS

Besides using drawing techniques, you may use other methods to communicate design ideas. For example, photographs and slides of objects can be used instead of drawings. Written or spoken words can be used to communicate an idea, as in the design briefs in the previous chapter, and as in printed advertisements.

Design ideas can sometimes be communicated through an audio or video tape.

Three-dimensional models also can be made to communicate design solutions. These are often used when the designer wants to show a working model.

Modeling techniques are useful problem-solving tools.

Communicating design ideas through photography. The line drawing D shows how the final composite was made. (Photos A, B, and C courtesy of DS America)

A

B

C

COMBINE TREES, WITH SKY FROM SECOND
PICTURE, USING DENSITY DIFFERENCE MASK.

MAKE THE MOUNTAIN APPEAR
IN NATURAL PERSPECTIVE.

REMOVE NUMBERS AND
GIVE NATURAL TONE.

CUT OUT AND MONTAGE CAR
RETAINING SHADOW UNDER
CAR.

REPLACE BRICK BACKGROUND BY
TREE ONE THROUGH CAR WINDOW.

D

Communicating design ideas through text. (Courtesy of Princeton Graphics Systems)

Communicating design ideas through video. (Photo courtesy of Sony Corporation of America)

A computer-generated 3-D model can be used instead of actual models to manufacture items such as a false tooth. (Courtesy of Structural Dynamics Research Corporation)

Modeling Design Solutions

When solving problems, we must develop and test alternative solutions. Sometimes, however, it is costly or dangerous or both to carry out such tests. To test alternatives without trying them, **models** are often used.

Suppose that there are several alternatives for the design of a large power plant. It will take years and millions of dollars to build such a plant. It's not possible, then, to build and test each alternative design. Instead, the planners use models. Models are used to test ideas without risking a great amount of time, capital, or public safety. There are five kinds of models.

1. **Charts** and **graphs** describe how an alternative solution might work.
2. **Mathematical models** show how an alternative will work by use of mathematical equations that predict performance.
3. **Sketches, illustrations, and technical drawings** show the ideas in picture form so they can be understood by others. Drawing a design often brings up ideas to improve it. The same thing often happens when you discuss the design with others.
4. **Working models** show how an alternative would work. A working model can be partly functional (only part of the idea is modeled) or fully functional. They can be made of the material that will be used, or of a different, more easily worked material. They can be full size or made to scale (larger or smaller than the alternative).
5. In **computer simulation**, a computer does mathematical modeling. The computer may display a picture of the idea on the screen. Computer simulation is most useful when a large number of calculations must be carried out.

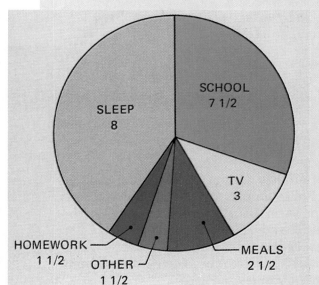

This pie chart describes the number of hours a student spends on various activities. (Reprinted from LIVING WITH TECHNOLOGY by Hacker and Barden, © 1988 by Delmar Publishers Inc.)

SCHOOL 7 1/2

SLEEP 8

TV 3

HOMEWORK 1 1/2

OTHER 1 1/2

MEALS 2 1/2

$$V = C \log_e \frac{M_o}{M_t}$$

V = ROCKET VELOCITY
C = EXHAUST VELOCITY
M_o = INITIAL MASS
M_t = MASS AT TIME "t"

This equation predicts the velocity of a rocket during flight.

Men creating a model of an automobile. (Photo courtesy of General Motors)

An example of a drawing used to convey ideas. (Reprinted from MECHANICAL DRAFTING by Madsen, Shumaker, and Stewart, © 1986 by Delmar Publishers Inc.)

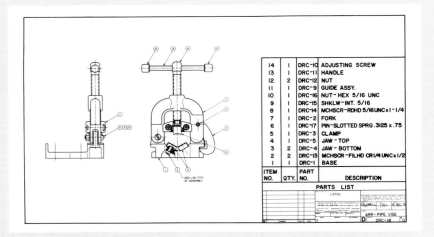

This model tests wind effects on the city of Boston, Massachusetts. Engineers used data from the model to design buildings that would not create harmful wind effects in the city. (Reprinted from TECHNIQUE magazine, Summer 1985, with permission. Copyright Data General Corporation)

A computer simulation used to model the "blind spot" experienced by the driver of a 120-ton truck. (Courtesy Komatsu Presser and Structural Dynamics Research Corporation)

CONNECTIONS: Technology and Science

■ There are a variety of types of models. Scientists often use models to explain their observations. For example, the model showing that matter is composed of tiny particles can be used to explain the differences among solids, liquids, and gases; in solids the particles are closest together, and in gases they are farthest apart. Models are quite useful in helping scientists to explain their observations and make predictions of future outcomes.

SUMMARY

Designing is a process people use to plan and produce processes, products, and systems that meet specified criteria. When technologists design, they must consider funtionality, quality, safety, ergonomics, and aesthetics.

Eight elements of communication are used to develop effective designs. These elements are shape, line, texture, color, proportion, balance, unity, and rhythm.

To communicate designs, people use graphics. These include sketching and freehand drawing, technical drawings, and computer-aided drafting. Sketches and freehand drawings are simple, give few details, and are done quickly.

Technical drawings are done with more detail and care than sketches. Instruments and tools are used to do technical drawings. Drawings are two- or three-dimensional. Two-dimensional drawings provide views of the front, side, top, or bottom of an object. Three-dimensional drawings are more realistic. Rendering and using color make drawings look more real. Rendering includes shading, adding shadows, and using texture.

Computer-aided drafting (CAD) is designing or drawing on the computer screen. A CAD system uses people, software, and hardware. CAD saves time, improves accuracy, and makes changing a design faster and easier.

Models are used to test ideas. Five kinds of models are: charts and graphs; mathematical models; sketches, illustrations, and technical drawings; working models; and computer simulations.

1. What is designing?
2. What are five criteria people must consider when designing?
3. Describe ergonomics and aesthetics and explain why they are important in designing for people.
4. Explain how texture and color can make designs more effective.
5. Make a technical drawing of a cube three inches on a side.
6. Trace the outlines of a car or an article of clothing in a magazine photograph. Render it to make it appear more realistic, using shading, shadows, and texture. Use color if you wish.
7. How has CAD affected the fields of drawing and design?
8. Why is modeling useful? Describe four modeling techniques.

(Courtesy Nike Inc.)

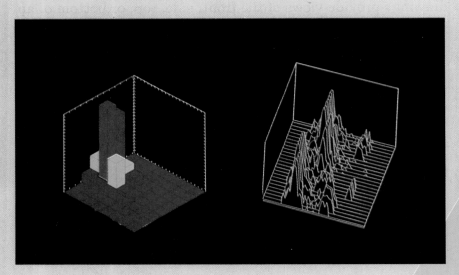

(Courtesy Structural Dynamics Research Corporation)

(Courtesy U.S. National Highway Transportation Safety Administration. "Vince & Larry" copyright 1985 by U.S. Department of Transportation. Characters used by permission.)

CHAPTER KEY TERMS

Across

2. A design consideration which reduces danger.
3. The quality of an object that is defined by its outline.
5. The specifications and requirements that a solution to a problem must fulfill.
8. A planned human activity intended to accomplish a desired result.
9. Involves the sizes within the design; a ratio.
12. This is used to define shape, to put boundaries around space, and form objects.
14. Refers to a product, system, or process fulfilling its intended purpose.
15. The product, system, or process must be designed to meet certain minimum standards of _____ .
16. Human factors engineering.

Down

1. Refers to the way the various parts of the design relate to one another.
4. Relates to the way the surface of an object looks and feels.
6. Has to do with the way something looks.
7. Limitations that form boundaries around the possible solutions to a problem.
10. The way in which all the parts of a design produce a general effect.
11. As in music, has to do with movement. The way the eye moves around the entire work.
13. One of the eight design elements that can produce psychological effects such as making an object appear colder or seem warmer.

Aesthetics
Analysis
Anthropometry
Balance
CAD
CADD
Color
Computer-aided drafting
Constraints
Design
Design criteria
Ergonomics
Functionality
Isometric
Line
Oblique
Perspective
Pictorial drawings
Proportion
Quality
Rendering
Rhythm
Safety
Shading
Shape
Sketching
Technical drawing
Testing
Texture
Unity

SEE YOUR TEACHER FOR THE CROSSTECH SHEET

ACTIVITIES

BEING A GRAPHIC DESIGNER

Problem Situation

The elements of good design are often used to communicate ideas to others. Architects, engineers, product designers, and graphic designers are just a few of the skilled people who incorporate the elements of good design into their work. Whether for a building, bridge or package, the elements of shape, line, texture, color, proportion, balance, unity, and rhythm help designers communicate effectively.

Design Brief

You are a graphic designer working for an advertising agency. You have been placed in charge of a new account and made responsible for developing a new line of wristwatches for a client. The watches will be used as promotional giveaways and must include the company's logo on the face. The company's product line is athletic shoes. You must stay within the following design criteria:

- your design must include the face of the watch and the watchband
- you must submit three different designs for the project
- the design may or may not include hour markings
- the company's logo must appear in each design

Design and develop a company logo and three watch designs. Incorporate into your design the eight elements of good design. Use a variety of materials in the development of your watch design presentation. Submit all sketches and drawings.

Suggested Materials

Drafting materials
Drawing papers
Color markers
Watercolors
Construction paper
Transfer type
Scissors
Rulers
Adhesives
Modeling materials

Procedure

1. Research some of the design qualities of famous modern logos. Identify the various principles of good design that are contained within each.
2. Develop a logo and company name that reflects the company's product line.
3. Develop several freehand sketches of possible designs. Incorporate the logo into the watch designs.
4. Select three of your best designs.
5. Using appropriate materials for your design, develop your three finished designs for presentation.

Classroom Connections

1. Design is the process of planning a desired result in the form of a product.
2. Designers must use their knowledge, experience, and creativity to develop new and inventive ideas.
3. The design elements of shape, line, texture, color, proportion, balance, unity, and rhythm help to communicate design ideas effectively.
4. Skills such as sketching and drawing allow the designer to better communicate a client's product in an advertisement.
5. Basic math skills such as measuring, addition, subtraction, and multiplication are necessary to calculate proportions and layout space.

Technology in the Real World

1. The use of specially designed computer software and hardware has changed how graphic designers prepare their work. Computer programs allow designers to draw, sketch, color, and shade designs on the computer screen. Designers can rotate and mirror geometric shapes, and they can also create irregular designs.

Summing Up

1. Briefly describe the qualities of each of the eight elements of design.
2. Show how lines and shading can be used to indicate the texture of an object.
3. Give examples of how the eight elements of design are incorporated into a design of a skyscraper.

ACTIVITIES

DESIGNING WITH PEOPLE IN MIND

Problem Situation

It seems each day new high-tech products find their way into our homes and workplaces. It is the responsibility of the product designer to make these products easy and comfortable to use. Products that are too hard to handle or operate are not matched well to the human user. Have you ever sat in a poorly designed chair that hurts the middle of your back? This chair shows poor ergonomics. Ergonomics, or human factors engineering, matches the human user to the design of the product.

Design Brief

You are a product designer working in the human factors department of a large television manufacturing company. You have been asked to design the remote control unit for the company's new line of televisions. The following are your design criteria:

- the user must be able to operate the remote control with one hand
- the remote must fit comfortably in an average size hand
- the remote must have a high-tech appearance to match the new line of televisions
- the remote must contain the following controls:
  - on/off
  - volume up/down
  - channel scan up/down
  - mute
  - individual channel buttons
  - clock set controls
- each control must contain a symbol identifying its use

Design and then build a full-size model of the remote control unit. Include in your presentation: sketches, mechanical drawings, and the final model.

Procedure

1. Gather information about the size, shape, and grip of people's hands. This can be accomplished by making impressions in clay and other materials.
2. Develop some freehand sketches, combining the elements of your design criteria with information you gathered about the human hand.

Suggested Resources

Drafting equipment
Drawing papers
Colored markers
Cardboard
Foam board
Transfer type
Scissors
Steel rule
Utility knife
Adhesives
Construction paper
Styrofoam
Balsa wood
Modeling tools

3. Select a design that meets these criteria and develop detailed mechanical drawings showing the shape, size, and placement of the buttons and controls on the remote.
4. Select appropriate modeling materials and build a full-size model of the remote from your detailed drawings.
5. Test your product design using the other students in your class for evaluation. Record their comments and suggestions.
6. Make modifications in your design if needed.

Classroom Connections

1. Ergonomics, or human factors engineering, deals with designing products that meet people's size and capabilities.
2. Designers must stay within the design criteria and constraints that limit their design options.
3. Designs are communicated to the people who will produce them through freehand drawings and detailed mechanical drawings.
4. Three-dimensional models can communicate design solutions effectively.

Technology in the Real World

1. Many computer-aided drafting (CAD) systems and computer-integrated manufacturing (CIM) systems can create three-dimensional views of products on a computer screen. These computer models can then undergo computer simulated testing and evaluation.
2. Automobile manufacturers develop full-scale models of new automobiles made from clay and plastic. These models are used for testing and evaluation of the design.

Summing Up

1. What are some of the graphic communication techniques a designer may use to communicate ideas to others?
2. What are the advantages that building a working model may offer in the design of a product?
3. What qualities does your remote control model have that demonstrate good ergonomic design?
4. What were some of the design constraints you had to work within when developing your remote model?

CHAPTER 4

TECHNOLOGICAL RESOURCES

MAJOR CONCEPTS

- Every technological system makes use of seven types of resources: people, information, materials, tools and machines, energy, capital, and time.
- Since there is a limited amount of certain resources on the earth, we must use these resources wisely.
- To choose resources wisely, people must understand the uses and limitations of each resource.
- When we choose resources, we must sometimes make trade-offs in order to reach the best possible solution.
- To solve technological problems, we choose resources from each of the seven categories of resources.
- Important factors in choosing resources are cost, availability, and appropriateness.

TECHNOLOGICAL RESOURCES

Resources are things we need to get a job done. Think about hamburgers. McDonald's restaurants have sold almost 100 billion of them. McDonald's uses resources to cook and serve its hamburgers.

What resources are needed? First, we need **people**. To make hamburgers, we need people to raise and butcher cattle, to grind up the meat, and to cook the hamburgers.

We need **information**. We must know how to feed and take care of cattle, how to keep the meat fresh, how to cook it. We need to know how to manage a restaurant, schedule workers in shifts, and keep an inventory of needed supplies.

We need **materials**. The materials we process to make a hamburger are meat, spices, and seasoning.

We need **tools**. Some of these tools are a stove and cooking utensils.

We need **energy**. It will take energy to cook the meat. McDonald's uses electricity, but at home we might use gas or charcoal to do the job.

We also need to buy the meat and the tools, and pay the people. We need money. Money is a form of **capital**.

Finally, **time** is a resource, too. Cooking a hamburger takes time.

You can see that making hamburgers requires the use of seven kinds of resources. They are: people, information, materials, tools and machines, energy, capital, and time. The same seven resources are needed to build a skyscraper or make a jet plane.

Every technological process involves the use of these seven basic kinds of resources. Whether it's plowing a field, cooking a hamburger, or using a telephone, resources from all seven categories are needed.

Every technological system makes use of seven types of resources: people, information, materials, tools and machines, energy, capital, and time.

People eat the hamburgers and create a demand for the product. (Courtesy of McDonald's Corporation)

We need information about how to keep cows healthy. (Courtesy of Cetus Corporation)

Tools are an important resource for preparing food. (Courtesy of McDonald's Corporation)

Problem Situation 1:   Wind chimes can be hung outside a home, or near an open window. They make lovely musical sounds when the wind blows.

Design Brief 1:   Design a set of wind chimes that are made from materials that have good acoustic properties.

Problem Situation 2:   Materials are chosen for certain products based upon their physical properties.

Design Brief 2:   Design a system for testing the compressive strength of various pieces of wood.

Problem Situation 3:   Clocks, sundials, and sand glasses are all used to measure time. Imagine that you are marooned on a desert island where there are plenty of sea birds' eggs but no timing device.

Design Brief 3:   Design and construct an aid that will enable you to measure a time span of about four minutes to cook an egg. (If possible the device should make a noise when the time is up!)

People's needs for exercise have caused the development of clothing that makes running easier. (Courtesy of NASA)

PEOPLE

Let's take a closer look at the seven technological resources. First, people are needed. Technology comes from the needs and wants of people. People create technology and people consume its products and services.

It is people whose needs bring about technology in the first place. People have many needs that are filled by technology, as you learned in Chapter 1. Sometimes, governments start programs that make technology grow. For example, the Soviet Union sent the first satellite, *Sputnik*, into space in 1957. The United States government decided to match this achievement. In 1958, NASA (National Aeronautics and Space Administration) was created to direct the space program. In 1969, U.S. astronauts landed on the moon.

It is people who use what they know and continue to learn more who make and develop technology. NASA scientists had to combine their knowledge with new ideas to come up with a space vehicle and a way to get it to the moon and back safely.

Of course, people provide the labor on which technology depends. Many workers are needed to provide the products and services we use every day.

People are also the consumers of technology. They eat the hamburgers, drive the cars, watch the television sets, and travel the roads and airways.

INFORMATION

Technology requires information. We need to know what to do and how to do it. Technology has grown quickly during the last few decades because of an explosion of information. Information is now doubling every five years. It is shared throughout the world because of new and better ways to communicate.

We use information in many different ways. A surgeon must know what tools to use during an operation. A farmer must know what corn will grow best in local soils. A factory worker must know how to operate a machine.

Everyone in our technological world uses information. Information begins as **data** (raw facts and figures). Data are then collected, recorded, classified, calculated, stored, and retrieved. They become information. **Data processing** is the act of turning data into information.

Information can be found in many places—in computer files, books, films, and museums, to name a few. But information is not valuable until we make use of it. We process information by collecting it, thinking about it, and applying it to meet our needs and wants.

(Courtesy of NASA)

MATERIALS

When people hear the word "resources," they think of materials first. Materials are a very important resource for technology. **Natural resources** are materials that are found in nature. These include air, water, land, timber, minerals, plants, and animals. Natural resources that will be used to make finished products are called **raw materials**.

Countries that are rich in natural resources have lots of raw materials. The United States is rich in some natural resources but must import others. We have a great deal of timber, oil, coal, iron, and natural gas. However, we must import most of

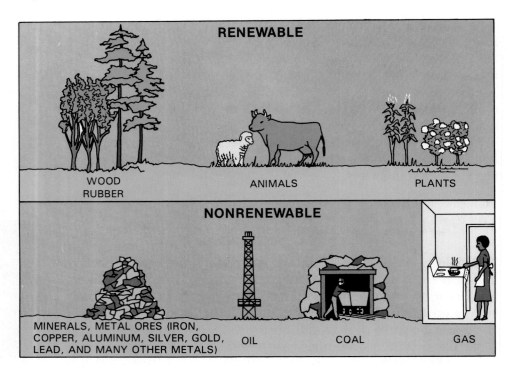

Renewable and nonrenewable raw materials.

our chromium, platinum, and industrial diamonds from South Africa. Our nickel comes from Canada and Algeria. Our cobalt is imported from Africa and Europe. Our aluminum comes from South America and Australia.

Raw Materials

There are two kinds of raw materials: **renewable raw materials** and **nonrenewable raw materials**. Renewable raw materials are those that can be grown and therefore replaced. Wood is a renewable raw material. Natural rubber comes from the tree *Hevea brasiliensis*. It is found in South America and the Far East. It, too, is a renewable raw material. Animals and plants are renewable resources.

Nonrenewable raw materials cannot be grown or replaced. Oil, gas, coal, and minerals are nonrenewable. Once we use up our supplies of these resources, there are no more.

Limited and Unlimited Resources

Since there is a limited amount of certain resources on the earth, we must use these resources wisely.

Some resources are available in great amounts, like sand, iron ore, and clay. Others are in short supply. When we can, we use plentiful materials instead of scarce ones. Fresh water is a resource that is scarce in some places. Some people think that the future may continue to bring a shortage of water.

New synthetic fibers provide bright colors and exciting styles for clothing. (Courtesy of DuPont Company)

The largest fabric structure in the world is the Haj Airline terminal in Saudi Arabia. (Courtesy of OC Birdair)

Sails made from synthetics (Mylar™ and Kevlar™) helped speed the yacht *Freedom* to victory in the America's Cup races. (Courtesy of DuPont Company)

Synthetic Materials

People have long used technology to make substitutes for some resources. Materials made in the laboratory are called **synthetic materials**. Many everyday materials are synthetics. Plastics like acrylic, nylon, and Teflon™ are not found in nature. They are made from chemicals. Industrial diamonds are synthetics. So are Dacron™, rayon, and fiberglass.

Many synthetics are less costly and more useful than natural materials. They can also be made stronger, lighter, and more long-lasting than the materials they replace. For example, we have glass that conducts electricity, plastics that last longer than metal, and fabrics that repel water. Synthetics can also be used in place of scarce materials, helping to save our natural resources.

TOOLS AND MACHINES

People have been using tools for more than a million years. Tools include hand tools and machines. Tools extend human capabilities. Some of them let us do jobs faster and better. Others let us do jobs we couldn't do at all without them.

Although some forms of animal life can use tools, only human beings can use tools to make other tools.

We use tools to fix things around the house. (Photo by Michael Hacker)

We use kitchen tools to prepare food. (Photo by Michael Hacker)

85

(Courtesy of Stanley Tools, Division of The Stanley Works, New Britain, CT 06050)

Hand Tools

Hand tools are the simplest tools. Some examples are screwdrivers, saws, hammers, and pliers. Human muscle power makes these tools work. They extend the power of human muscle.

Machines

Machines change the amount, speed, or direction of a force. Early machines were mechanical devices. They used the principles of the six **simple machines**: lever, wheel and axle, pulley, screw, wedge, and inclined plane. Early machines used human, animal, or water power.

Many modern machines have moving mechanical parts. Some, such as televisions and stereos, use electrical energy. Some machines use electricity to move mechanical parts (for example, those that have electric motors). These machines are called **electromechanical devices**. A robot is such a device.

Automatic machines do not need people to operate them. They must only be started and watched by workers to make sure they are working properly.

Electronic Tools and Machines

Some electronic tools are used for testing electrical circuits. The computer is an electronic tool. Computers are used to pro-

An electric sewing machine is an example of an electromechanical device. (Photo by Michael Hacker)

All this early juke box needs is for the human to put two cents in the slot. Quite a bargain! (Courtesy of the Smithsonian Institution)

This welding robot can follow a seam between two pieces of metal. (Courtesy of General Electric Research and Development Center)

This computerized system controls the flow of over one-half million gallons of fuel oil through pipelines in Texas. (Courtesy of DuPont Company)

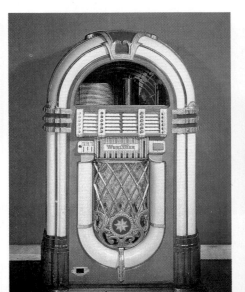

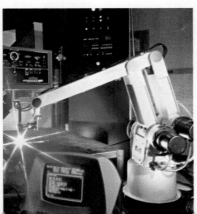

Six Simple Machines

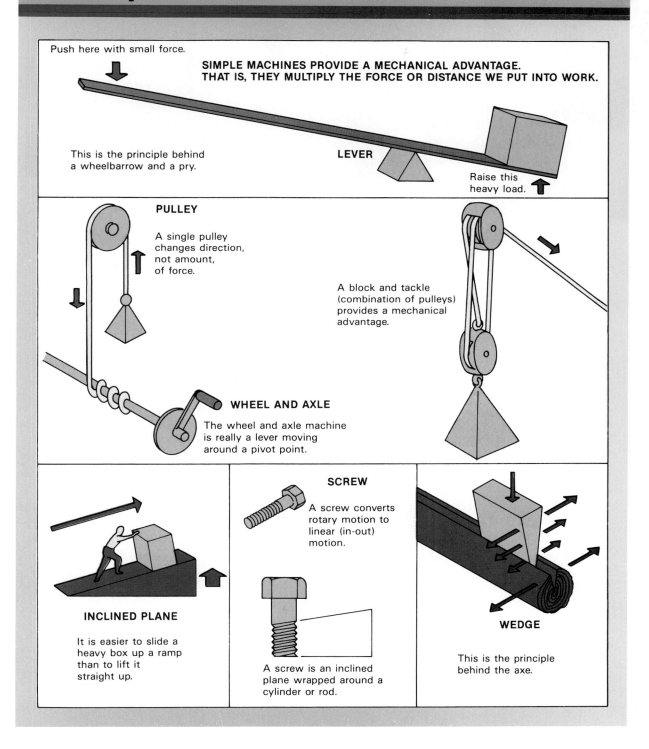

Push here with small force.

SIMPLE MACHINES PROVIDE A MECHANICAL ADVANTAGE. THAT IS, THEY MULTIPLY THE FORCE OR DISTANCE WE PUT INTO WORK.

This is the principle behind a wheelbarrow and a pry.

LEVER

Raise this heavy load.

PULLEY

A single pulley changes direction, not amount, of force.

A block and tackle (combination of pulleys) provides a mechanical advantage.

WHEEL AND AXLE

The wheel and axle machine is really a lever moving around a pivot point.

SCREW

A screw converts rotary motion to linear (in-out) motion.

INCLINED PLANE

It is easier to slide a heavy box up a ramp than to lift it straight up.

A screw is an inclined plane wrapped around a cylinder or rod.

WEDGE

This is the principle behind the axe.

Hot laser pulses cause a metal surface to soften and vaporize. (Courtesy of General Electric Research and Development Center)

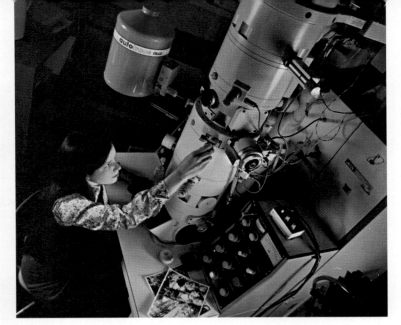

This powerful microscope can magnify the surface of a piece of metal as much as 140,000 times. (Courtesy of General Electric Research and Development Center)

cess information. They are also used to run factory machinery. Computers are used in many machines, saving greatly on energy and labor.

Optical Tools

Some optical tools extend the power of the human eye. Lenses magnify objects, making them easy to see and study. Microscopes and telescopes are optical tools. Today, optical tools help us learn more about the genes that direct human growth and functions.

Another optical tool is the **laser**. Lasers send very strong bursts of light energy. The light energy can be used to measure, cut, and weld materials. It can also be used to send messages over long distances. Lasers are accurate tools. They can be used to cut through pieces of metal, or repair damage in a human eye.

ENERGY

The United States uses a huge amount of energy. Energy is used to make products, to move goods and people, and to heat, cool, and light the places where people work and live. The United States has only about 6 percent of the world's population. But it uses about 35 percent of the energy used in the world. Some energy resources are in great supply. Others are in limited supply and can get used up.

Renewable and Nonrenewable Energy Sources

Renewable energy sources are those that can be replaced. Human and animal muscle power are renewable. So is the energy we get from burning wood, and from the sun. Gravitational energy and geothermal energy are renewable.

Nonrenewable (finite) energy sources are those that can be used up. Some of them are coal, oil, natural gas, and nuclear fission. Most of the energy we use comes from nonrenewable energy sources.

SOURCES OF ENERGY

Many forms of energy start with the sun. The source of all food is plants, which use sunlight and carbon dioxide for growth. People and animals get their energy from the food they eat. Coal, oil, and gas come from decayed plant and animal matter. The heating of air masses by the sun causes winds.

(Courtesy of Tennessee Valley Authority)

Where the United States Gets Its Energy

Most of our energy comes from non-renewable resources. Because of this, we are experimenting with new sources of energy like fusion, solar, wind, and water.

Cogeneration is a way to use energy from the byproducts of technology. Heat generated in factories is used to heat water and make steam, and the steam is used to make electricity. Wood chips from paper mills and other factory wastes are burned to create heat energy that is turned into electricity.

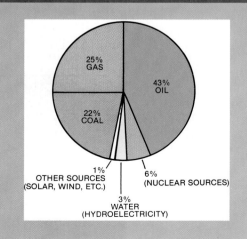

25% GAS

43% OIL

22% COAL

1% OTHER SOURCES (SOLAR, WIND, ETC.)

6% (NUCLEAR SOURCES)

3% WATER (HYDROELECTRICITY)

There are six different kinds of energy sources. They are: human and animal muscle power, solar energy, gravitational energy, geothermal energy, chemical energy, and nuclear energy.

Human and Animal Muscle Power

Early people used their own muscle power to get things done. Later, people learned to work together to do tasks that required larger amounts of energy.

People soon learned to use donkeys, oxen, and other animals to do work. Harnessed to machines, these animals could plow the land, pump water, and do other jobs. In Egypt, Greece, Rome, and Mexico, there are large stone buildings that were built thousands of years ago. Human and animal power hauled and lifted the stone blocks from which these buildings were made. In some parts of the world, human and animal power are still important sources of energy.

CONNECTIONS: Technology and Mathematics

- A small amount of mass can produce a large amount of energy. This is because the amount of energy (E) produced is equal to the mass (m) times the square of the speed of light (c). (To square means to multiply a number by itself.) How much energy would 1 gram of mass produce?

$$E = mc^2$$
$$E = 1(186,000)^2$$
$$E = 1(186,000)(186,000)$$
$$E = 34,596,000,000 \text{ units of energy}$$

- A cord of wood (a pile $4' \times 4' \times 8'$) can provide about the same amount of heat as 200 gallons of oil. About how many gallons of oil would a pile that is twice as long, twice as wide, and twice as high be equal to? The *volume* of each pile can be calculated.

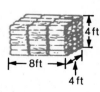

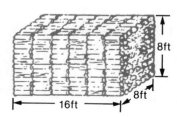

$$V_1 = l \times w \times h \qquad V_2 = l \times w \times h$$
$$V_1 = 8 \times 4 \times 4 \qquad V_2 = 16 \times 8 \times 8$$
$$V_1 = 128 \text{ cubic ft.} \qquad V_2 = 1024 \text{ cubic ft.}$$

The second pile has a volume 8 times as great as the first pile, and will provide about the same amount of heat as 1600 gallons of oil.

In China, human power is an important source of energy. Bicycles outnumber private automobiles by almost a million to one. (UN Photo 152,715/John Isaac)

Solar Energy

Every fifteen minutes, the sun produces enough energy to meet the world's energy needs for one year. We can now use and store **solar energy** by converting it into electricity. Solar cells, or **photovoltaic cells** or **photocells**, are used to do this job. Solar energy is renewable and nonpolluting.

NASA is developing this Power Extension Package (PEP), which will use a large array of solar cells to convert sunlight into electrical energy. (Courtesy of NASA)

An aerial view of a solar electricity-generating plant. (Courtesy of Southern California Edison)

The Solar Challenger's only power source was sunshine. About 16,000 photovoltaic cells converted the sun's rays into electricity to power the motor. (Courtesy of DuPont Company)

(Courtesy of the U.S. Department of Energy)

This Darrieus windmill looks like an eggbeater. Wind blowing from any direction makes the blades turn. (Courtesy of the Department of Energy)

Wind energy is another form of solar energy. Wind is air that moves because of the sun's unequal heating of the earth's surface. The wind has been used as a source of energy for thousands of years. Its earliest use was to power sailing ships. In some places (like the Far East), sailing ships are still used to transport goods and people, and to harvest food from the oceans.

People have long used machines that harness the wind to pump water and grind grain. Today, windmills are used in some places to produce electricity.

Gravitational Energy

Gravity makes water flow downhill. In the past, water wheels used the energy of falling water to grind grain and mill flour. Today, falling water turns **turbines** at hydroelectric plants on dams. These turbines produce **hydroelectricity**. Gravity also causes tides in oceans and rivers. People are experimenting with **tidal energy** to learn whether it, too, could be a useful source of energy.

Geothermal Energy

Rock beneath the earth's crust is so hot that it can flow. In some places, it flows to the surface, heating water in the cracks in rocks. "Old Faithful" at Yellowstone National Park is a geyser of hot water and steam that results from geothermal energy.

Steam from geothermal sources could be used to turn turbines and produce electricity. A tiny amount of our energy today is geothermal, but new technology could make it an important energy source in the years to come.

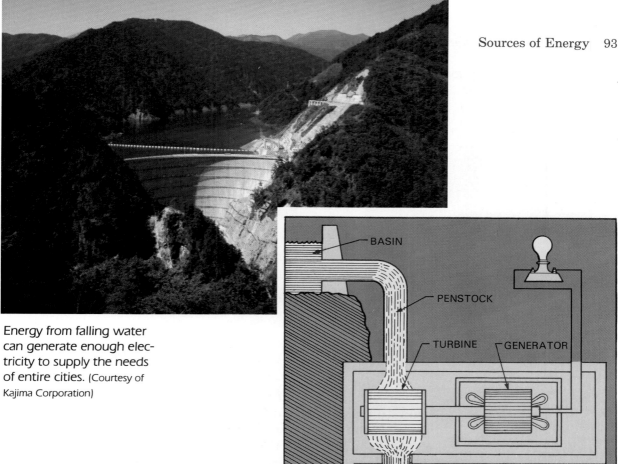

Energy from falling water can generate enough electricity to supply the needs of entire cities. (Courtesy of Kajima Corporation)

Water falling through a penstock turns the blades of a turbine. The turbine is connected to a generator, which produces electricity.

Chemical Energy

Energy is produced as a result of chemical change. For example, when wood burns in air, a chemical change takes place. Atoms of wood combine with oxygen. Heat energy is produced.

Chemical energy sources are wood, coal, oil, and gas. A cord of wood (a pile 4 ft. by 4 ft. by 8 ft.) can provide the same amount of heat as about 200 gallons of oil. Some kinds of wood are better than others for firewood. Hardwoods are better than softwoods. Oak is a very good hardwood for burning.

Wood was burned as the best source of heat energy until the Middle Ages. Around A.D. 1600, the supply of firewood ran out. The forests had been destroyed. As wood became scarce, coal took its place. The United States is rich in coal, with about 250 billion tons. This is about half the world's supply. As oil and gas are being used up, the demand for coal is growing.

The formation of coal

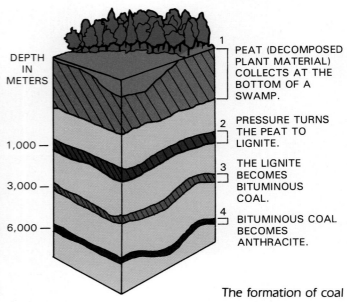

DEPTH IN METERS

1 PEAT (DECOMPOSED PLANT MATERIAL) COLLECTS AT THE BOTTOM OF A SWAMP.

2 PRESSURE TURNS THE PEAT TO LIGNITE.

1,000 —

3 THE LIGNITE BECOMES BITUMINOUS COAL.

3,000 —

4 BITUMINOUS COAL BECOMES ANTHRACITE.

6,000 —

The formation of coal

Train carrying coal from mines in Pennsylvania. (Photo by Jeremy Plant)

The change to coal brings problems with it. Coal is not as easy to transport as oil or gas. It cannot move through pipelines. It must be moved by boat, truck, or railway. Mining of coal on the land's surface (strip mining) can destroy the landscape. In addition, burning coal puts pollutants such as sulfur dioxide into the air.

But work is being done to solve these problems. Coal can be crushed and mixed with water, creating a slurry which can then be pumped through pipelines. Landscapes can be restored after strip mining is done. Coal-burning power plants are finding ways to cut down on pollutants. They clean the coal before it is burned. They also clean the smoke given off, using devices called scrubbers in their smokestacks.

Oil and gas are two **fossil fuels**. They come from decayed plant and animal remains. Millions of years ago, this material settled into layers that were then covered by sand and soil. As time passed, many deep layers of material built up, and the sand and soil were turned to rock. The heat and pressure of these great depths turned the plant and animal remains to oil and gas.

Sound waves are used to find oil deep underground. Satellite photographs also show where oil might be discovered. Once oil is located, wells are drilled and oil and gas are carried through pipelines to refineries. There they are made into fuels.

Oil, or "black gold," is the energy source we depend on the most. Half our energy comes from oil. Oil is a good source of energy. It can be stored and transported easily. It can be made into many different fuels, from heating oil to gasoline.

Pipeline carrying oil from tankers to a refinery. (*Photo by Jonathan Plant*)

Natural gas is used for home heating and cooking, and in the glass industry. Gas-fired furnaces melt glass in large quantities.

Other energy sources such as coal and hydroelectric power can take the place of oil. They cannot, however, be carried around in the form of liquid fuel, as can fuels made from oil. Oil will probably remain the most important energy source for cars and trucks for many years to come.

CONNECTIONS: Technology and Science

■ Lasers are energy converters. The word "laser" stands for Light Amplification by Stimulated Emission of Radiation. In a laser, electrons in the atoms of laser material emit light of a certain wavelength.

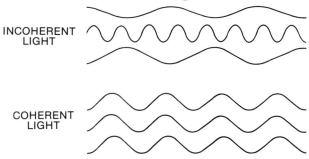

The emitted light waves are coherent; that is, they have the same direction, phase, and wavelength. The laser light is concentrated.

■ Two methods of taking advantage of solar energy for heating are active and passive solar systems. The active system involves using a collector panel to collect and concentrate the solar radiation. The panel generally contains a fluid that transfers the heat to a storage area. The heat from the storage area can then be circulated throughout the building.

Passive systems rely on the house itself, its orientation to the sun, landscaping, and construction to heat the building. For example, windows are located on the southern side of the house since this is the side that receives the most solar radiation during the day.

This floating oil-drilling platform in the North Sea is producing an average of 70,000 barrels of oil per day.
(Courtesy of DuPont Company)

Nuclear Energy

Scientist Albert Einstein proved that energy can be changed into matter and matter can be changed into energy. His famous equation is $E = mc^2$. E stands for energy. The m stands for mass (the amount of matter). The symbol c stands for the speed of light (186,000 miles per second). The equation shows that even a tiny amount of matter can be changed into a huge amount of energy.

Nuclear energy comes from just such a change. There are two ways to produce nuclear energy: fission and fusion. In **fission**, an atomic particle called a neutron hits an atom, and the atom divides into two. The two parts have less mass than the whole atom had. The missing matter has turned to energy. The explosion produced by an atomic bomb comes from fission. Huge amounts of energy come from a small amount of matter.

Nuclear power plants use fission to produce electricity. One problem with this energy source is that fission produces radio-

CONNECTIONS: Technology and Social Studies

- As goods are produced for the marketplace, society needs to become more and more concerned about how they are manufactured. Such problems as limited raw materials and waste management must be solved if we are to continue to advance as a society.
- Nuclear energy has created a great deal of controversy since it came into use as a source of power. Questions of safety, cost effectiveness, and the disposal of nuclear waste are issues that still have not been answered to everyone's satisfaction.
- When other sources of energy, besides water power, became available it brought about a significant change in where people lived. No longer did they have to settle by rivers and other waterways; towns and villages could develop almost anywhere.

The atomic bomb produces incredible amounts of energy from nuclear fission. (Courtesy of Los Alamos National Laboratory)

active material. This material is dangerous to human health. Ways to safely store this radioactive waste are being studied.

In **fusion**, atoms join together. Huge amounts of energy make two atoms combine into one. But the new atom has less mass than the two atoms separately. Some matter has been changed to energy. It is possible that someday fusion will provide our energy needs. There are many problems to overcome before this can happen. One is that temperatures of about 900 million degrees Fahrenheit are needed to start the fusion process.

CAPITAL

Capital is one of the seven technological resources. To build homes or factories, to make toasters or automobiles, to move people or goods, capital is needed. Any form of wealth is capital. Cash, stock, buildings, machinery, and land are all capital.

A company needs capital to operate. To raise capital, a company may sell stock. Each share of stock has a certain value. When people buy the stock, this money goes to operate or expand the business. These investors become part owners (shareholders) in the company. Shareholders hope that the company will do well, and their stock will become more valuable. The company may turn back some of its profit to investors in the form of **dividends**.

Companies also borrow money from banks. Banks charge **interest**. This means that the amount of money paid back is more

Capital resources are necessary for any technological act. (Courtesy of PPG Industries)

than the amount borrowed. A company borrows money with the hope that its profits will help pay the loan and the interest.

TIME

Early people measured time by the rising and setting of the sun and the change of seasons. Later clocks were used to measure time periods shorter than a day.

When most people lived by farming, time was measured in days. In the industrial era, time became more important. It began to be measured in hours, minutes, and seconds. In today's information age, tasks are done in fractions of seconds. Computers process data in nanoseconds (billionths of a second). Time has become an increasingly important resource in our information age.

HOW TO CHOOSE RESOURCES

To choose resources wisely, people must understand the uses and limitations of each resource.

When choosing resources to solve technological problems, we have to consider various factors. Among them are the cost, the availability, and the appropriateness of the resources.

COST OF RESOURCES

Since companies like McDonald's are in business to make a profit, the cost of resources used is very important. It costs money to hire fast-food workers, and to buy land on which to build restaurants. It is expensive to transport materials, produce energy, and buy tools and machines.

In some industries, like the automotive industry, decisions must be made about whether to hire human workers or to purchase robots. Robots never complain about getting tired, never call in sick, and produce high-quality work. However, they cost a lot of money to buy. The company must decide which resources will be cheaper and better in the long run. When we choose resources, we must sometimes give up desirable things (like low costs) to get some other desirable thing (like higher quality). Giving up one thing to get another is called making a **trade-off**.

When we choose resources, we must sometimes make trade-offs in order to reach the best possible solution.

Another decision involving cost might involve a company's deciding where it is best for them to build a new factory. Is it better for them to build the plant in the United States, where workers earn high salaries, or to build it in a foreign country where labor costs will be cheaper? For example, factory workers in Haiti work for wages of less than $5.00 per day.

Time Measurement Throughout the Ages

When people lived by farming, they needed to know when to plant seeds. They did not need the precise time of day. A sundial or burning rope clock was adequate. A ring of huge stones at Stonehenge, England, is believed to have been built in ancient times to be used as a calendar. Stars and planets seen through spaces between the stones told the time of year.

During the Middle Ages, monks invented the water clock for telling time. In the industrial age, people used mechanical clocks to tell time.

In today's information age, it is sometimes necessary to measure even small fractions of a second. Tools used for these measurements include the quartz crystal clock, the atomic clock, and the oscilloscope.

Stonehenge (Photo by Susan Warren)

An oscilloscope (Courtesy of Tektronix, Inc.)

SUNDIAL

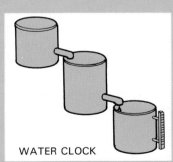

WATER CLOCK

An atomic clock (Courtesy of Hewlett-Packard Company)

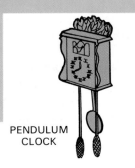

PENDULUM CLOCK

Companies must decide whether it is better to invest in expensive machinery or to hire human labor.

AVAILABILITY OF RESOURCES

We also make choices based upon how available resources are. During the energy crisis in 1973, people waited in lines for hours at gasoline stations. Gasoline prices doubled and speed limits on U.S. highways were lowered to 55 miles per hour to save fuel. Automobile makers changed, too. They began to manufacture smaller cars that used less fuel. Choices were made based upon availability of energy resources.

Resources are unevenly distributed throughout the world. In some countries, certain resources are scarce. In other countries, those same resources are plentiful. Usually scarce resources cost more than plentiful resources. Sometimes we must use resources that are rare and costly for special purposes. Jet planes used by the armed forces are made from a metal called titanium. The United States doesn't have much titanium. We must import it from other countries like Australia and India. Because of this, titanium is very expensive to use.

APPROPRIATENESS OF RESOURCES

We must also determine which resources are most appropriate. That is, which are best suited to the task at hand. For example, if McDonalds were to set up several restaurants in France, they might look for managers who speak both English and

The Lockheed SR-71 spy plane is made out of titanium. (Courtesy of Lockheed-California Company)

When metal was in short supply during World War II, people made bicycles out of wood. (Courtesy of Smithsonian Institution)

French. A person who did not speak both languages well would not be an appropriate resource.

When choosing expensive machinery, companies must consider how appropriate it is for the place in which it will be used. For example, if a company was hired to do farming in an area without technically trained people, would they want to use modern tractors? Modern tractors are complicated pieces of equipment. They require trained service people to keep them in repair. In some countries, there are many unskilled workers and relatively few technicians. Although tractors can plow much faster than people and animals, human and animal labor would be a much more appropriate resource for such areas.

We choose energy resources according to how appropriate they are. In a country like Israel, oil prices are very high and there are many sunny days. Almost every house has a solar hot water heater on the roof. Solar energy is an appropriate resource in such a climate.

Important factors in choosing resources are cost, availability, and appropriateness.

PROPERTIES OF MATERIAL RESOURCES

Material resources have different **properties**. Properties include how strong and hard they are, what they look like, and whether they rust or break down.

Glass is used for windows because it lets light in. Plastic is used for dishes because it is strong and is easily cleaned. Copper is made into electrical wire because it conducts electricity. We make phonograph needles from pieces of diamond because diamonds are hard. The needle will last for a long time. We make clothing out of nylon because it is lightweight and looks good.

Mechanical Properties

Force can bend or break materials. Some materials bend more than others. A fiberglass fishing rod will bend a great deal without breaking, but wood will not. Materials that can bend or deform without breaking are called **ductile materials**. Pots and pans are made from a flat sheet of metal. Pressure on part of the metal forms it into the shape of the pot. The metal must be able to stretch and get thinner without breaking. In making wire, a thick rod is pulled through a small hole to make it thinner. This process, called **drawing wire**, works because the metal used is ductile. Materials that can be twisted, bent, or pressed into shape have high ductility.

The opposite of ductile is brittle. A **brittle material** breaks instead of deforming. Window glass is brittle.

There are two kinds of ductile materials. An **elastic material** can bend and then come back to its starting shape and size. Rubber bands, springs, and fishing rods have high elasticity. A **plastic material** can be bent and will stay bent. The material we know as plastic got its name because of that property. When we heat a piece of plastic and bend it, it will stay that way after it cools. Modeling clay and some metals are among materials with high plasticity.

Some materials are stronger than others. A material's strength is how well it keeps its shape when force is applied to it. **Tension** is a pulling force that acts on a piece of material. When we pull on a spring, it is under tension. Tensile strength refers to a material's ability to withstand tension. **Compression** is the opposite of tension. It is a pushing or squeezing force. Squeezing a sponge or walking on a carpet are examples of compression. Compressive strength refers to a material's ability to withstand compression. **Torsion** is when a material is twisted. When we twist a garden hose, the hose undergoes torsion. The twisting force itself is called **torque**. When we use a wrench to turn a bolt, we apply torque. A **shear** force acts on a material the way a pair of shears (scissors) works. One part goes in one direction and the other part goes in the opposite direction.

Toughness is the ability of a material to absorb energy without breaking. Leather is tough.

Properties of Materials

MECHANICAL PROPERTIES

ELASTICITY
The rod returns to its original shape.

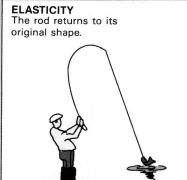

PLASTICITY
A coil of clay will stay bent.

BRITTLENESS

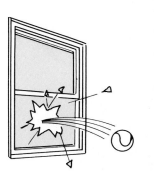

DUCTILITY

HARDNESS

TOUGHNESS
The frame of the car will withstand impact.

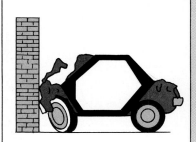

STRENGTH
ABILITY TO WITHSTAND . . .

Torsion

Compression

Tension

Shearing Action

OPTICAL PROPERTIES

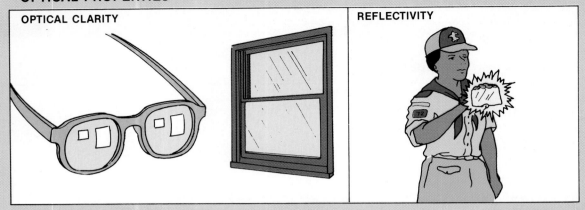

OPTICAL CLARITY

REFLECTIVITY

THERMAL PROPERTIES

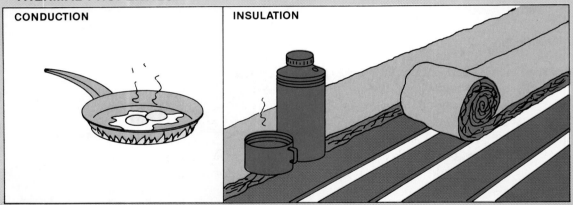

CONDUCTION

INSULATION

ELECTRICAL AND MAGNETIC PROPERTIES

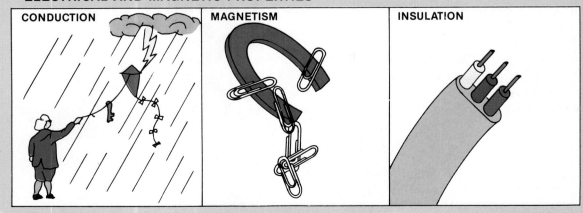

CONDUCTION

MAGNETISM

INSULATION

A very important property is hardness. **Hardness** is a material's ability to resist being scratched or dented. Diamonds are the hardest materials. Some metals are also very hard. Metal tools must be hard to resist wear. Tungsten carbide is a hard material. The teeth of saw blades are made from it.

Electrical and Magnetic Properties

Conductors are materials through which electricity moves most easily. Most good conductors are metals, the best of which is silver. The next best is copper. Most wire is made from copper, because silver is so costly.

Insulators are materials that do not conduct electricity easily. Insulators can be useful. Copper wire is covered with plastic or rubber for safety.

Magnetic materials are those that are attracted by a magnet. Iron, steel, nickel, and cobalt are magnetic. Copper, wood, glass, and leather are not.

Thermal Properties

In Greek, "therm" means heat. **Thermal** properties have to do with how well materials conduct heat. Some materials conduct heat easily, others do not.

Metals are good heat conductors. Copper and aluminum are two of the best. Other materials such as rubber and fiberglass are poor heat conductors. We call them insulators. A building's outside walls are insulated to keep heat in during the winter and out during the summer.

Optical Properties

Optical properties have to do with how a material transmits or reflects light. Light passes easily through clear window glass. Glass used for scientific devices often must be even clearer. Plastic is lightweight and will not break. It is therefore used for contact lenses and eyeglasses. Some pure kinds of glass are used to carry light signals for communications systems. Metals that reflect light well are used in headlights and flashlights.

MAINTENANCE OF RESOURCES

We must keep in mind how well the resources we use will stand up over time. We might use silver to make an item because of the way it looks. Do we want to have to polish it every few

Disposal of Resources

Instead of fixing something when it breaks, we often throw it away. We have cups, razors, and flashlights that are made to be thrown out after use. Food and drink comes in disposable bottles, boxes, and cans.

We must think about what happens to the resources we choose after use. Some are stored in landfills. Others are burned or thrown into the oceans. Burning pollutes the air, and ocean dumping pollutes the water. We must try to choose resources that will decompose (break down) and become part of the earth again. Those that will not decompose, like plastic or glass, should be **recycled** (reused).

Material Decomposition Time

MATERIAL	TIME TO DECOMPOSE
Orange peels	1 week to 6 months
Paper containers	2 weeks to 4 months
Paper containers with plastic coating	5 years
Plastic bags	10 to 20 years
Feathers	50 years
Plastic bottles	50 to 80 years
Aluminum cans with flip-top tabs	80 to 100 years
Plutonium	24,390 years (half-life)
Glass bottles	Indefinite

To dispose of radioactive waste, technologists cover it with lead and then bury it in salt formations far underground. (Courtesy of Rockwell International Corp.)

Recycling is an important source of aluminum supply. (Courtesy of Aluminum Company of America)

Waste disposal is a major technological problem. (Courtesy of Sperry Corp.)

(Courtesy of International Business
Machines Corporation)

weeks to keep it shiny? Maintenance is an important factor.
Machines and tools must be kept in working order. This takes
trained repair people and costly parts. Maintaining information resources is often a problem. Should we rebind old library
books or buy new ones? Should we put our files into a computer or go on keeping written records?

CULTURAL VALUES

Our values also help us choose which resources to use. Some
people think that animals should not be used for research.
Others think it is all right because human lives may be saved
by such research.

SUMMARY

Every technological activity involves the use of seven resources.
They are: people, information, materials, tools and machines,
energy, capital, and time.

People's needs drive technology. Humans design and create
technology using their knowledge and intelligence. People can
make policies that promote technological growth, and people
use the products and services of technology.

Information is needed to solve problems and to create new
knowledge. Information comes from raw data, which are processed by collecting, recording, classifying, calculating, stor-

*To solve technological prob-
lems, we choose resources
from each of the seven categories of resources.*

(Courtesy of DuPont Company)

(Courtesy of Boise Cascade Corporation)

(Courtesy of Southern California Edison)

ing, and retrieving the facts. People turn information into knowledge by giving it meaning.

Materials found in nature are called raw materials. Raw materials can be made into useful products. Renewable raw materials are those that can be grown and therefore replaced. Nonrenewable raw materials are used up and cannot be replaced. Synthetic materials are human-made materials. They can be produced with useful characteristics natural materials do not have.

Tools extend the capabilities of people. Hand tools extend the power of the human muscle. Optical tools extend the power of the eye. Computers extend the power of the brain. Machines are tools that change the amount, speed, or direction of a force. Most modern machines have moving parts. Machines that use electrical energy to move mechanical parts are called electromechanical devices. Automated machines can operate without much human control. Electronic tools, particularly the computer, save huge amounts of time, energy, and labor.

Energy sources, like materials, are either renewable or nonrenewable. Renewable energy sources include human and animal muscle power, water, wind, geothermal energy, and solar energy. Nonrenewable energy sources include oil, gas, coal, and nuclear energy.

Capital is any form of wealth. Capital can be cash, shares of stock, buildings, machinery, or land.

Time is an important resource in the information age. Electronic circuits carry huge amounts of data in billionths of a second.

We must make choices about resources. To make these choices, we must have information about the properties, uses, and limitations of resources. Most often, we'll have to make trade-offs.

We must consider cost when choosing resources. We must pay workers, move materials, pay for energy, and buy tools and machines.

Availability is also important. Sometimes, one material can be substituted for another that we can't find. One energy source can be substituted for another. Solar energy can be used instead of oil in some places.

Resources that are appropriate to a certain job must be chosen. We would not want to choose machines if we could not find people to run and fix them.

Materials are chosen for their properties. There are many mechanical properties. Elasticity is a material's stiffness. Plasticity is the ability to be bent and stay bent. Strength is the resistance to tension, compression, torsion, and shear force. Toughness is the ability to absorb energy. Hardness is the ability to resist scratching and denting. Materials have electrical properties (how well they conduct electricity). Materials have magnetic properties (whether they are attracted by a magnet). Materials have thermal properties (how well heat flows through them). Materials have optical properties (how well they transmit or reflect light).

How easy it is to maintain and dispose of resources must be kept in mind in choosing them. Our cultural values are also important in choosing resources.

Feedback

1. What are the seven resources common to all technologies?
2. Define a machine in your own words. Then use your definition to determine whether the following items are machines:
   a. a baseball bat
   b. software for a video game
   c. a radio
   d. a ramp for wheelchairs
   e. a hand-operated drill
   f. a wrench
3. Should the United States build nuclear-powered plants to generate electricity? Explain your answer.
4. Wood is a renewable resource. Does that mean that we can cut down all the trees we need? Explain your answer.
5. What are two advantages of synthetic materials over natural materials?
6. How are modern hydroelectric power plants similar to ancient water wheels?
7. What is a disadvantage of using windmills to generate electricity?
8. Name three renewable and three nonrenewable sources of energy.
9. If a person wanted to start a company, how might he or she arrange to get capital?
10. How do knowledge and information differ?
11. Suppose we choose to locate a toy-making factory in Korea instead of the United States because labor is cheaper. What trade-offs would we have to make?
12. What are three factors that influence our choice of resources?
13. You plan to purchase a screwdriver for electrical work. From what material should the handle be made? Why?
14. Give an example of how our values influence our choice of resources.
15. Why must people understand the uses and limitations of resources in order to choose them wisely?

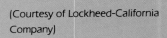

(Courtesy of DuPont Company)

(Courtesy of Lockheed-California Company)

(Courtesy of Southern California Edison)

CHAPTER KEY TERMS

Across

3. A force that pulls on a piece of material.
4. _____ refers to a material's ability to resist being scratched.
5. Refers to a material's ability to conduct heat.
8. A _____ material can be deformed without breaking.
10. A good one permits electric current to flow easily.
11. Tools that multiply force.
13. This resource has six sources.
14. This happens when a twisting force is applied.
18. This will stretch without breaking.
20. Energy produced by fission or fusion.
21. A pair of scissors produces this kind of force.
22. A resource used to purchase other resources.

Down

1. A good one does not conduct electricity.
2. Coal, oil, and gas.
6. Electricity from water power.
7. To reuse.
9. Limited in supply.
10. A force that squeezes a material.
12. A resource that extends human capabilities.
15. There are seven kinds of these needed for technology.
16. Energy from heat inside the earth.
17. Energy from the sun.
19. Materials made by humans are _____.
22. Substance formed from decomposed plant and animal life under great pressure.

Brittle
Capital
Coal
Compression
Conductor
Ductile
Elastic
Energy
Finite
Fossil fuels
Geothermal
Hardness
Hydroelectricity
Insulator
Machines
Nuclear energy
Recycle
Resources
Shear
Solar
Synthetic
Tension
Thermal
Tools
Torsion

SEE YOUR TEACHER FOR THE CROSSTECH SHEET

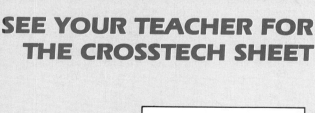

ACTIVITIES

WIND POWERED SYSTEMS
Problem Situation

For thousands of years people have used wind to move ships. Wind systems are energy converters that change wind energy into another form. Windmills that pump water or grind grain are changing wind energy into mechanical energy. A windmill used to generate electricity changes wind energy to mechanical energy, which is used to operate a generator that produces electrical energy.

There are advantages and disadvantages to using wind-powered electrical generators. The wind is free and wind generators don't pollute the air. However, they are expensive to construct and only operate efficiently when there is a steady wind. Wind generators are being used successfully in a number of places, including the San Francisco area. Researchers are also trying to develop less expensive and more efficient wind-powered generating systems.

Design Brief

Using the wind produced by a fan, construct a device to lift as much weight as possible and operate a DC motor to generate electricity. The windmill you design should have a horizontal shaft.

Procedure

1. Using the safety procedures described by your teacher, construct three sides for a tower to support your wind turbine using the 6" pine strips and craft sticks. Use masking tape to hold the pieces together until the glue dries.
2. Put the three sides together to form a triangle-shaped tower. Glue together using white glue. Use masking tape to hold the structure together until the glue dries.
3. Drill a 5/16" hole 1 inch from the end of both 1/8" × 1 1/4" × 4" strips. Mount these on the tower to support the horizontal shaft.
4. Cut an 8" piece of 1/4" dowel for the horizontal shaft.
5. Use reference materials to get ideas for the turbine design. Select one design and construct it using modeling materials made available by your teacher. Attach the turbine blades to the 1/4" dowel.
6. Test your turbine using a 20" square fan as the wind source. Make necessary improvements.
7. Attach a string long enough to reach the floor to the horizontal shaft. Use a paper clip on the end to attach to the aluminum can.

Suggested Resources

20" box fan
DC motor
Galvanometer
24 craft sticks
String
Gram scale
Paper clip
Masking tape
White glue
6 pieces of 1/8" × 3/4" × 6" pine
2 pieces of 1/8" × 1 1/4" × 4" pine
12-ounce aluminum can (top removed)
Assorted modeling materials

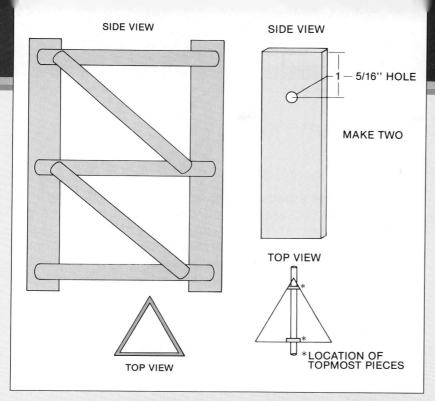

SIDE VIEW

SIDE VIEW

1 — 5/16" HOLE

MAKE TWO

TOP VIEW

TOP VIEW

*LOCATION OF
TOPMOST PIECES

8. Test your wind turbine to determine how much weight it can lift from the floor to the bench top.

9. Modify your turbine design to try to increase its lifting power. Retest and record the maximum weight your turbine is able to lift.

10. Remove the string and design a way to mount the DC motor so that it will be driven by the horizontal shaft when wind is supplied by the fan. Measure the motor's electrical output with the galvanometer.

Classroom Connections

1. We need energy converters because we do not always want to use energy in its original form.

2. Generators convert mechanical energy into electrical energy. Motors convert electrical energy into mechanical energy. If the shaft of a motor is turned, it acts like a generator.

3. Winds are caused by uneven heating of the earth's surface by the sun. Bodies of water and mountains also affect local winds.

4. Scientists estimate that the sun's energy will continue for 50 billion years.

Technology in the Real World

1. The burning of fossil fuels to produce electricity contributes to acid rain, a form of air pollution.

2. Wind turbine generators will not work everywhere. Also, some people object to them for aesthetic reasons.

Summing Up

1. Give three examples of energy conversion.

2. Explain what you did to improve the way your wind turbine works.

3. Why is it important to continue to develop alternate energy systems?

113

ACTIVITIES

IT'S ABOUT TIME

Problem Situation

Throughout history people have been interested in measuring time. More than 4,000 years ago sundials were used. Since these were useless at night and on cloudy days, people worked to develop ways of measuring time that did not require sunlight. Candles were marked so that intervals of time could be counted as they burned. Burning rope was also used. Equally spaced knots indicated units of time as the rope was burned.

Water clocks were developed in Greece and Rome. These kept time by having water drip from one container to another. Hourglass "clocks" made from two connected globes, one empty and one filled with sand, were also common. Columbus used a 30-minute hourglass on his ship. Today we use similar devices to time the cooking of eggs and the length of telephone calls.

The first mechanical clocks were very large and were driven by falling weights. Accurate clocks were introduced in the 1500s and 1600s when springs and pendulums were used as control mechanisms. Today it is possible to measure intervals smaller than one billionth of a second.

Design Brief

Design and construct a timekeeping device that measures at least three minutes. Calibrate the device so that it indicates regular intervals. For example, a five-minute clock should indicate one, two, three and four minutes.

Procedure

1. Your teacher will identify the materials and equipment that may be used for this activity.
2. Use reference materials available in the Technology Education laboratory and the school library to gather information about timekeeping devices.
3. Sketch at least three different devices.
4. Next to each sketch, identify the materials that could be used to construct the device.

Suggested Resources

Plastic and metal
 containers
Tape
Glue
Chalk
Markers
Straws
Salt
Sand
Tubing
Other materials made
 available by your
 teacher.

5. For each device write a brief paragraph describing how you expect to construct it.
6. Choose one device to construct. Begin work after you obtain teacher approval. NOTE: Follow all safety instructions given by your teacher.
7. Using the safety procedures described by your teacher, construct your timekeeping device. Remember that it must measure at least three minutes and should indicate regular intervals.
8. Test and improve your device until it works properly.

Classroom Connections

1. A nanosecond is one billionth of a second; a microsecond is one millionth of a second; a millisecond is one thousandth of a second.
2. A pendulum is a free-swinging object which has back and forth movement.
3. Velocity is the speed of an object as it travels between two points. Miles per hour in a given direction is an example of this kind of measurement.
4. Three dimensions—length, width and height—are used to describe the size of objects. In his Theory of Relativity, Albert Einstein identified time as the fourth dimension.

Technology in the Real World

1. Technology has reduced the time required for many activities. A journey of 1,000 miles that took weeks a century ago can now be completed in two hours.
2. Electronic timing can measure the speed of athletes to one thousandth of a second.

Summing Up

1. Draw a time line that places five timekeeping devices in chronological order.
2. Describe several ways in which time influences your daily routines.
3. Describe several problems that you had to solve during construction on your time-keeping device.

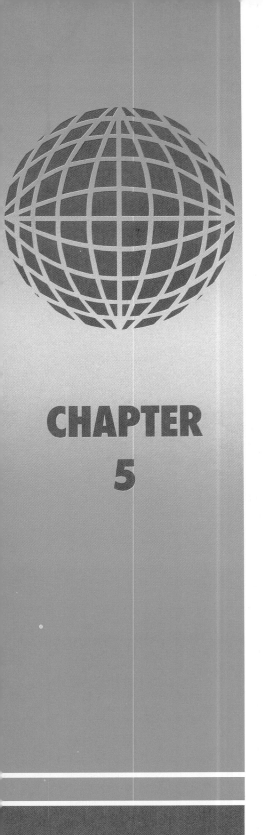

CHAPTER 5

PROCESSING RESOURCES

MAJOR CONCEPTS

After reading this chapter, you will know that:

- Technological systems convert resources into end products. These end products can be materials, information, or energy.
- The actual conversion of resources occurs within the technological process.
- The processing of resources is usually directed by people, since it is people who design and develop technological systems.
- There are several ways people can process materials. These processes include forming, separating, combining, and conditioning.
- Computers process and communicate information. They also control systems that process materials and energy.
- The end result of a conversion process must be monitored if we want to be sure the desired goals are achieved.

INTRODUCTION

To make resources more useful, people process them. To **process** something means to change it. For example, wood can be made into a table or the handle of a shovel or paper. Cotton can be made into thread, then woven into fabric to make clothing. Animal hides can be cut and sewn to make shoes, handbags, and coats. Plants and animals can be processed to make food.

Information is processed data. Energy can also be processed. For example, an electric motor changes electrical energy into the mechanical energy that runs a model train.

Technological systems convert resources into end products. These end products can be materials, information, or energy.

Problem Situation 1: Safety goggles are essential in the school and home workshop but there are two problems associated with their use
1. they are easily misplaced, and
2. they are often found in a dirty, dusty condition.

Design Brief 1: Design a way of storing a pair of goggles for use with a drill press that overcomes these two problems.

Problem Situation 2: Boiled eggs roll around on a plate and many traditional egg cups are both unhygienic and unstable.

Design Brief 2: Using acrylic sheet or a similar material, design and construct a support for one or more boiled eggs. The support must stack, be easily cleaned, and must be stable.

Problem Situation 3: If food is not preserved it will spoil. One way to preserve food is by drying.

Design Brief 3: Design a food drying system that can be used to dry fruits.

PROCESSING MATERIAL RESOURCES

The actual conversion of resources occurs within the technological process.

The processing of resources is usually directed by people, since it is people who design and develop technological systems.

Primary raw materials are those that are taken from the earth and changed into a form that can be made into useful products. For example, trees are harvested and cut into boards. Iron ore, limestone, and coke are processed into standard-size steel slabs, sheets, and rods. Rubber is heated with sulfur in a process called **vulcanization**, making it able to withstand great temperature changes. **Primary industries** such as the timber and mining industries process primary raw materials. These raw materials become basic **industrial materials**.

We use all the other resources when we process a material. For example, tools are used when lumber is cut into long boards. Sometimes wood is dried in an oven called a **kiln**, which uses energy. In most cases, processing a material adds value to it.

Some basic industrial materials (Courtesy of Commercial Metals Company)

The aluminum industry processes bauxite ore into many different products. (Courtesy of Aluminum Company of America)

Processing Resources

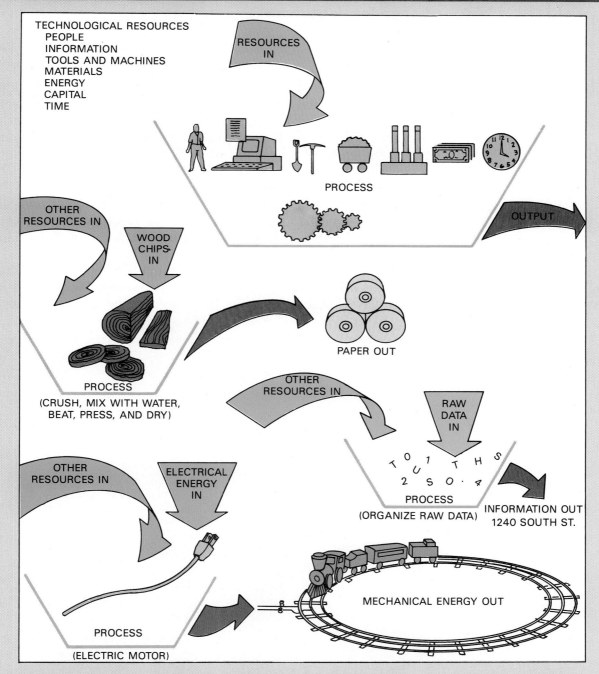

TECHNOLOGICAL RESOURCES
  PEOPLE
  INFORMATION
  TOOLS AND MACHINES
  MATERIALS
  ENERGY
  CAPITAL
  TIME

RESOURCES IN

PROCESS

OUTPUT

OTHER RESOURCES IN

WOOD CHIPS IN

PAPER OUT

PROCESS
(CRUSH, MIX WITH WATER, BEAT, PRESS, AND DRY)

OTHER RESOURCES IN

RAW DATA IN

PROCESS
(ORGANIZE RAW DATA)

INFORMATION OUT
1240 SOUTH ST.

OTHER RESOURCES IN

ELECTRICAL ENERGY IN

MECHANICAL ENERGY OUT

PROCESS
(ELECTRIC MOTOR)

The technological process combines all the technological resources to produce an output.

FORMING PROCESSES

Forming a material means changing its shape without cutting it. When we bend a metal rod into a coat hanger shape, we are using a forming process.

Casting

Casting is one way of forming a material. Castings are made from **molds**. When you walk on damp beach sand, your feet make footprints, or molds. If you melted metal, poured it into a sand mold, and let it cool, you would have a casting of your foot. The casting process is used in making ice cubes. Pour water into an ice cube tray, and when it freezes, it forms ice cube castings.

Molds can be one-piece molds or they can be made of several pieces of material. The footprint is a one-piece mold. Another one-piece mold is a cake pan. Batter takes the shape of the pan and the cake forms a casting.

A two-piece mold can be used to cast ceramic (clay) objects. First, we make a two-piece mold from plaster. We pour in a liquid clay called **slip**. We let the clay set for a few minutes,

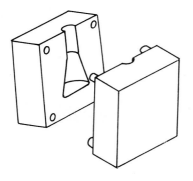

A two-piece mold for casting ceramic objects.

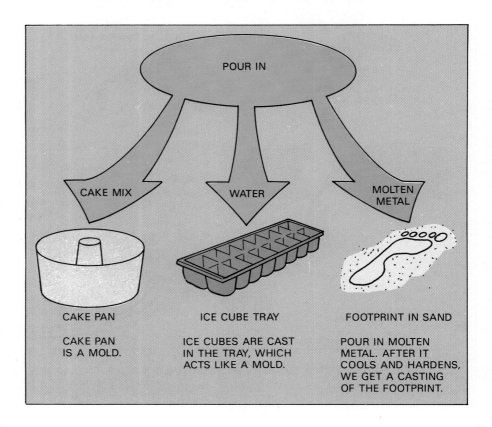

POUR IN

CAKE MIX   WATER   MOLTEN METAL

CAKE PAN   ICE CUBE TRAY   FOOTPRINT IN SAND

CAKE PAN IS A MOLD.

ICE CUBES ARE CAST IN THE TRAY, WHICH ACTS LIKE A MOLD.

POUR IN MOLTEN METAL. AFTER IT COOLS AND HARDENS, WE GET A CASTING OF THE FOOTPRINT.

Some common one-piece molds.

then pour out the extra slip. The plaster absorbs water from the slip and leaves a thin wall of clay inside the mold. When the clay is dry, the mold is opened and the casting is removed.

Pressing

Pressing is much like casting. Material is poured into a mold. Then a plunger with its own shape is lowered, forcing the material to spread out to fill the mold. The material takes the shape of the mold on the bottom and of the plunger on the top. The plunger is lifted and the object is removed. Waffles are made this way.

Powdered metal can be pressed to make objects. The metal powder is placed in the bottom of the mold and pressed by the plunger into a solid mass. The object is then heated to bond the particles. This is called **sintering**.

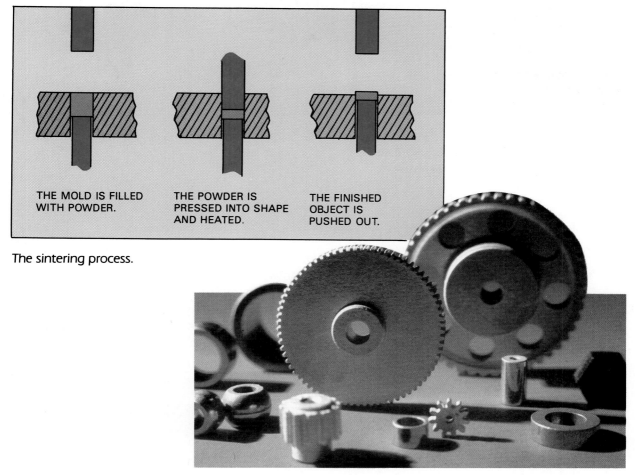

THE MOLD IS FILLED WITH POWDER.

THE POWDER IS PRESSED INTO SHAPE AND HEATED.

THE FINISHED OBJECT IS PUSHED OUT.

The sintering process.

Finely ground metal powders can be sintered into strong, lightweight objects.
(Courtesy of Aluminum Company of America)

Forging

A **forging** is a metal part that has been heated (not melted) and hammered into shape. Blacksmiths used to hammer metal to make horseshoes. Today, forging is done by huge machines. A piece of metal is placed in the lower half of a mold. A powerful ram presses the metal downward into the mold with as much as 2,500 tons of force.

In days past, hand forging was a common method of forming tools and other useful items. (Courtesy of The Citizens & Southern National Bank of South Carolina)

A large forge press (Courtesy of Cerro Metal Products)

Extruding

Extruding is a way of forming by squeezing soft material through an opening, as when we squeeze a tube of toothpaste. The material takes the shape of the opening. For example, if a toothpaste tube had a square end, the paste would come out as a rectangular block. Extruding can save work because the finished product often needs little more in the way of processing.

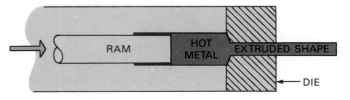

In extrusion, hot metal is forced through a hole called a die. The metal takes the shape of the die.

Blow Molding and Vacuum Forming

Blow molding and **vacuum forming** are used in forming plastics. In both, a thin plastic sheet is heated to soften it. In blow molding, air blows the sheet into a mold. Plastic bottles are made this way. In vacuum forming, the sheet is drawn into a mold by a vacuum. The plastic takes the form of whatever it is drawn against. Vacuum forming is used to plastic-wrap some items. A plastic sheet is drawn against the object being packaged. This is known as blister packaging.

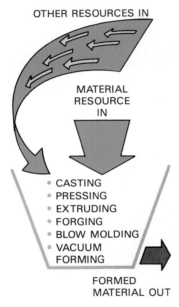

OTHER RESOURCES IN

MATERIAL
RESOURCE
IN

- CASTING
- PRESSING
- EXTRUDING
- FORGING
- BLOW MOLDING
- VACUUM
  FORMING

FORMED
MATERIAL OUT

Some forming processes.

SEPARATING PROCESSES

When we use knives, saws, or scissors we are **separating** one piece of material from another. Separating is the same as **cutting**. Tools are used in many separating processes, including **shearing**, **sawing**, **drilling**, **grinding**, **shaping**, and **turning**.

Shearing

Shearing is using a knife-like blade to separate materials. We can use one blade, like a knife, or two blades, like scissors (sometimes called shears). When we shear material, the blade compresses it. When the force on the material is high enough, the material breaks along the line of the cut. The sharper the blade, the greater the force on the area being cut.

Six Types of Separating Machines

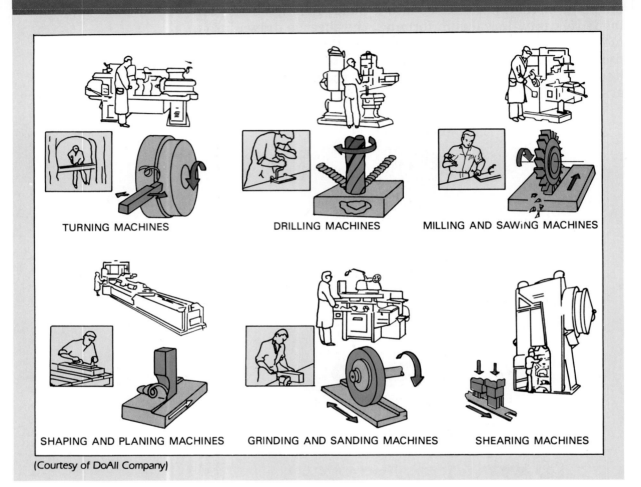

TURNING MACHINES

DRILLING MACHINES

MILLING AND SAWING MACHINES

SHAPING AND PLANING MACHINES

GRINDING AND SANDING MACHINES

SHEARING MACHINES

(Courtesy of DoAll Company)

Sawing

Sawing is cutting material with a blade that has teeth. Each tooth chips away tiny bits of material as the saw cuts. There are two ways of sawing wood. **Ripping** is used to cut wood in the direction of the grain. **Crosscutting** is used to cut across the grain. Hand rip and crosscut saws have from six to ten teeth per inch. Metal is cut by hand with a saw called a **hacksaw**, which has a very hard steel blade. The blade has about eighteen teeth per inch.

Saw blades can be made in circular shapes or in continuous bands. Some power saws, such as radial arm saws and table saws, use circular blades. The blade spins rapidly, making many cuts each minute on the material being sawed.

Sawing and Drilling

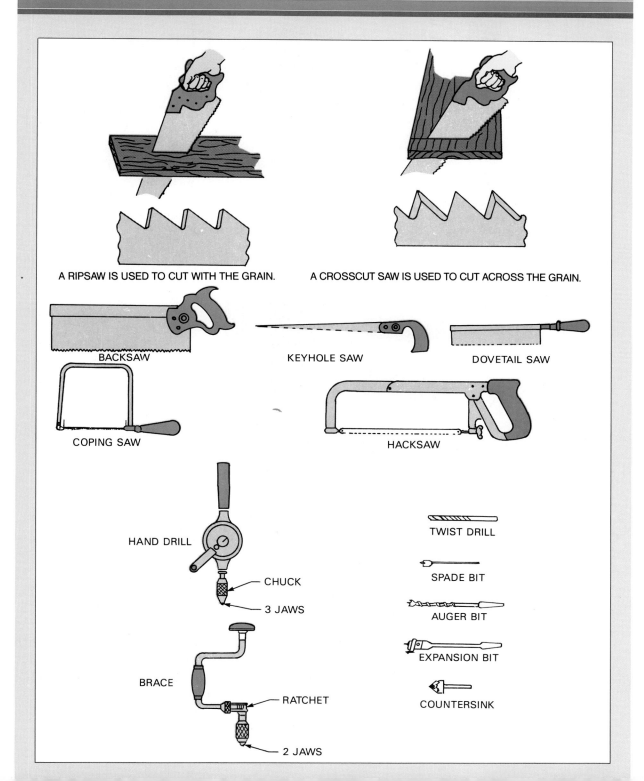

A RIPSAW IS USED TO CUT WITH THE GRAIN.

A CROSSCUT SAW IS USED TO CUT ACROSS THE GRAIN.

BACKSAW

KEYHOLE SAW

DOVETAIL SAW

COPING SAW

HACKSAW

HAND DRILL

CHUCK

3 JAWS

BRACE

RATCHET

2 JAWS

TWIST DRILL

SPADE BIT

AUGER BIT

EXPANSION BIT

COUNTERSINK

A drill press (Courtesy of Delta
International Machine Corporation)

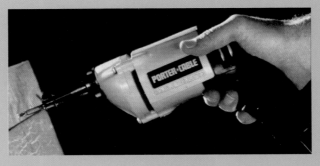

An electric hand drill (Courtesy of Porter-Cable Corporation)

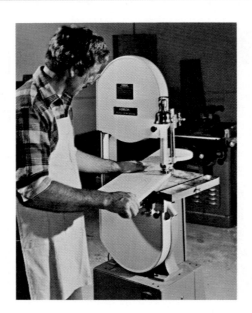

Operator using a band saw
(Courtesy of Delta International
Machine Corporation)

Drilling

Drilling is used to cut holes in materials. A pointed tool with a sharp end is turned quickly by a hand or electric drill. Drilled holes can be as small as 1/10,000 of an inch or as large as 3½ inches in diameter.

Grinding

Grinding is done using tiny pieces of very hard material called **abrasives**. These abrasive particles can be glued onto a flat sheet to make sandpaper or emery cloth. The particles may also be formed into grinding wheels. As the wheel turns, the particles rub against and cut away bits of the materials being ground.

Often, we sharpen tools by grinding them. Grinding removes a very small amount of material at a time.

Polishing is a form of grinding. Polishes most often contain abrasive powder. When you rub with a polish, you are removing a tiny bit of the surface. This makes the surface smooth, clean, and shiny. Toothpaste is an abrasive. We clean our teeth by using an abrasive to remove plaque.

Notice the safety shields over this grinding machine and the safety glasses worn by the operator. Sparks and tiny particles of material can fly into the eye unless proper safety precautions are taken. (Courtesy of Delta International Machine Corporation)

Shaping

Shaping is used to change the shape of a piece of material. Chisels and planes are hand tools used for shaping. Shaping tools have cutters with chisel-like edges that chip away material. Jointers, planers, and shapers are shaping machines.

Wood lathes are used to make cylindrical objects, such as legs for tables and chairs.
(Courtesy of Delta International Machine Corporation)

Turning

Turning tools are also used to shape materials. A turning tool is different from other shaping tools in that the tool itself does not move. Instead, the material moves. A lathe spins the material. The cutting tool is held against the spinning material to cut grooves or reduce the diameter. Wood chair legs and stair railings are made this way.

Other Separating Processes

Materials can be separated by means other than cutting. This can be done **chemically**, as happens when water is separated into hydrogen and oxygen. We can get salt from salt water by letting the water evaporate. **Filtering** is used to separate solids and liquids in a mixture. The vegetables in a can of soup can be separated from the broth by pouring the soup through a strainer. **Magnets** can separate magnetic materials from nonmagnetic materials.

COMBINING PROCESSES

Sometimes materials are **combined**. We **fasten** one material to another. We **coat** materials with a protective finish. Using two or more materials, we make a **composite** that has special properties we need.

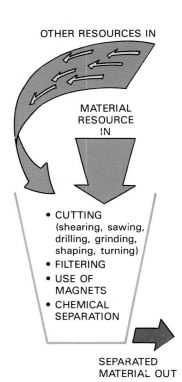

OTHER RESOURCES IN

MATERIAL RESOURCE IN

• CUTTING
  (shearing, sawing, drilling, grinding, shaping, turning)
• FILTERING
• USE OF MAGNETS
• CHEMICAL SEPARATION

SEPARATED MATERIAL OUT

Some separating processes

Mechanical Fasteners

We can fasten materials together mechanically using nails, screws, and rivets. Different fasteners are used with different materials. For example, nails are made to be used with wood.

Long ago, nails were made by hand and were very costly because it took a long time to make them. People did not often use nails in building houses. Today, machines make nails by

---

CONNECTIONS: Technology and Mathematics

■ In 1950 about 30 out of every hundred workers were employed in manufacturing industries compared to 16 out of every hundred today. We can calculate the percent of decrease.

$$\% \text{ of decrease} = \frac{\text{amount of decrease}}{\text{original number}} \times 100$$

$$\% \text{ of decrease} = \frac{30 - 16}{30} \times 100$$

$$= \frac{14}{30} \times 100$$

$$= 0.466 \times 100$$

$$= 46.6\%$$

The number of workers employed in information processing increased from 17 per hundred in 1950 to 60 per hundred today. To find the percent of increase, calculate the amount of increase compared to the original number.

$$\% \text{ of increase} = \frac{60 - 17}{17}$$

$$= \frac{43}{17}$$

$$= 2.53$$

$$= 253\%$$

■ Drilled holes can be made as small as 1/10,000″ or as large as 3½″ in diameter. How large would the radius (r) be of a hole that is 3½″ in diameter (d)? How large would the area (A) of the hole be?

$r = \frac{1}{2} \times d$	$A = \pi r^2$
$r = \frac{1}{2} \times 3\frac{1}{2}''$	$A = (3.14)(1.75'')^2$
$r = \frac{1}{2} \times 7\frac{1}{2}''$	$A = (3.14)(1.75'')(1.75'')$
$r = \frac{7}{4}'' = 1.75''$	$A = 9.62''$

the thousands, making them cheaper. Because the price of nails has dropped, they are used by many people to build houses.

To hold two pieces of wood together well, nails should be driven in at right angles to the grain. They should be long enough so that two-thirds of a nail's length is in the second piece of wood. A nail can split very hard wood. This can be avoided by drilling holes smaller than the diameter of the nail. These holes, called **pilot holes**, make it easier for the nail to go into the wood.

There are five common kinds of nails: brads, finishing nails, casing nails, common nails, and box nails. Nails come in many different sizes. The size of nails is measured in **penny weight**. Penny weight used to refer to the amount of money it cost to buy one hundred nails. An eight-penny nail (abbreviated 8d) used to cost eight cents per hundred. Now penny weight refers only to nail length.

Screws can be used to draw pieces of materials tightly together. Screws hold better than nails. They are also easier to remove.

Different kinds of screws are made for use in metal and wood. Wood screws are threaded about two-thirds of the way to the head. **Sheet metal screws**, used for fastening sheets of metal, are threaded all the way. Pilot holes are needed for sheet metal screws. A sheet metal screw can't pierce a piece of metal, so a pilot hole is drilled. They make their own thread as they are threaded into a piece of sheet metal.

Machine screws and bolts do not have pointed ends. The thread is the same diameter from tip to head. These fasteners are held in place by a **nut**. **Washers** are used between the nut and the material being fastened. Flat washers keep the material from being damaged by the nut. Lock washers keep the nut from loosening when the screw is vibrated, such as when a lawn mower is in use.

Riveting is often used on aircraft. Rivets hold pieces of sheet metal together. One end of a rivet is already formed. The other end is formed after it is placed through the two pieces of material being fastened, and hammered into shape.

Fastening Materials with Heat

Soldering is joining metals by melting metal between them. Solder is an alloy made from lead and tin. It melts at about 450° F. Soldering irons or guns are used to melt the solder. This method is most often used to attach wires in electronic circuits.

Welding is another way to use heat to join metals. Metals to be welded are heated (to 6000°-7000° F) so they join. The heat comes from a welding torch. The torch heats by burning gas and oxygen (gas welding) or by using electrical currents (arc welding).

Types of Fasteners

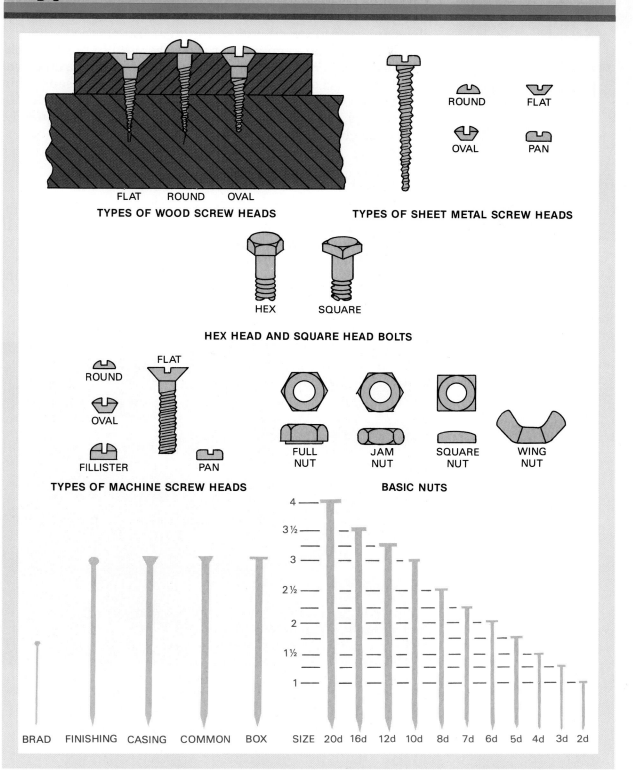

TYPES OF WOOD SCREW HEADS
FLAT   ROUND   OVAL

TYPES OF SHEET METAL SCREW HEADS
ROUND   FLAT
OVAL   PAN

HEX HEAD AND SQUARE HEAD BOLTS
HEX   SQUARE

TYPES OF MACHINE SCREW HEADS
ROUND
OVAL
FILLISTER
FLAT
PAN

BASIC NUTS
FULL NUT   JAM NUT   SQUARE NUT   WING NUT

BRAD   FINISHING   CASING   COMMON   BOX   SIZE   20d   16d   12d   10d   8d   7d   6d   5d   4d   3d   2d

Pennsylvania's Washington County courthouse has been restored to look as beautiful as it did when it was built at the turn of the century. The project required 450 gallons of paint, 15 gallons of stain, and 26 quarts of custom colorants. (Courtesy of PPG Industries)

Gluing Materials

Gluing is another way to fasten materials. Glue forms chemical bonds between itself and the material being glued. A good glue joint is stronger than the materials it fastens. Different glues are used for wood, metal, ceramic, and plastic.

Until a few years ago, copper pipe was used in most plumbing jobs. Today, plastic pipe called PVC (made from a chemical called polyvinyl chloride) is used. Special glues are used to join pieces of PVC pipe.

Hot glues, put on with glue guns, are strong and set quickly. They are being used to fasten airplane parts together.

Coating Materials

To protect or beautify a material, we can coat it. Paint, stain, and wax are some of these coatings. Glazes are used to coat ceramics. Metals can be coated with other metals by **electroplating**. Gold-plated jewelry and silver-plated tableware are made this way. Aluminum is coated through **anodizing**. A thin oxide coating forms on the metal. In **galvanizing**, steel is coated with zinc to keep it from rusting.

This van has a body manufactured from composite materials. (Courtesy of General Motors Corporation)

The Electroplating Process

Electroplating combines electricity and chemistry. The part to be plated is connected to the negative terminal of a battery. The metal that does the plating is connected to the positive terminal. In copper plating, a liquid plating solution of copper sulfate is used.

The solution breaks down into charged particles called ions when electricity passes through it. The copper ions are positive and move to the negative terminal. They are replaced in the solution by ions from the copper metal. The process can be used to plate many different kinds of metals.

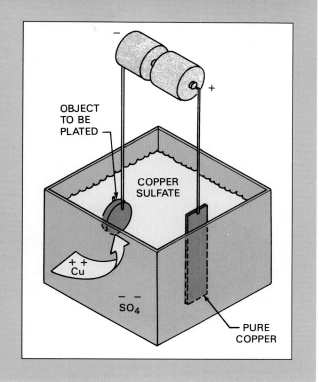

OBJECT TO BE PLATED

COPPER SULFATE

$+ + Cu$

$- - SO_4$

PURE COPPER

Making Composite Materials

The ancient Egyptians learned how to make composites. They added straw to the clay they used for making bricks, making the bricks stronger. Today we have many composite materials. These are materials formed from several different kinds of materials. A composite has properties that are better than those of the materials it is made from. Plywood is a composite. Layers of wood are crisscrossed and glued, forming a strong wood panel.

Fiberglass is a composite that can be stronger than steel, though it weighs much less. Fiberglass is used to make the bodies of some automobiles. Other composites are used to make the bodies of airplanes.

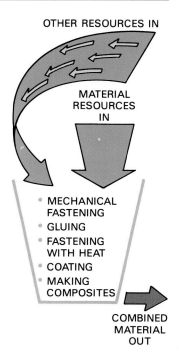

OTHER RESOURCES IN

MATERIAL RESOURCES IN

- MECHANICAL FASTENING
- GLUING
- FASTENING WITH HEAT
- COATING
- MAKING COMPOSITES

COMBINED MATERIAL OUT

Some combining processes

CONDITIONING PROCESSES

Conditioning a material changes its internal properties. In magnetizing a piece of steel, for example, we cause iron molecules to line up in one direction.

Heat-treating can also cause changes in a material. When steel is heated cherry-red and quickly cooled in water, the steel becomes harder. If it is heated again, not quite as hot, and

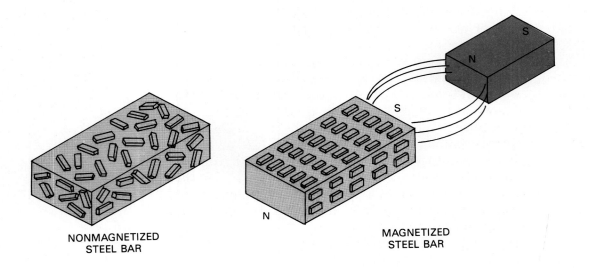

NONMAGNETIZED
STEEL BAR

MAGNETIZED
STEEL BAR

There are several ways people can process materials. These processes include forming, separating, combining, and conditioning.

cooled, it becomes tough. This process is called **tempering**. Heating steel until it is cherry-red and cooling it slowly makes softer steel. This is known as **annealing**.

When clay is fired in a kiln, it becomes much harder and stronger. When a piece of metal is hammered, it becomes harder. This is **mechanical conditioning**. When we mix plaster with water, a chemical reaction occurs. Heat is given off and the plaster hardens. This is **chemical conditioning**. In each case, the change takes place inside the material itself.

Other examples of conditioning are:

■ light exposing photographic film;
■ chemicals developing photographic paper;
■ baking a cake;
■ cooking an egg;
■ boiling water;
■ melting ice;
■ placing a metal in liquid nitrogen to make the metal a superconductor;
■ using radiation to shrink a tumor;
■ making butter from cream;

- sending electricity through the filament wire of a light bulb; and
- making wine from grape juice.

PROCESSING INFORMATION RESOURCES

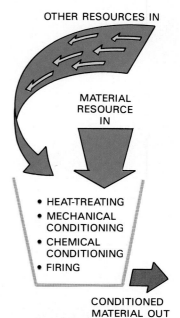

Some conditioning processes

Information makes today's world run. For all people, information is necessary for living. For people such as computer programmers, bankers, teachers, doctors, and stockbrokers, information is a major resource.

Information is data that has been processed. **Data** by themselves are raw facts and figures. Data alone don't tell us very much. They are just numbers, letters, and symbols. To make sense out of raw data, we must convert them into meaningful information. A series of numbers and letters like 12125551212 means nothing unless we understand what the data stand for. We must give meaning to the data. Once we know that the data stand for a telephone number, 12125551212 begins to have some meaning. When we know that the telephone number is that of directory assistance in New York City, the data have become meaningful information.

Data must be organized so that they make sense. **Data processing** is the act of converting raw data into information. We processed data when we learned that 1-212-555-1212 is a telephone number.

Information can be used by people who understand it. When people interpret information, they create knowledge. **Information processing** is the act of converting information into new,

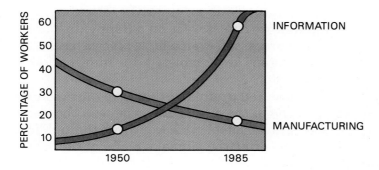

In 1950, about 30 percent of the workers in this country worked in manufacturing industries. Today, only about 16 percent of the workers work in such industries. In 1950, only about 17 percent of the workers worked in jobs that process information. Today, over 60 percent of the working population in the United States works in information jobs.

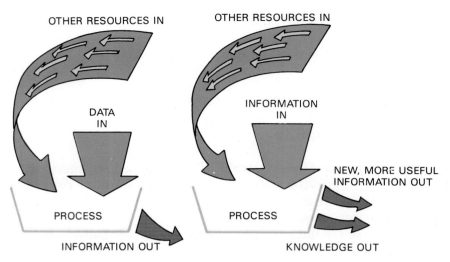

Data and information processing.

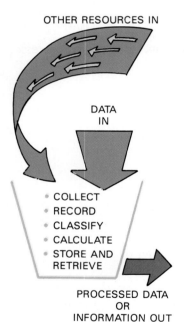

Steps in processing data

more useful information or into knowledge. When we learned that 1-212-555-1212 is the telephone number for New York City directory assistance, we processed information.

Five steps are needed to change data into information.

First, the data must be **collected**. If we are electing a class president, we first count the votes. This is collection.

Second, we must **record** the data. We write down on the blackboard how many votes each person received.

Third, we **classify** the data. We must determine who received each vote.

Fourth, we must **calculate** the number of votes each person received. We add the votes. Calculating provides new data. In our case, it tells us who has the most votes and has won the election.

Fifth, we must **store and retrieve** the data. We can store and retrieve it from our minds or from the blackboard. If we're using a computer, we can store it on a floppy disk or tape.

Once data are processed even a little bit, we have some information. As more steps are completed, the information becomes more useful.

To make information useful to others, we must **communicate** it. We let everyone in the class know who won the election. We might tell the whole school who won over the public address system. We might print the winner's name in the school newspaper. Communicating information can be done in many ways.

Computers process and communicate information. They also control systems that process materials and energy.

A traveling executive communicating information by means of a hand-held computer and a public telephone. (Courtesy of RCA)

CONNECTIONS: Technology and Science

■ Matter has physical and chemical properties that affect how it reacts to change. A physical change occurs when physical properties such as size, shape, or phase change without changing the identity of the substance. For example, crushing a rock or melting an ice cube are physical changes.

  Chemical changes occur when substances are changed into different kinds of matter. Burning oil and digesting food are examples of chemical changes.

■ Matter can be classified as elements, compounds, and mixtures. The simplest forms of matter that cannot be broken down into simpler substances by ordinary methods are elements. Compounds are made of two or more elements that are chemically combined. Water (H_2O) made up of the elements hydrogen and oxygen, and carbon dioxide (CO_2) composed of carbon and oxygen are compounds. Mixtures consist of elements and/or compounds that are not chemically combined. The components in a mixture keep their own properties. Air is a mixture, made up of oxygen, nitrogen, carbon dioxide, and water vapor.

PROCESSING ENERGY RESOURCES

Energy is the ability to do work. When we move something, we do work. If we lift a box weighing 50 pounds a distance of 1 foot, we have done 50 foot-pounds of work.

Heat energy is measured in Btus (British thermal units). A Btu is the amount of heat energy needed to raise the temperature of 1 pound of water 1° Fahrenheit. Heat energy can be changed to work. For example, 1 Btu is equal to 778 foot-pounds of work. That means it takes 1 Btu of heat energy to move a 778-pound weight 1 foot.

Potential and Kinetic Energy

Suppose we lift an object off the ground. If we let it fall, it could do work. It could flatten something or trip a lever. Because the object has the potential (the ability) to do work, we say it has **potential energy**. Potential energy is energy an object has because of its position with respect to another object. Potential energy depends on the weight of the object and the distance through which it could move. An object on a table has potential energy with respect to the floor. It could fall to the floor and do work. It has no potential energy with respect to the table top.

Stored energy is also potential energy. A battery stores energy, so it has potential energy. A firecracker, a compressed spring, and a stretched rubber band all have potential energy.

If we drop an object, its potential energy is used up. It is changed to **kinetic energy**. Kinetic energy is the energy of things that are moving. A moving automobile has kinetic energy. This energy must be absorbed when a car crashes. Large barrels filled with sand are placed in front of bridges by the road to do this job. How much kinetic energy an object has depends on how fast it is moving and on its mass.

Potential energy can be changed to kinetic energy and vice versa. Water held behind a dam has potential energy. It can fall and turn a turbine. When this happens, its potential energy is changed to kinetic energy. Likewise, kinetic can be changed to potential energy. When we compress a spring, we use kinetic energy. We are storing potential energy. When we let go of the spring, its potential energy changes to kinetic energy.

The box on the table has potential energy. It could fall and do work. For example, it could fall on the lever and make the person on the seesaw go for a ride.

Principles of Energy Conversion

In Chapter 4, six sources of energy were listed. **Human and animal muscle power** is still in use today. People in developing nations rely on this kind of energy. **Solar energy** gives us wind, heat, and light energy. **Chemical energy** comes from burning wood, coal, oil, and gas. **Gravitational energy** comes from tides and falling water. Heat deep inside the earth makes **geothermal energy**. **Nuclear energy** is made by turning matter into energy.

The sun is the original source of much of the energy on earth. We cannot create more energy. We can only change matter into energy, or we can change one form of energy into another.

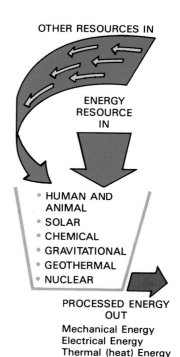

OTHER RESOURCES IN

ENERGY RESOURCE IN

- HUMAN AND ANIMAL
- SOLAR
- CHEMICAL
- GRAVITATIONAL
- GEOTHERMAL
- NUCLEAR

PROCESSED ENERGY OUT
Mechanical Energy
Electrical Energy
Thermal (heat) Energy
Light Energy
Sound Energy

Processing energy resources

Energy Conversion in an Automobile Engine

In an automobile engine, gasoline burns (chemical energy), producing heat (thermal energy), which is changed to mechanical energy.

This is done in four steps. In step one, the piston moves down, pulling a mixture of gasoline and air into the cylinder. In step two, the piston moves up and compresses the fuel. At this instant, the spark plug fires, igniting the gas-air fuel mixture. The heat formed makes the gases expand, pushing the piston down. This is the third step. In step four, the piston moves back up and the exhaust valve opens. Burned gases are forced out. The cycle begins again with the first step.

Gasoline is a source of potential energy in the engine. As it burns, it is changed to heat energy. The heat energy causes the engine parts to move. As the parts move, they have kinetic energy.

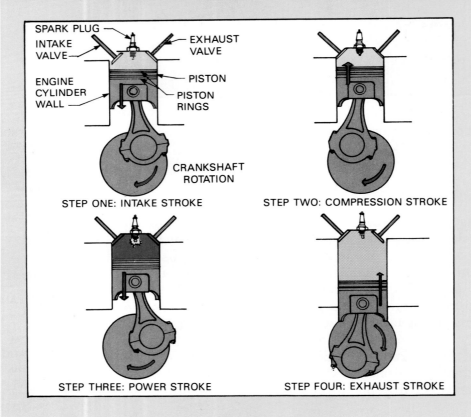

STEP ONE: INTAKE STROKE

STEP TWO: COMPRESSION STROKE

STEP THREE: POWER STROKE

STEP FOUR: EXHAUST STROKE

Energy Converters

Energy converters are used to change one form of energy into another. We use them because we do not always want to use energy in the form it comes in. Cars use gasoline's chemical energy by changing it to heat and then to mechanical energy. This kind of change is very common. Automobile, diesel, and rocket engines all work this way.

We can also change mechanical energy into electrical energy. Electrical generators work this way. The electricity we use in our homes and cars is made by changing mechanical energy.

Power companies burn coal or oil (chemical energy) to heat water and make steam (thermal energy). The steam turns turbine blades (mechanical energy) that are connected to electrical generators (electrical energy).

Hydroelectric power comes from changing the energy of falling water (mechanical) to electrical energy. Falling water turns turbine blades (mechanical energy), and the turbines turn the electric generators (electrical energy).

Generators change mechanical energy to electrical energy. Electric motors do the opposite. They use electricity to provide motion.

Electrical energy can be changed to chemical energy. Copperplating, described earlier, changes electrical energy to chemical energy.

Some Energy Converters

CHEMICAL TO THERMAL	THERMAL TO MECHANICAL	MECHANICAL TO ELECTRICAL	ELECTRICAL TO LIGHT
home heating furnace	steam-powered turbine	electrical generator (dc)	light bulb
gasoline engine	gasoline engine	electrical alternator (ac)	fluorescent tube
rocket engine burning coal, oil, or wood	rocket engine	crystal used in microphones and phonograph cartridges	lasers and light-emitting diodes (LEDs)

The end result of a conversion process must be monitored if we want to be sure the desired goals are achieved.

Photocells change light to electrical energy. They are made from materials that give off electrons when light falls on them.

Nuclear power plants change matter to energy. When an atom of uranium 235 is struck by another atomic particle, the atom is split. Some of its mass is changed into energy. Such a change can create huge amounts of energy. From each gram of matter, 25,000,000 kilowatt-hours of energy could be produced.

SUMMARY

We process resources to make them more useful and valuable. All seven technological resources are needed in processing. Primary raw materials are processed into industrial materials. From industrial materials, products are manufactured. Materials are processed by forming, separating, combining, and conditioning them.

Forming a material means changing its shape without cutting it. Forming processes include casting, pressing, extruding, blow molding, and vacuum forming. Separating processes separate one piece of the material from another. Cutting is a separating process. We can cut materials by shearing, sawing, drilling, grinding, shaping, and turning. Combining materials means putting two or more materials together. We can combine materials by fastening, coating, or making composites. Conditioning materials means changing their internal properties.

CONNECTIONS: Technology and Social Studies

- The constant and rapid changes in the labor market have a large impact upon the work force. It is unlikely that a person will hold the same job for his/her entire life. Changing careers have become the norm rather than the exception.
- Can the United States continue to use more than its share of the earth's energy resources? It would appear that the other nations of the world will demand their fair share. Greater efforts to develop and use increased amounts of renewable energy sources has to be the wave of the future.
- Selling stock in a corporation is a major means of getting the necessary capital to begin or expand a business. This type of investment can benefit everyone. For the investor it is an opportunity to make money while the company executives get the capital they need to continue their business.

Conditioning processes include magnetizing, heat treating, mechanical working, and chemically processing a material.

Information processing has become very important in the modern world. Many jobs depend on information. Information is data that have been processed. Data are raw facts and figures. Data processing is converting raw data into information. Information processing is converting information into new, more useful information.

We can turn matter into energy or change one form of energy to another. Energy is the ability to do work. Potential energy is energy an object has because of its location. It is also stored energy. Kinetic energy is energy that moving objects have.

We can change the six kinds of energy into other kinds that are more useful to us. Devices that do this changing are called energy converters. We can change chemical energy into thermal energy and thermal energy into mechanical energy. A gasoline engine makes these changes.

Feedback

(Courtesy of PPG Industries)

1. This chapter discussed processing three resources: materials, information, and energy. Which of the seven technological resources can be processed? Which cannot? Why?
2. What is the output of a system that processes materials?
3. What is the output of a system that processes information?
4. What is the output of a system that processes energy?
5. What part do people play in the resource conversion processes?
6. What actually occurs within a technological process?
7. Why must the end results of a conversion process be monitored?
8. What is the role of computers in processing materials, information, and energy?
9. What are some ways materials may be processed?
10. What processes would you use to make hamburgers from raw meat?
11. How do sawing, drilling, and grinding differ?
12. How would you fasten a metal bracket to a wooden shelf?
13. Why would you want to weld two pieces of metal together rather than use screws or glue?
14. How does coating wood with wood finish differ from the way you would finish a metal surface?
15. What happens to the internal structure of a piece of steel when it is magnetized?
16. List five examples of conditioning materials.
17. Make a chart showing jobs that involve manufacturing products and jobs that involve processing information.
18. Suppose you are going to buy a car. List the steps used in data processing and explain how you would use each step in deciding which car to buy.
19. List six sources of energy.
20. Describe five examples of how energy can be converted from one form to another.

(Courtesy of Delta International Machine Corp.)

(Courtesy of Delta International Machine Corp.)

CHAPTER KEY TERMS

Across

4. Raw materials are processed into these.
6. A means of joining metals with heat and a low-temperature alloy.
8. The process of fastening, coating, and making composite materials.
10. Combining the seven technological resources.
11. Data that are processed.
12. Changing the internal properties of a material.
14. A device that changes one form of energy into another.
15. Can be done by mechanical means, heat, or glue.
16. Hardening, tempering, and annealing metal.

Down

1. Pouring liquid into a mold.
2. Raw facts and figures.
3. A cutting process using blades with teeth.
5. Materials found in nature.
7. To distribute information to others.
9. You pour a casting into this.
13. The process of separating one piece of material from another.

Casting
Combining
Communicate
Composites
Conditioning
Cutting
Data
Drilling
Energy
  converter
Extruding
Fastening
Forming
Grinding
Heat treating
Industrial
  materials
Information
Mold
Processing
Primary raw
  materials
Sawing
Separating
Soldering

SEE YOUR TEACHER FOR THE CROSSTECH SHEET

(Courtesy of Cerro Metal Products)

DESIGNING A SOLAR FARM

Problem Situation

For many years, one goal of scientists and engineers has been to find alternative methods of providing electrical power. The use of solar energy to create steam for the turbines has proven very promising. Hundreds of mirrored reflectors covering acres of land have created the first solar farms. These reflectors track the sun's path across the sky, aiming its heat energy at a pressured tank of water standing atop a tower. The water boils, creating steam that powers a small turbine generator. The result: electricity from free and abundant fuel, the sun.

Design Brief

In this activity you will design tracking solar reflectors for a solar farm, staying within the following criteria:

- reflector diameter = 6' − 0"
- tower height = 20' − 0"
- reflector style = flat or parabolic
- manual tracking of 180 degrees horizontally and 45 degrees vertically

Build a scale model of a tracking reflector. When you and your classmates are finished, set up a miniature solar farm. Measure the temperature rise of a coffee can filled with water over a two-hour period.

Suggested Resources

Prototyping materials such as: wood, metal, plastic, etc.
Mylar for reflector
Material processing equipment

Procedure

1. The tower and reflectors will be built at a scale of ½" = 1' − 0".
2. Make a few freehand sketches of possible tower designs. Be sure to stay within the constraints of the problem.
3. Make a few sketches of pivoting mechanisms that will allow the reflector to track according to the problem specifications.
4. Select a design and appropriate materials.
5. Using the safety procedures described by your teacher, build and test your prototype.

Classroom Connections

1. Materials are processed by separating, combining, and forming techniques.
2. The layout of a parabolic curve can be calculated using the principles of geometry. See diagram.

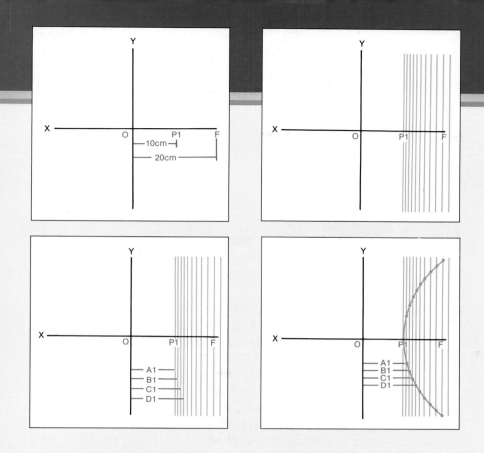

3. The law of Conservation of Energy states that we can only change energy from one form to another, or store energy. We can neither create nor destroy energy.

Technology in the Real World

1. The power of the sea and its tides is being used to turn specially designed turbine generators. Tidal energy, as it is called, may help ease our dependence on fossil fuels.
2. The use of superconductors in wires and generators creates an almost resistance-free pathway for electricity to flow.
3. Nuclear fusion is a safer and cleaner method of nuclear energy production than nuclear fission. Recent developments in this field may bring fusion technology to reality some day.

Summing Up

1. What advantages does a parabolic reflector have over a flat plate reflector?
2. In what parts of the country would solar farm technology be most practical?
3. What are some other applications of solar energy technology?
4. List some of the material processing techniques you used in the construction of your solar reflector and tower.

ACTIVITIES

TECHNOLOGY TESTING LABORATORY

Problem Situation

The study of metals is called **metallurgy**. Scientists use microscopes and other instruments to determine the internal structure of metals. Technologists are concerned about the way metals behave in use.

The chemical properties of a metal show how it is affected by its environment. One example of this is the ability of steel to resist rusting. Engineers study the corrosion resistance of steel because corrosion is the leading cause for the failure of steel parts such as automobile fenders and the cables of suspension bridges.

The corrosion resistance of steel can be increased several ways. Steel can be dipped in molten zinc to provide a protective coating. This process is called galvanizing. Combining steel with other metals can produce corrosion-resistant alloys. Stainless steel, an alloy made by adding chromium (and sometimes nickel) is very corrosion resistant.

Design Brief

In this activity you will investigate the corrosion resistance of steel. Protected and unprotected samples will be exposed to plain water and salt water. You will record your observations and evaluate the effectiveness of several coatings.

Suggested Resources

Eight #10 common nails
Two #10 galvanized nails
Ten test tubes
Materials to construct test tube racks
Table salt
Exterior enamel
Primer

Procedure

1. Coat four common nails by dipping in primer. Allow to dry overnight.
2. Apply enamel to two of the primed nails. Also apply enamel to two unprimed common nails. Allow two days for drying.
3. Construct two test tube racks that hold five test tubes each. NOTE: Use the materials and equipment provided by your teacher. Follow all safety procedures.
4. Prepare the salt water solution by dissolving two ounces of salt in a cup of warm water.
5. Fill five test tubes with salt water and five with plain water.
6. Set up the two test tube racks as shown in the diagram. In each rack you should have five different kinds of nails: one plain common nail, one primed common nail, one common

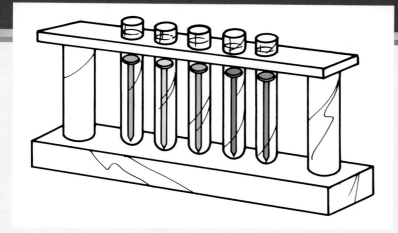

nail with enamel only, one common nail with primer and enamel, and one galvanized nail. Place the racks in a safe location as specified by your teacher.

7. Prepare a log to record your observations for ten days.
8. Write a brief statement describing what you see in each of the ten test tubes each day.
9. After ten days, write a paragraph summarizing the results of your investigation. Follow your teacher's directions for cleaning the test tubes and disposing of the other materials.

Classroom Connections

1. Corrosion resistance is the ability to resist oxidation.
2. Oxidation is the interaction of oxygen with a material.
3. The corrosion (oxidation) of iron and steel is called rusting.
4. The oxide coating that forms on aluminum prevents additional corrosion.

Technology in the Real World

1. Millions of dollars are spent each year to repair corrosion damage. Air pollution and road salt contribute to this problem.
2. Manufacturers are developing new ways to prevent oxygen and moisture from reaching metal. Automobile bodies are often dipped into protective coatings.

Summing Up

1. Why is it important to prevent the corrosion of metals?
2. List three ways of increasing the corrosion resistance of steel products.
3. Which coating(s) used in your investigation provided the best protection? Why?

CHAPTER 6

TECHNOLOGICAL SYSTEMS

MAJOR CONCEPTS

After reading this chapter, you will know that:

- People design technological systems to satisfy human needs and wants.
- All systems have inputs, a process, and outputs.
- The basic system model can be used to analyze all kinds of systems.
- In a technological system, a technological process combines resources to provide an output in response to a command input.
- Feedback is used to make the actual result of a system come as close as possible to the desired result.
- Systems often have several outputs, some of which may be undesirable.
- Subsystems can be combined to produce more powerful systems.

SYSTEMS

A **system** is a means of getting a desired result. A technological system does this through technology. For example, an automobile is a technological system for traveling from one place to another. A radio lets us listen to music or news. A computer lets us do calculations quickly. A system can be huge, such as the space shuttle. It can also be small, like a pocket calculator.

Technological systems are all alike in one way. Each has **inputs**, a **process**, and **outputs**.

People design technological systems to satisfy human needs and wants.

Modern technological systems are all around us, touching our lives in many ways. (Courtesy of (clockwise from sailboat) Pearson Yachts; Perini Corporation; AT&T; NASA, Cray Research Inc.)

All systems have inputs, a process, and outputs.

151

Problem Situation 1:   A cam is used to change circular motion to vertical or horizontal motion.

Design Brief 1:   Design and construct a pull toy that uses a cam to get amusing movements.

Problem Situation 2:   Car thieves may be discouraged by cars with alarm systems and engine cut-out switches.

Design Brief 2:   Design and construct a system that includes both these features; use the car horn as the alarm. The system should activate when any door opens. The car owner should be able to turn the system on and off.

Problem Situation 3:   Black asphalt roof shingles often absorb the summer sun's heat to create a very high temperature in the house.

Design Brief 3:   Design a temperature-controlled system for cooling the roof of a house; the system should utilize the process of evaporation.

Problem Situation 4:   At manufacturing sites, it is often necessary to sort products according to color.

Design Brief 4:   Design and construct a mechanism which will automatically sort colored and clear marbles of the same size.

THE BASIC SYSTEM MODEL

The basic system model can be used to analyze all kinds of systems.

All systems include inputs, a process, and outputs. Feedback is added to provide a better way of controlling the system. The **basic system model** can be used to describe any technological system. A **system diagram** can be drawn to show how these parts work together in a system.

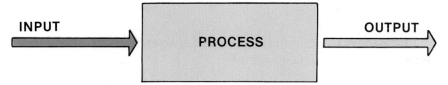

Diagram of a basic technological system.

Inputs

The **input** is the command we give a system. It is also the **desired result**. When we turn on a television set, we are giving it a command. That command is: "Give us picture and sound."

Let's look at another example. A car moves when we tell it to by stepping on the gas. The input command (or desired result) might be: "Go 30 miles an hour."

Other system inputs are the resources needed by the system (see Chapter 4). The seven kinds of resources used by technological systems are people, information, materials, tools and machines, energy, capital, and time.

In a technological system, a technological process combines resources to provide an output in response to a command input.

CONNECTIONS: Technology and Science

- An ecosystem is a complex system of interactions of living things with each other and their physical environment. Forests, ponds, streams, marshes, and fields are some ecosystems. Living things include plants, animals, and decomposers. The physical environment includes minerals, water, and gases, which are continuously being recycled. Energy, mainly sunlight, enters the ecosystem. Sunlight, as well as carbon dioxide and water, is important to plants' food-making capabilities. This food in turn provides the energy and materials needed by animals. When the plants and animals die, decomposers return the material and energy back to the environment so it can be reused.

- An electric circuit can be thought of as a simple system. The components of the system include an energy source, such as a battery; a load, such as a light bulb; a switch; and connecting wires. The battery provides the source of electrical energy which leaves the negative end and flows through the wires, the closed switch, and the light bulb. The light bulb will remain lit as long as the parts are connected and working.

The Process

The **process** is the action part of a system. It combines the resources and produces results.

In an automobile, the process involves both the car and the driver. The seven technological resources are used in the process. Energy is stored in the gasoline. The machine is the car. People (the driver), information, time, materials, and capital work together to make the car go 30 miles an hour.

Outputs

The **output** is what is produced. It's the **actual result**. We hope that the output matches the command input. That is, we hope that our car will go 30 miles an hour when we step on the gas. Most systems have more than one output.

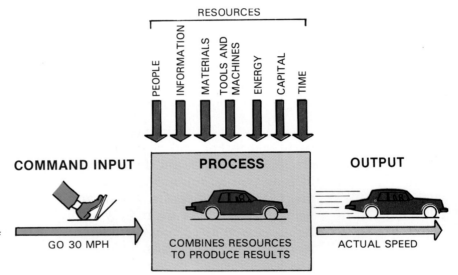

The process combines the seven technological resources to produce the desired result. The output of the system is the speed that the car actually goes.

Feedback

How does the driver know when the car is going 30 miles an hour? The driver checks the speedometer. The speedometer gives the driver **feedback**. Feedback is information about the output that can be used to change the output. When the car's speed reaches 30 miles an hour, the driver lets up on the gas. The driver then pushes only hard enough to keep the car going at 30 miles an hour.

The speedometer is a **monitor**. A monitor gives feedback about output. It lets us compare the actual result to the result we want. We can **control** the system, if needed, to get the out-

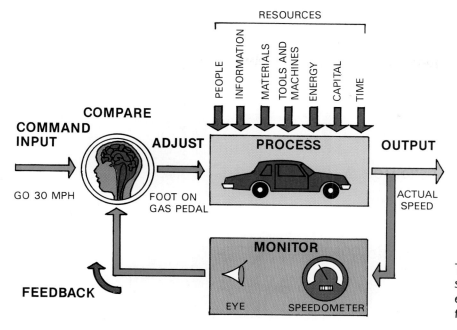

The combination of the speedometer, the driver's eye, and the driver's brain forms the feedback loop.

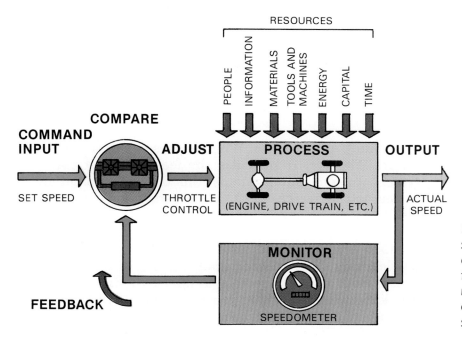

If a car has an automatic speed control system, the driver is removed from the feedback loop. The speed is monitored automatically and compared to the desired speed set by the driver.

put we want. Systems with feedback are sometimes called **control systems** or **feedback control systems**.

An example of a technological control system is a sump pump. A sump pump is used to pump water out from under a house before the basement floods in times of heavy rainfall. It can be turned on and off by a person. When the water gets

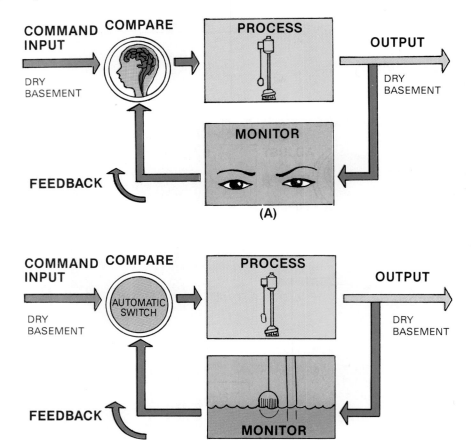

The system diagram looks the same, whether a person provides feedback and control of the pump (A) or the feedback and control are supplied automatically by a float and switch (B).

Feedback is used to make the actual result of a system come as close as possible to the desired result.

high, the pump is turned on. When it goes down, the pump is turned off. The person provides the feedback that controls the system. In some sump pumps, this is done automatically. A float turns the switch on or off. A person isn't needed for the job.

A control system is being used to help you learn. Your teacher uses homework and tests as monitors to see how well you are learning. Your teacher grades and returns homework and tests. Both of you use this feedback to find out where you are doing well, and where you need more work or perhaps some help.

Systems that have feedback are called **closed-loop systems**. Feedback "closes" the loop from input to output. Some systems don't use feedback. They are called **open-loop** systems. A person who wears a blindfold while trying to draw a picture of a dog is an example of an open-loop system. Without feedback, the person cannot compare the output to the input. When the blindfold is removed, the person can do the job. Now we have a closed-loop system.

Instant Feedback, Medieval Style

In the Middle Ages, teaching machines were used to train knights. A knight on horseback would charge a wooden figure mounted on a pivot. If he struck the figure in the center, it would fall over. If not, the figure would swing around and hit the knight with a club. It was instant feedback and instant learning.

Even our own bodies contain systems. For example, our bodies maintain a temperature of about 98.6° Fahrenheit. The input command to the body's temperature regulation system is the desired temperature, 98.6° Fahrenheit. Maintaining that temperature involves the action of the muscles, skin, blood, and the body "core." The output is the actual body temperature.

The body has many other control systems. There are systems that regulate sugar level, heartbeat, oxygen collection, and other important activities. These control systems keep our body conditions just about the same, even though outside conditions may be changing.

You can perform a feedback experiment using a pencil. Place the pencil on a table. Close your eyes, turn around once, and

IN HOT WEATHER, EVAPORATING SWEAT COOLS THE BODY DOWN. IN COLD WEATHER, SHIVERING WARMS THE BODY UP.

Sweating and shivering are part of the body's feedback control system that maintains a constant temperature.

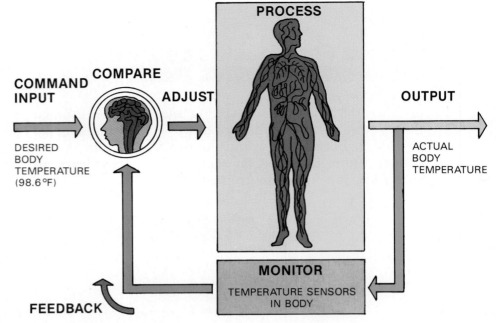

A human system

try to pick up the pencil without opening your eyes. You probably can't do this easily. You have no visual feedback (you can't see the pencil). You may find that if you feel around you can get the pencil. You will be using tactile (touch) feedback. You use tactile feedback when you use your fingers to pick up or hold something.

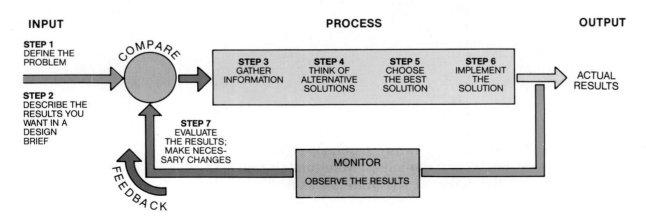

CONNECTIONS: Technology and Mathematics

■ Using the formula $D = R \times T$ (distance = rate × time), you can find any one of the variables, D, R, or T, when you are given the other two variables. For example, if you are traveling on a highway and your speedometer reads 50 miles per hour (R), you can calculate your distance (D) after 4 hours (T) of driving.

$$D = R \times T$$
$$D = 50 \text{ miles per hour} \times 4 \text{ hours}$$
$$D = 200 \text{ miles}$$

Your speedometer is broken, but you notice that you have traveled 10 miles in 15 minutes. How fast have you been traveling?

$$D = R \times T$$

$$\frac{D}{T} = R$$

$$\frac{10 \text{ miles}}{15 \text{ minutes} \times \dfrac{1 \text{ hour}}{60 \text{ minutes}}} = R$$

$$\frac{10 \text{ miles}}{¼ \text{ hour}} = R$$

$$40 \text{ miles per hour} = R$$

■ A thermometer tells you the output of your body's temperature regulation system. If the thermometer has a Fahrenheit scale, normal body temperature will read 98.6°. What will the thermometer read if it has a Celsius scale?

37° C   98.6° F

$$C = \frac{5}{9}(F - 32°)$$

$$C = \frac{5}{9}(98.6° - 32°)$$

$$C = \frac{5}{9}(66.6°)$$

$$C = \frac{333°}{9}$$

$$C = 37°$$

MULTIPLE OUTPUTS

Systems often have several outputs, some of which may be undesirable.

A system may produce several outputs. They can be desirable, undesirable, expected, or unexpected. This may be true even if we designed the system to produce only one desired output. A coal-burning power plant is designed to produce electricity. However, it also produces heat, smoke, ash, noise, and other outputs.

Sometimes extra outputs are useful. Heat produced by a power plant can be used to heat nearby buildings. Other outputs, such as noise or smoke, may be unwanted. We may have to take steps to reduce or eliminate them, even if we get less electricity when we do so.

When designing systems, a person must consider unexpected outputs as well as expected outputs. Sometimes, designs must be changed. We may lose some of the desired output when we reduce the unwanted outputs.

CONNECTIONS: Technology and Social Studies

- In an economic system, manufacturers receive feedback regularly from the consumer. This is done in the marketplace where people decide what they will buy. Whether a product sells or not tells the producer what future course of action they need to pursue.
- As a result of improved systems of communication, the ability of people to interact with each other has become almost instantaneous. Fax machines, space satellites and computers have reduced the amount of time people need to wait to communicate with each other.

Four Kinds of Output

Outputs from a system can be of four types. A power plant's outputs could include all four.

1. The expected, desirable output from a power plant is electricity. The added output of heat is also expected. If something useful is done with it, it is also desirable.
2. The expected, undesirable output from the power plant is noise and smoke.
3. An unexpected, desirable output was found at one power plant. The plant discharged some of its heat into a river, warming the water. Tropical fish flourished in the river near the plant, creating an attraction.
4. An unexpected, undesirable output of some plants is acid rain. Acid rain is caused by the pollution power plants produce.

This power plant produces all four kinds of output.

SUBSYSTEMS

Systems are often made up of many smaller systems called **subsystems**. When you are trying to understand a large system, you might find it helpful to break it into subsystems. You can study each of them separately. Suppose you want to look at a transportation system that carries goods by truck from Los Angeles to New York City.

Subsystems can be combined to produce more powerful systems.

You could break down the large system into smaller ones. Some of the subsystems would be the vehicle system, the management system, and the communication system. Each of these could be broken down further into more subsystems. You would create a subsystem tree.

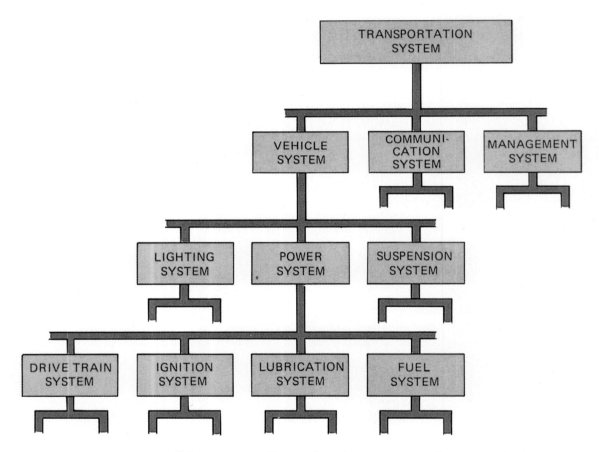

Subsystem tree diagram for a large transportation system. Each subsystem can be broken down into smaller subsystems. (From Hacker & Barden, *Living With Technology,* Copyright 1988 by Delmar Publishers Inc.)

SUMMARY

Technological systems are made by people to satisfy human needs or wants. A system is a method of achieving the results that we desire.

All systems have inputs, processes, and outputs. The input is the command we give a system. The input command is also called the desired result. The process is the action part of a system. It combines the resources and produces results. The output is the actual result delivered by the system.

Feedback is used to make the actual result of a system come as close as possible to the desired result. Feedback is made up of a monitor that observes the actual result, a comparator that compares the actual result with the desired result, and a controller that changes the process to make the output closer to the desired result.

Systems with feedback are called closed-loop systems. Systems that do not have feedback are called open-loop systems. Open-loop systems cannot be controlled as well as closed-loop systems.

Large systems are often made up of smaller subsystems. Examining each subsystem by itself can be useful in understanding a large, complex system.

Systems often have multiple outputs. When we design a system, we must think about any possible undesirable outputs. We may have to modify a system design to reduce or eliminate the undesirable outputs.

Feedback

1. List five important parts of a technological system.
2. What does the process part of the system do?
3. Why is feedback important in a system?
4. Give an example of
   a. feedback you've received recently at school.
   b. feedback you've received recently from a friend.
   c. feedback you can receive when riding a bicycle.
   d. feedback in a technological system.
5. Give an example of feedback in a body system.
6. Name some subsystems that make up a large railroad system.
7. Using the basic system model, model the operation of a nuclear power plant. Indicate the input (desired result), the process, the output (actual result), monitoring, and comparison.
8. The automobile has become a very important system of transportation. Identify an output resulting from the development of the automobile that is:
   a. expected and desirable.
   b. expected and undesirable.
   c. unexpected and desirable.
   d. unexpected and undesirable.

CHAPTER KEY TERMS

Across
2. Information we get by monitoring the output of a system.
4. The command you give a system.
5. Could be expected, unexpected, desirable, or undesirable.
6. The action part of the system.
7. A system that does not have a feedback loop.
9. This system includes a feedback loop.
11. The part of a system that adjusts the process.
12. People design technological _____ to help achieve desired results.

Down
1. We compare the actual results to these.
3. The output of a system.
8. A smaller system within a larger system.
10. Sensing the output of a system.

Actual results
Basic system model
Closed-loop system
Control
Desired results
Feedback
Input
Monitoring
Open-loop system
Output
Process
Subsystem
Systems

SEE YOUR TEACHER FOR THE CROSSTECH SHEET

ACTIVITIES

MODEL ROCKET-POWERED SPACECRAFT

Problem Situation

To explore space, NASA developed and launched several generations of unmanned and manned spacecraft. The Space Shuttle, a reusable vehicle, is one of man's most complex technological achievements. Dozens of systems and thousands of subsystems are on board the Space Shuttle. For the safety of the crew, there are many back-up systems. The failure of a single system can cancel a launch or cut short a mission.

A model rocket is a system that is similar to rockets launched by NASA. Subsystems of the model rocket include the structure, ignition, propulsion, guidance, and recovery systems. All these systems must work together for a successful launch of a model rocket.

Design Brief

Construct a model rocket using the directions in the procedure below or from a kit. Launch and recover the rocket. As you build the rocket, learn about each of the subsystems so you will be able to explain how they work together.

Suggested Resources

Paper mailing tape—2" wide and 20" long

One wooden dowel rod —¾" in diameter and 12" long

Two pieces of wooden dowel rod—¾" in diameter and ½" long

One paper soda straw— 2" long

One strip of tagboard— 1½" × 10" long

Three pieces of heavy thread (crochet thread is good)—each 10" long

One sheet of light plastic (bags from drycleaners are a good source)— 12" × 12"

Six pieces of masking tape—½" × 5"

Two cork stoppers—#4 and #6 sizes

Soft elastic cord—⅛" wide and 10" long

One model rocket engine (Estes A8-3 or B6-4)

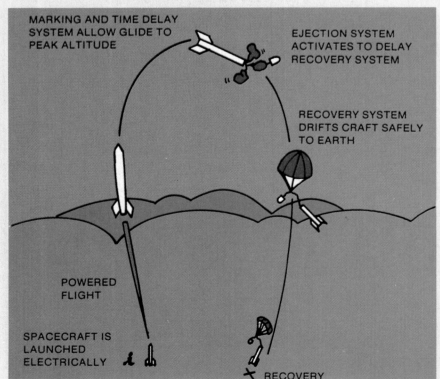

MARKING AND TIME DELAY SYSTEM ALLOW GLIDE TO PEAK ALTITUDE

EJECTION SYSTEM ACTIVATES TO DELAY RECOVERY SYSTEM

RECOVERY SYSTEM DRIFTS CRAFT SAFELY TO EARTH

POWERED FLIGHT

SPACECRAFT IS LAUNCHED ELECTRICALLY

RECOVERY

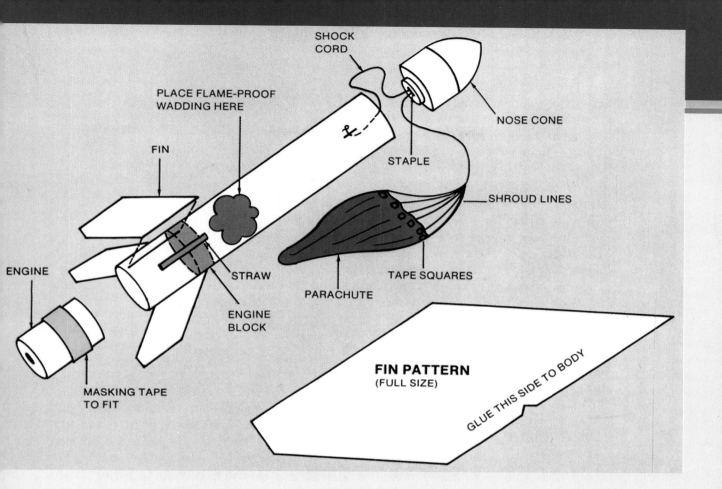

PLACE FLAME-PROOF WADDING HERE

FIN

ENGINE

MASKING TAPE TO FIT

STRAW

ENGINE BLOCK

SHOCK CORD

STAPLE

NOSE CONE

SHROUD LINES

TAPE SQUARES

PARACHUTE

FIN PATTERN
(FULL SIZE)

GLUE THIS SIDE TO BODY

Procedure

1. A few safety rules must be followed:
   a. Engines must only be ignited electrically and by remote control. Each package of engines contains instructions.
   b. Rockets must never be fired indoors or in a congested area. A launch rod must be used, and no one should stand closer than ten feet from the launch area.
   c. Rockets should never be recovered from power lines or dangerous places.
   d. All vehicles should be tested for flight stability before being flown for the first time.
2. To make the body tube, tear or cut a sheet of notebook paper in half so that you have two sheets, each 4¼" × 11".
3. Carefully roll one piece of paper around the 12" dowel rod and glue it together with white glue. You have a paper tube 11" long. Do not glue the tube to the rod, and do not remove the rod.
4. Cut one end of the mailing tape at an angle of 45 degrees.
5. Wet the tape (do not soak it) and carefully spiral the tape around the paper tube on the dowel. The angled edge should be started along the top edge of the paper tube. Be careful

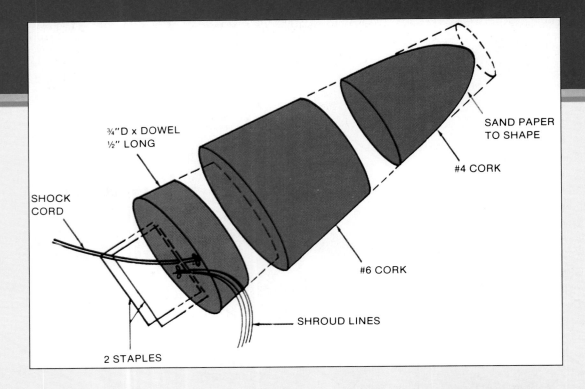

¾"D x DOWEL
½" LONG

SHOCK
CORD

SAND PAPER
TO SHAPE

#4 CORK

#6 CORK

SHROUD LINES

2 STAPLES

not to glue the tape to the rod. As you reach the lower end of the tube, slide the rod up inside the paper tube and trim the tape along the edge of the paper. Remove the finished body tube from the rod to dry.

6. To make the nose cone, glue the two corks together and to one of the ½" dowel pieces as shown. When dry, shape the nose cone with sandpaper.

7. Cut the plastic into an 8" hexagon and attach the shroud lines (crochet thread) at the corners with the masking tape squares. This is the parachute.

8. Drill a ¼" hole in the remaining dowel piece, and glue it firmly inside the body tube, 2½" from the bottom of the tube.

9. Using the fin pattern, cut three fins of tagboard. Glue them near the end of the body tube. When dry, reinforce the joints with more white glue.

10. Glue the straw to the body tube between two fins.

11. Attach the elastic shock cord to the top of the body tube. Staple the free end of the shock cord and the parachute shroud lines to the nose cone.

12. Before you launch your spacecraft, test it for flight stability. Place an engine in the craft and balance it at mid-point on the end of a string. If it flies straight when you swing it around in a circle, it will fly straight when you fire it. If it does not fly straight, add weight to the nose and try again.

SAFETY NOTE: Prepare and fly your rocket only as directed by your teacher and using the instructions that came with the engine. Use flame-proof material for wadding.

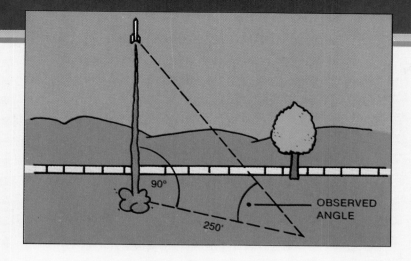

Classroom Connections

1. People design technological systems to satisfy human needs and wants.
2. In a technological system, a technological process combines resources to provide an output in response to a command input.
3. Newton's third law of motion states that for every action there is an equal and opposite reaction.
4. To overcome the gravitational pull of the earth, a rocket's engine must accelerate the rocket to a speed of 8 km/second. This is called **escape velocity**.

Technology in the Real World

1. NASA developments that have applications outside the space program are called spin-offs. Today, people benefit from dozens of spin-offs in the areas of medicine and electronics.
2. In the future, private companies will manufacture products in space. The zero-gravity environment of space is particularly useful for producing certain kinds of very pure drugs, glass, and crystal structures.

Summing Up

1. Describe the role of each subsystem of your model rocket.
2. Explain how the rockets launched by NASA and model rockets are similar. Explain how they are different.
3. List three subsystems that are required on the space shuttle but not on a model rocket.

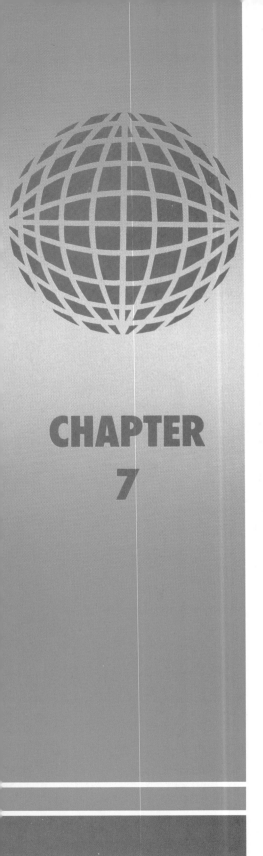

CHAPTER 7

ELECTRONICS AND COMPUTERS

MAJOR CONCEPTS

- The use of electronics has completely changed our world in the last hundred years.
- Electric current is the flow of electrons through a conductor.
- Electronic circuits are made up of components. Each component has a specific function in the circuit.
- An integrated circuit is a complete electronic circuit made at one time on a piece of semiconductor material.
- Computers are general-purpose tools of technology.
- Computers use 1s and 0s to represent information.
- Computers have inputs, processors, outputs, and memories.
- Computers operate under a set of instructions, called a program. The program can be changed to make the computer do another job.
- Computers can be used as systems or can be small parts of larger systems.

ELECTRONICS

The use of electronics has changed our world during the last hundred years. People have learned to use electricity to work with information. Electrical signals can carry information quickly over wires or through the air by radio. Electronics also lets people communicate with machines, making machines even more useful.

The use of electronics has completely changed our world in the last hundred years.

Electronics in Our World

The electric light has extended our day. (Courtesy of General Signal Corporation, Stamford, CT)

Electric appliances in the kitchen allow us to keep our food longer and prepare it more quickly. (Courtesy of Maytag)

Other electric appliances have reduced the amount of time we spend on work, giving us more free time. (Courtesy of Sears Roebuck & Company)

Electronic entertainment has changed the way we spend our free time. (Courtesy of RCA)

171

Problem Situation 1:   Portable, compact devices to measure and determine average wind speed are not readily available for home users of wind power.

Design Brief 1:   Design and construct a wind speed indicator (anemometer) that is simple, effective, and durable and that might be used by anyone wishing to assess wind speed. The anemometer must use a voltmeter or ammeter as the reading instrument.

Problem Situation 2:   Houses built in low-lying areas often have the problem of water in the basement.

Design Brief 2:   Design and construct a model of a system that will sound an alarm and turn on a pump when the water level in the basement is too high.

Problem Situation 3:   At a disco, four-colored transparent sheets of plastic are to be rotated in front of a spotlight for effect.

Design Brief 3:   Design a mechanism to rotate the disks so that they will stop at 90° intervals for 5 seconds before moving on to the next color.

THE SMALLEST PIECES OF OUR WORLD

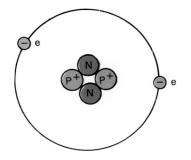

A helium atom has two protons and two neutrons in its nucleus. The atom also has two electrons in a shell (orbit) around the nucleus.

All materials are made up of tiny particles called **atoms**. Materials made up of atoms of only one kind are called **elements**. Iron and carbon are examples of elements. Only 104 elements are known to exist. An atom is the smallest part of an element that can exist and still have all the properties of that element. Atoms are so small that you cannot see them, even with the most powerful optical microscope.

Atoms of different kinds may be combined to make materials called **compounds**. When a compound is formed from two or more elements, it has properties all its own. It may be nothing like the elements it is made of. For example, sodium, a poisonous metal that reacts violently when it touches water, makes a compound with chlorine, a poisonous gas. The compound is sodium chloride, or common table salt. Because elements can be combined in many ways, millions of compounds can be formed from the 104 elements.

Atoms are made up of even smaller particles. These particles do not have the properties of the element. They have their own properties. Atoms have a center part called a **nucleus**. The nucleus is made up of **protons** and **neutrons**. Smaller particles called **electrons** circle the nucleus very rapidly.

The number of protons in an atom determines the kind of element it is. For example, atoms with 13 protons are aluminum atoms. Atoms with 29 protons are copper. Other elements you know are oxygen and nitrogen, which make up the air we breathe, and metals such as gold, nickel, and lead.

Protons and electrons have electric charges. Protons are positive and electrons are negative. Neutrons have no charge. In most natural atoms, the number of protons equals the number of electrons. The positive charges equal the negative charges. The atom, therefore, has no charge.

Particles with opposite charges are attracted to each other. Particles with the same charges repel each other (push each other away). Negative electrons are attracted to their nucleus because of the protons' positive charge. Their rapid motion around the nucleus keeps them from falling into it.

Objects with different charges attract each other (A). Objects with similar charges repel each other (B).

ELECTRIC CURRENT

In some atoms, the electrons are held tightly to the atom. In other atoms, some of the electrons are easily pulled away. They may move from one atom to another. Materials whose atoms give up some electrons easily are called **conductors**. Materials that hold tightly to their electrons are called **insulators**. In a conductor, electrons can move from one atom to another. In an insulator, each atom's electrons are tightly held, and electrons are not free to move between atoms.

Electrons can flow through a thin wire or a piece of solid material, or even through the air, as in a lightning bolt. The flow of electrons is called **current**. The measure of current flow is the **ampere** or **amp**. One amp is equal to about six billion billion electrons flowing past a point in one second.

When you want to move a chair, you have to push on it. You have to exert a force. In the same way, to get a current to flow, a force has to be applied. This force, called an **electromotive** force, is measured in **volts**. Without voltage, no current will flow.

Electric current is the flow of electrons through a conductor.

A battery provides electromotive force. A battery has a positive terminal and a negative terminal. When the terminals are connected to the opposite ends of a wire, electrons start moving. They are attracted to the positive terminal and repelled from the negative one. This sets up a current, or flow of electrons.

Electrical Current Flow

The flow of current through a wire is much like the flow of water through a pipe. In the electric circuit shown in (A), electromotive force comes from the battery. In the water pipe shown in (B), the force comes from a person pushing a piston.

In both cases, a greater force will make a larger current flow. More water will flow through a larger water pipe, as shown in (C). More current will flow through a larger wire, because a larger wire has less resistance.

Ohm's Law is an equation that describes the flow of electrical current.

$$\text{Current (amps)} = \frac{\text{Voltage (volts)}}{\text{Resistance (ohms)}}$$

If voltage gets larger, current gets larger. If resistance gets larger, current gets smaller.

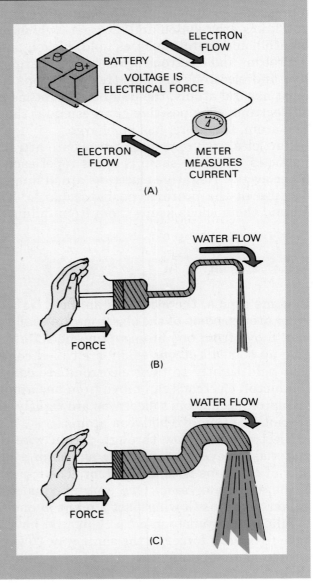

ELECTRON FLOW

BATTERY
VOLTAGE IS ELECTRICAL FORCE

ELECTRON FLOW

METER MEASURES CURRENT

(A)

WATER FLOW

FORCE

(B)

WATER FLOW

FORCE

(C)

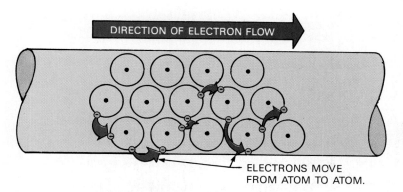

DIRECTION OF ELECTRON FLOW

ELECTRONS MOVE
FROM ATOM TO ATOM.

In a metal wire, electrons are free to move from one atom to the next. Electric current is the flow of electrons through the wire.

Some conductors are better than others. For a given voltage (force), more current flows through a good conductor than a poor one. **Resistance** is the opposition to a flow of current. It is the measure of how good a conductor is. A material with high resistance is a poor conductor. A material with low resistance is a good conductor. The unit of resistance is the **ohm**.

ELECTRONIC COMPONENTS AND CIRCUITS

Electronic **components**, or parts, control the flow of electricity (electrical current). They carry out many useful tasks. Components are connected together in different ways to form **circuits**. A circuit is a group of components connected together to do a specific job. Designers show plans for circuits using drawings called **schematics**. In these drawings, each component has its own **symbol**.

One of the simplest components is the **resistor**. A resistor has a known resistance value. It is used to control current flow. Resistors come in a wide range of values, from less than one ohm to tens of millions of ohms.

Electronic circuits are made up of components. Each component has a specific function in the circuit.

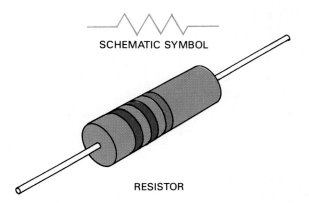

SCHEMATIC SYMBOL

RESISTOR

On many resistors, color-coded bands indicate the resistance in ohms.

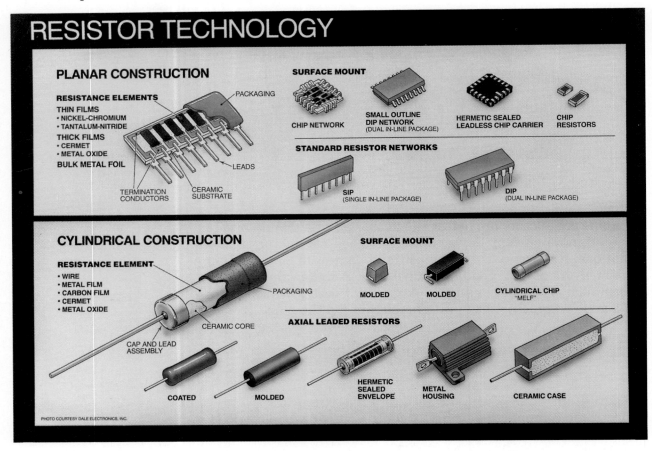

RESISTOR TECHNOLOGY

PLANAR CONSTRUCTION

RESISTANCE ELEMENTS

THIN FILMS
• NICKEL-CHROMIUM
• TANTALUM-NITRIDE

THICK FILMS
• CERMET
• METAL OXIDE

BULK METAL FOIL

PACKAGING

LEADS

TERMINATION CONDUCTORS

CERAMIC SUBSTRATE

SURFACE MOUNT

CHIP NETWORK

SMALL OUTLINE DIP NETWORK (DUAL IN-LINE PACKAGE)

HERMETIC SEALED LEADLESS CHIP CARRIER

CHIP RESISTORS

STANDARD RESISTOR NETWORKS

SIP (SINGLE IN-LINE PACKAGE)

DIP (DUAL IN-LINE PACKAGE)

CYLINDRICAL CONSTRUCTION

RESISTANCE ELEMENT
• WIRE
• METAL FILM
• CARBON FILM
• CERMET
• METAL OXIDE

PACKAGING

CAP AND LEAD ASSEMBLY

CERAMIC CORE

SURFACE MOUNT

MOLDED

MOLDED

CYLINDRICAL CHIP "MELF"

AXIAL LEADED RESISTORS

COATED

MOLDED

HERMETIC SEALED ENVELOPE

METAL HOUSING

CERAMIC CASE

PHOTO COURTESY DALE ELECTRONICS, INC.

Resistors come in many shapes and sizes. They are used in many different types of circuits. (Courtesy of Dale Electronics, Inc., Columbus, NE)

Semiconductors are materials that are neither good insulators nor good conductors. The most common semiconductor material is **silicon**. One kind of component made using semiconductors is the **diode**. A diode lets current flow in one direction but not the other.

One of the most important electronic components today is the **transistor**, which was invented in 1947. A transistor is a resistor that lets a small amount of current control the flow of a much larger amount of current. Transistors are used to control electric motors. They can also be used to control the storage of a small amount of electric charge used to represent information, as in a computer. The transistor is very small. It is a square wafer a few thousandths of an inch on a side. It is packaged in a larger metal or plastic container to make it easy to handle.

Other components are like transistors and resistors. A **thermistor** has a resistance that changes with the temperature.

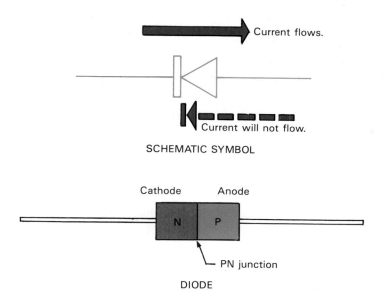

Current flows.

Current will not flow.

SCHEMATIC SYMBOL

Cathode   Anode

N   P

PN junction

DIODE

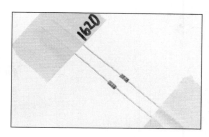

Small diodes are used in digital circuits and in radio detection circuits. Large diodes are used in circuits that supply power or control large currents. (Courtesy of Radio Shack, a division of Tandy Corporation)

A diode is made when a P-type semiconductor and an N-type semiconductor meet at a PN junction. Electron current will flow across the junction in one direction, but not in the other.

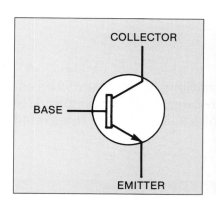

COLLECTOR

BASE

EMITTER

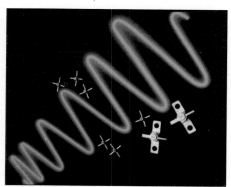

A small amount of base current in a transistor can control a large amount of collector current. Transistors are used for making small signals larger (amplification) and for controlling electrical devices. (Photo courtesy of Hewlett-Packard Company)

Thermistors can be used to make electronic thermometers. They can also be used as part of the control system of refrigerators or ovens. A **photoresistor** has a resistance that changes with the amount of light hitting it. Photoresistors can be used to turn on lights when it gets dark. They also can be used to measure light.

Many other components are used in building circuits. Batteries, switches, motors, and generators are a few of these.

Printed Circuits

Circuits are groups of components connected together to do a specific job. Components are often connected by wires.

Soldering is a method of joining two wires together. A metal called solder is melted on them, forming a connection. Solder has low resistance. It makes a good connection.

Care is taken to make sure wires in a circuit do not accidentally touch. This would set up an unintended flow of current (a short circuit). To prevent this, wires often have a covering of insulation. The insulation prevents short circuits.

Early circuits used large components that were connected to each other by several wires. Each wire was soldered by hand. Components became smaller over the years, and hard to solder. Also, a way was needed to make the same circuit over and over again, without mistakes.

The **printed circuit board** solved both problems. A printed circuit board is a thin board made of an insulating material, such as fiberglass. On one or both sides, a thin layer of a good conductor, often copper, is bonded right on the board. Patterns etched in the copper form paths for electricity. Holes for mounting components are drilled in the board. The components are then soldered to the conducting paths on the board. The conducting paths are photographically placed on the board, so many boards can be made with exactly the same circuit. Once the components are mounted on the board, they can all be soldered at once by an automatic soldering machine.

Integrated Circuits

One of the most important inventions of the twentieth century is the integrated circuit. An **integrated circuit** provides a complete circuit on a tiny bit of semiconductor. Integrated

An integrated circuit is a complete electronic circuit made at one time on a piece of semiconductor material.

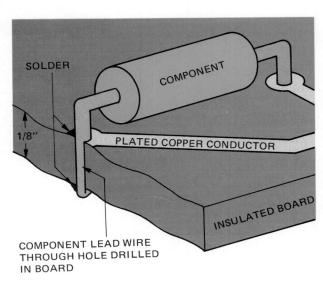

A typical component mounted on a printed circuit board. (From Barden & Hacker, COMMUNICATION TECHNOLOGY, copyright 1990 by Delmar Publishers Inc.)

SOLDER

COMPONENT

1/8"

PLATED COPPER CONDUCTOR

INSULATED BOARD

COMPONENT LEAD WIRE THROUGH HOLE DRILLED IN BOARD

This printed circuit board has many components mounted in a tight space. (Courtesy of Metheus Corporation)

A soldering iron melts the solder, joining the wires to the component terminals. Soldering provides a good mechanical and electrical connection. (Photo courtesy of Cooper Industries, Inc., copyright Tom Watson Photography)

circuits are often less than one-tenth of an inch long by one-tenth of an inch wide. This **chip** contains components and conducting paths.

A chip is designed by an engineer, who makes a drawing of it several hundred times larger than it will be. The drawing is photographically reduced, forming a **mask**. The mask is used to put patterns on a wafer of semiconductor material. Many identical circuits are made at once on a round wafer several inches across.

Computers are built using large integrated circuits. Computers that took up rooms of space twenty years ago now fit on a desk top because of integrated circuits. Chips have replaced large, bulky circuits in many other systems, as well.

Chips use less material, take up less space, and are cheaper to manufacture. Thus, the cost of electronic products came down quickly when chips came into use. If the same thing had happened over the last ten years in the automobile industry, a Rolls-Royce would now cost $500 and get 1,500 miles per gallon.

Analog and Digital Circuits

Information can be represented by electricity in several ways. One way is to have a voltage change based on the information it represents. A voltage could represent a person's speech. It would get larger as the person talked louder. In this case, the voltage is an **analog** of, or similar to, the person's speech. It is called an **analog signal**.

An electronic circuit that works with analog signals is called an **analog circuit**. Voltages in such a circuit change smoothly, as do the things they represent, such as a person's voice.

Sometimes information must be very accurate or must be sent over long distances. Under these conditions, analog cir-

Integrated circuits, like transistors, are packaged in larger plastic or metal containers for protection and ease of handling. Often, the more complex the integrated circuit, the more input and output connections it needs. This requires the package to be larger. (Courtesy of Hitachi America, Ltd., Semiconductor & IC Division)

Size and Complexity of Integrated Circuits

Integrated circuits have become more and more complex since they were invented by Jack Kilby in 1958. Single circuits can perform more and more complex tasks. One measure of a chip's complexity is the number of transistors it uses. People often talk about four types of integrated circuits:

1. SSI—Small Scale Integration: several dozen transistors.
2. MSI—Medium Scale Integration: up to several hundred transistors.
3. LSI—Large Scale Integration: up to several thousand transistors.
4. VLSI—Very Large Scale Integration: one hundred thousand or more transistors.

Each of the squares on the round wafers in the photograph is a complete integrated circuit. Wafers are usually two to four inches in diameter and contain dozens or hundreds of integrated circuits.

(Courtesy of Matsushita/Panasonic)

CONNECTIONS: Technology and Science

■ Electricity can exist as static electricity, electricity at rest, or current electricity. Static electricity involves the accumulation of an electric charge on an object. Lightning is an example of a discharge of electricity resulting from a buildup of static charge.

Current electricity involves the flow of electric charge in a circuit. It may be direct current (dc) in which the charge flows in one direction only, or it may be alternating current (ac) in which the direction of flowing charges changes rapidly. Batteries produce direct current. A generator, in which a conductor moves within a magnetic field, makes alternating current.

■ Circuits are pathways along which electrical energy can move. Any break in a circuit results in the current stopping. Circuits may be series or parallel. In a series circuit only one pathway exists for the current. In a parallel circuit more than one pathway is present. Therefore a break in one part of a parallel circuit will not stop the flow of electrical energy in the other path.

cuits are not good enough. In these cases, **digital technology** is used.

In digital circuits, information is first coded into a series of 0s and 1s. A voltage above a certain value is coded as a 1. A voltage below that value is coded as a 0. Each 1 or 0 is called a **bit**, short for **binary digit**. Binary refers to the number system that has only two numbers, 0 and 1.

COMPUTERS

One field that electronics and integrated circuits have changed drastically is the field of computers. For thousands of years, people have used machines to help with calculations. The Chinese abacus is still used today. Napier's bones, invented in 1617, was used to multiply. The first mechanical adding machine was invented by Blaise Pascal in 1645. It used sets of wheels, moved by a needle, to add numbers and give a sum.

The first all-purpose calculator was developed in the mid-1800s by Charles Babbage. It could be instructed, or programmed, to do sets of calculations. Ada Lovelace, Babbage's co-worker, was the world's first computer programmer. A modern computer programming language, ADA, is named after her.

Herman Hollerith devised a way of automating the U.S. Census in 1890. Information was coded as holes punched into cards. The information could be tabulated by machine. Hollerith's method was a complete system. It had a machine to punch the cards (input). It had a tabulator for sorting the cards (processor). It had a counter to record the results (output). A sorting box rearranged the cards for reprocessing (feedback). In 1911, Hollerith's Tabulating Machine Company became a part of a

The Pascaline was a mechanical adding machine invented by Blaise Pascal in 1645. (Courtesy of The Computer Museum, Boston, MA)

(Courtesy of Buick Motor Division)

(Courtesy of Goldome)

(Courtesy of Whirlpool Corp.)

company that later became International Business Machines Corporation (IBM). IBM is the largest computer company in the world today.

Integrated circuits made computers smaller and cheaper. Computers have been used since the late 1940s. But not until computers could be put onto one chip did they come into household use.

Computers and microcomputers are used everywhere. Microcomputers are used in electronic games, including video games. They are used in automatic bank teller machines (ATMs) and cash registers. They are used in home appliances, such as micro-

---

CONNECTIONS: Technology and Mathematics

■ The computer uses the binary (base 2) system to represent information. The base 2 system is a place value system similar to our decimal system, which uses powers of ten. In our decimal system the number 172 can be expressed as follows:

$$172 = (1 \times 100) + (7 \times 10) + (2 \times 1)$$

Standard Form   Expanded Notation

$$= (1 \times 10^2) + (7 \times 10^1) + (2 \times 10^0)$$

In the binary system 172 is written "10101100". Each digit represents a power of two.

128	64	32	16	8	4	2	1
2^7	2^6	2^5	2^4	2^3	2^2	2^1	2^0
1	0	1	0	1	1	0	0

172 is the sum of $128 + 32 + 8 + 4$.

■ Large numbers can be written in different ways. In this chapter, the amp, which is a measure of current flow, is described as equal to about six billion billion electrons flowing past a point in one second. In standard form this number would be written 6 000 000 000 000 000 000. This is difficult to read because it has so many zeros. Scientists use a special notation, called scientific notation, to write large numbers.

six billion billion = 6 000 000 000 000 000 000 =   6 $\times$ 10^{18}

a number   a
from   power
1 to 10   of 10

A 640 kByte computer can store 640,000 bytes of information. In scientific notation, this number is 6.4×10^5.

wave ovens and VCRs. They are also widely used in automobiles.

The computers described above are small, even though they might be powerful. But there are also very large computers that are used for large, complex jobs. Big computers record income tax information sent in each year by taxpayers. They help engineers design new airplanes.

WHAT IS A COMPUTER SYSTEM?

A computer does its work according to a list of instructions, called a **program**. The program can be changed at any time. A computer is therefore a general-purpose tool. A programmer makes it do a given job by providing instructions, but can change the instructions to make it do a different job. The computer is under **program control**.

Computers are general-purpose tools of technology.

Most of today's computers are digital, using 1s and 0s to represent information. Any number can be represented by a binary number, a group of 1s and 0s.

Computers use 1s and 0s to represent information.

Bits are organized into groups of eight to make them easier to work with. These groups of eight bits are called **bytes**. Each byte can represent one of 256 different characters (numbers, letters, punctuation, or other information). Data can be repre-

The ENIAC

The ENIAC (Electronic Numerical Integrator And Computer) used 18,000 vacuum tubes. It was ten feet tall, three feet deep, and 100 feet long. The vacuum tubes gave off light and heat that attracted moths. The moths became trapped in the wires and moving parts of the computer. People had to clean the moths out, a process known as "debugging." Modern computers don't have problems with moths. But people say they are "debugging" when they find and fix problems in a computer.

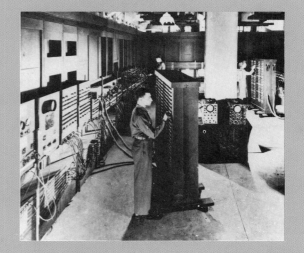

(Courtesy of Sperry Corporation)

DECIMAL	BINARY
0	0000 0000
1	0000 0001
2	0000 0010
3	0000 0011
4	0000 0100
5	0000 0101
6	0000 0110
7	0000 0111
248	1111 1000
249	1111 1001
250	1111 1010
251	1111 1011
252	1111 1100
253	1111 1101
254	1111 1110
255	1111 1111

This chart shows some conversions from decimal (Base ten) numbers to binary (base two) numbers. Eight-bit binary bytes are shown.

sented by bytes, kilobytes (one kByte = one thousand bytes), or megabytes (one MByte = one million bytes).

The Computer Processor

All computers have some parts in common. The first is the **processor**. The processor controls the flow of data, its storage, and what the computer does with the data. The processor reads the program and changes the instructions into actions. The actions might be to add two numbers or to store a number or letter.

The **power** of a processor refers to how fast it is. Personal computers can carry out hundreds of thousands of instructions in a second. Large business computers can carry out millions of instructions per second (MIPS). Very fast computers can handle hundreds of millions of instructions per second.

Memory

The place where the program is stored is called the **memory**. The memory also stores the information being worked on. Most modern computers use integrated circuit memory circuits. A tiny chip can store more than one million characters.

The **random access memory** or **RAM** stores the program and the information currently being worked on. When a computer is referred to as a 640 kByte computer, the RAM can store 640 kilobytes. Personal computers have RAMs of 256 kBytes (lap-

(Courtesy of Hewlett-Packard Company)

(Courtesy of Radio Shack, a division of Tandy Corporation)

(Courtesy of Prime Computer, Inc.)

These three computers have very different main memory sizes. The small lap-top computer has 640 kilobytes of memory; the desk-top personal computer has 4 megabytes of memory; the large superminicomputer has up to 16 megabytes of memory.

top portables) to several MBytes (larger desktop computers). Very large computers have RAMs of many megabytes.

A computer must have another, much larger memory. This other memory, **storage** or **secondary storage**, stores information for use later. Storage is very large so that many different kinds of information or large amounts of the same kind of information can be stored.

Secondary storage includes floppy disks, hard disks, and magnetic tape. Floppy disks are thin, flexible magnetic storage disks that can be easily inserted into or removed from the computer. Hard disks are thick, rigid magnetic storage disks that are permanently mounted in the computer.

On all three, data are stored magnetically. The surface of the disk or tape (called the **medium**) is coated with a thin layer of iron oxide, a magnetic material. A tiny electromagnet, or **head**, is placed near the tape or disk as it moves. A voltage applied to the head magnetizes tiny bits of iron oxide. Information is stored as magnetic fields. This storage process is called "writing to disk" or "writing to tape." Information can also be taken back from the disk or tape by the head. The magnetic fields in the disk or tape are changed back into electrical impulses. This is called "reading from disk" or "reading from tape."

This fixed disk unit contains several disks stacked one on top of another. There is a separate head for each disk. (Courtesy of Pertec Peripherals Corporation)

Secondary storage can hold an unlimited amount of information. When one disk or tape is full, it can be replaced by another. A floppy disk used in some personal computers can store more than one MByte of information, equal to more than 600 pages of typed text. A small hard disk, which is also used in personal computers, stores up to 20 MBytes, or 10,000 pages of typed text. There are hard disk drives containing several large disks that can store hundreds of MBytes.

Tape storage is much the same as with floppy disks, except that the information is stored on tape that looks similar to audio reording tape. The tape is often held in cassettes for ease of loading, unloading, and storage.

A single optical disk, 8 inches in diameter, can store as much information as 15,000 sheets of paper. The disk and its drive can retrieve the information within 0.5 second. (Courtesy of Matsushita/Panasonic)

The **optical disk** is now also being used for memory storage. Optical disks are used to store data in audio compact disks (CDs) and video disks (see Chapter 8). In computers, they are able to store billions of bytes of information, or gigabytes (GBytes).

Computer Input

Information stored in and processed by the computer must be exchanged with people or other machines. This exchange is called **input/output** or **I/O**. Many I/O methods are in use because of the many ways computers are used.

Computers have inputs, processors, outputs, and memories.

Information provided to a computer is called **input**. You are probably familiar with computer input through a **keyboard**.

Bar Code Readers

The bar code reader is an input device used in supermarkets. It reads labels on boxes and bottles of store products.

A special code, called the Universal Product Code (UPC), is used. The UPC is made up of vertical lines of different thicknesses. At the check-out counter, a device called an **optical scanner** picks up light reflected from the white spaces between the bars of the UPC. The black-and-white stripes are changed to on/off pulses of electricity.

The computer uses the code to tell what the product is, how much it costs, and whether to charge tax for it. The computer might also keep a record of how many of that product have been sold and how many should be ordered from the supplier.

GENERAL PRODUCT CATEGORY I.D.

DOUBLE-CHECK CODE

BEGIN/END CODE

BEGIN/END CODE

MANUFACTURER I.D. NUMBER AND BINARY CODE

PRODUCT I.D. NUMBER AND BINARY CODE

The Universal Product Code (UPC) uses thick bars, thin bars, and blank spaces to encode information about a product. (From THE WAY THE NEW TECHNOLOGY WORKS by Ken Marsh)

A keyboard is similar to a typewriter keyboard. People use a keyboard to program a computer. A keyboard can also be used to provide data to a computer. The data are then processed through a program already stored in the computer. Communicating with a computer by means of a keyboard is called an **interactive** process. Input through the keyboard is followed quickly by output from the computer.

Tapes, hard disks, and floppy disks can also provide input to a computer. Tapes and disks can hold a program or data. The computer can move, or load, the data into its main memory. People can write new programs on their computers, storing them on tape or disk. These tapes or disks can then be copied and sold to others.

A device called an optical character reader can transfer input from the printed page directly to the computer. It changes letters and numbers into a code of bytes. In this way, many pages of text can be put into a computer's secondary storage for later use.

Input can also be in the form of human speech. Special devices can recognize some spoken words (up to a few hundred) and change them into bytes. This technology is still being improved. Someday it may be an important source of computer input.

Computer Output

Computer output appears in many forms. You probably have seen the most common, the **video monitor**, or **CRT screen**. CRT stands for cathode ray tube. A cathode ray tube changes electrical signals to light images. CRTs are used in televisions to produce the picture you watch. They are also used in test equipment called oscilloscopes.

Video monitors can be used to display text, like that on this page, or graphics (pictures), or both at the same time. They can be black-and-white or green-and-white, or full-color.

Often, a keyboard is used with a monitor. The result is a system that can provide input to the computer (through the keyboard) and display output from the computer (through the monitor). The combination is called a **terminal**. Sometimes, other input devices are added to the terminal such as the **mouse**, the **light pen**, and the **touch-sensitive screen**.

A **printer** is another common computer output device. A printer records output on a piece of paper. Such a paper record is called **hard copy**. Two kinds of printers used with computers are **dot matrix printers** and **daisy wheel printers**. Dot matrix printers are used to make drawings as well as print letters and numbers. Because these printers use dots, not lines, the pictures and letters are sometimes not very clear.

A mouse combines output with input. (Courtesy of Apple Computer, Inc.)

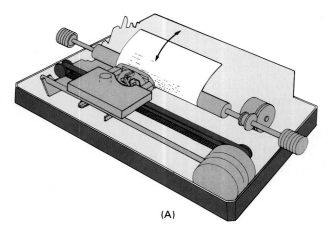

(A)

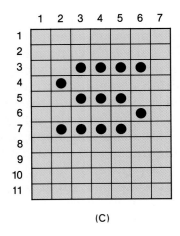

(C)

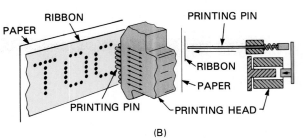

(B)

(A) A dot matrix printer has a print head that contains a group of pins. (B) The pins can be moved by electromagnets to strike a ribbon, making impressions on paper. (Parts A and B from Brightman & Dimsdale, USING COMPUTERS IN AN INFORMATION AGE, copyright 1986 by Delmar Publishers Inc.) (C) The letter "s" is made by activating a number of pins in the right pattern. The matrix shown here is 11 × 7. Other matrix sizes are used.

Daisy wheel printers, on the other hand, print clearly. But they cannot print pictures. Daisy wheel printers are usually more costly than dot matrix printers. They are also slower.

Large computers use printers that can print much faster than dot matrix or daisy wheel printers. They can print whole lines or pages at a time. Another kind of printer used with both large and small computers is the **laser printer**. The laser printer can print both pictures and text very quickly.

Like a printer, a plotter marks on paper. Plotters are output devices that use one or more pens to make drawings. Plotters are often used in systems that make, change, and store drawings, such as CAD, or computer-aided design systems.

Audio output takes the form of tones, beeps, music, and voice. Tones and beeps signal that the end of a page has been reached. They can tell the operator that an instruction is unknown to the computer. Music and voices can be made inside the computer using a device called a **synthesizer**. A synthesizer can produce a wide range of tones, volume, and even drum-like sounds. Voice synthesizers reproduce speech. They are used by telephone companies to give customers telephone numbers or the time of day.

Computers can be connected to each other. They can also be connected to terminals far away, a process called **data commu-**

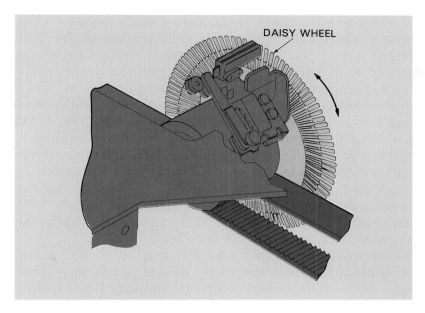

DAISY WHEEL

A daisy wheel contains many "petals." Each petal has a raised character on it. A daisy wheel printer positions the correct character, and strikes it. This pushes the character onto an inked ribbon, which leaves an impression on the paper. (From Brightman & Dimsdale, USING COMPUTERS IN AN INFORMATION AGE, copyright 1986 by Delmar Publishers Inc.)

nications. A special device called a **modem** is often used. Modems must be used in pairs. One modem converts the 1s and 0s into tones of two different pitches (frequencies) that can be sent on the telephone lines. The modem at the other end changes the tones back into 1s and 0s. An inexpensive modem can send the information on one typed page in seconds.

COMPUTERS, LARGE AND SMALL

Computers come in all sizes. They range from a chip less than one inch square to a room full of equipment. Computers also have all sizes of capability. Some are slow processors, while

This small desktop laser printer produces very high-quality printed copies. (Courtesy of Hewlett-Packard Company)

A plotter may be used to make multicolored drawings. Photo courtesy of Houston Instruments, a division of AMETEK)

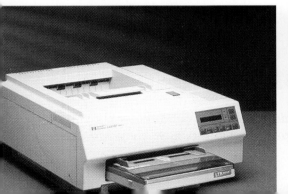

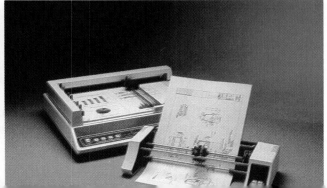

A microcomputer (Courtesy of NCR Corporation)

others do jobs very quickly. Some have a small memory while others have a very large memory. A user must choose the right computer for the job. You can think of computers as falling into one of four categories. They are: microcomputer, minicomputer, mainframe computer, and supercomputer.

Microcomputers are found in appliances, automobiles, and personal computers. They can be as small as one chip. More often, they are a group of integrated circuits including a microprocessor chip. Their memories can be small (one kByte) or large (several MBytes, as in some "supermicrocomputers"). Microcomputers handle data or instructions 8, 16, or 32 bits at a time (1, 2, or 4 bytes at a time). They may be called "8-bit machines" or "16-bit machines," or "32-bit machines," depending on how many bits at a time they handle.

Minicomputers are slightly larger than microcomputers. Minicomputers are often shared by several people in a small company, or one department of a large company. They handle 16, 24, 32, or more bits at a time. They come with large disks or tapes for secondary storage.

Mainframe computers are the large computers used by large companies, government agencies, and universities. They are used to make out payroll checks, keep personnel records, keep track of orders, or keep lists of warehoused items. These large computers handle data and instructions 32, 36, 48, and 64 bits at a time. They may have very large secondary storage devices such as hard disks and tapes. Mainframes can carry out millions of instructions per second.

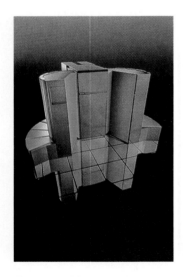

This Hitachi supercomputer is very fast. It is used for artificial intelligence experiments. (Photo by Paul Shambroom, courtesy of Cray Research, Inc.)

Supercomputers are the fastest and largest computers. They are most often used for research, for analyzing huge amounts of data, or for other very large jobs. Supercomputer speed is measured by the number of multiplications or divisions (floating point operations) per second (FLOPS) the computer can carry out. (A floating point operation might be $1.23 \times 2.6 = 3.198$.) The largest supercomputers can carry out billions of floating point operations per second (GFLOPS). Supercomputers are costly to buy and use.

Computers are becoming more powerful all the time. The power of a supercomputer from twenty years ago is now available in a desktop personal computer. This trend toward more power in a smaller package may continue through the year 2000 and beyond.

COMPUTER SOFTWARE

Computers can be used for many different jobs. How they are used depends on input/output devices and how they are programmed. A computer program is called **software**. There are three important kinds of computer software. They are: operating systems, applications programs purchased for use, and applications programs written by the user.

The computer's **operating system** lets a user control the computer and its components. It also makes the components available to other kinds of software. An operating system is chosen to fit the computer and the job it must do. Sometimes more than one operating system is available for a particular computer. The user must choose the right one for a job. Examples of operating systems used in personal computers are MS-DOS and OS/2.

An applications program tells the computer the steps to follow to carry out a specific task. These programs include computer games, word processors, and car engine control programs. Applications programs are stored on tape or disk. They can be bought for use with personal or mainframe computers.

A user must make sure a program will work (is **compatible** with) the operating system. Many programs can be used with several different operating systems. They can be used on different computers.

Sometimes no program exists to do a job. In this case, a program will have to be written. This may be easy and quick,

Computers operate under a set of instructions, called a program. The program can be changed to make the computer do another job.

```
10 INPUT "What is your name ";N$
20 INPUT "Please tell me a number";A
30 INPUT "Please tell me another number";B
40 PRINT "Thank you,";N$;", the product of";A;"and";B;"is :";A*B;"."
50 PRINT "The Quotient of";A;"and";B;"is:";A/B;"."
60 PRINT "The sum of";A;"and";B;"is :";A+B;"."
70 PRINT "The difference between";A;"and";B;"is :";A-B;"."
80 END
```

A) This simple BASIC program instructs the computer to record the name of the operator and to ask for two numbers. The computer will then multiply, divide, add, and subtract the numbers.
(Program courtesy of R. Barden, Jr.)

```
RUN
What is your name ? Mary
Please tell me a number? 8
Please tell me another number? 2
Thank you,Mary, the product of 8 and 2 is : 16 .
The Quotient of 8 and 2 is: 4 .
The sum of 8 and 2 is : 10 .
The difference between 8 and 2 is : 6 .
```

B) The actual exchange between the operator and the computer is shown here. The entries in red are made by the operator. All others are output from the computer. The entry "RUN" starts the program.

or it may be a long, hard effort. Some programs take years to write. An applications program is written in one of many programming languages. Each language has features that make it best for writing certain kinds of programs.

One common programming language is **BASIC** (**B**eginner's **A**ll-purpose **S**ymbolic **I**nstruction **C**ode). It is used by students, businesspeople, and hobbyists. It is fairly easy to learn and use. BASIC can be used on most small personal computers as well as on large mainframe computers.

Pascal is a language that is becoming more and more popular. It is often the language students learn after they learn BASIC. Pascal is named after Blaise Pascal, who invented the first mechanical adding machine.

Some languages have been invented to help computers "think." **Artificial intelligence** is the imitation of human thought by computers. Programming in most computers uses complete information to arrive at clear answers. In artificial intelligence, information may not be complete. Answers may, or may not, be correct. The computer is able to learn from its mistakes.

COMPUTERS AND THE SYSTEM MODEL

It is helpful to think of the computer in terms of a system model. The terms used in describing how a computer works are system terms. Input devices provide command inputs and

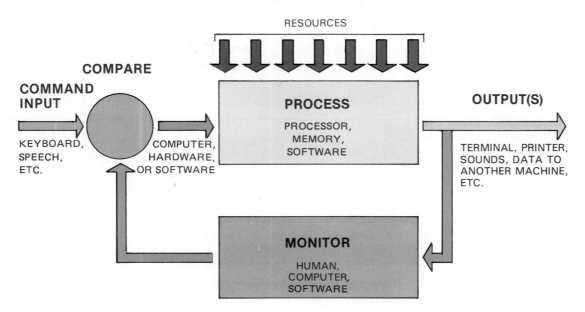

The general system diagram can be readily applied to computers.

CONNECTIONS: Technology and Social Studies

■ Advances in electronics have had an impact upon the lifestyles of people all over the world. The telephone enabled people to talk to each other quickly and over long distances. The invention of the radio, and subsequently television, found people staying at home for their entertainment.

■ With more and more daily tasks being handled by computers and machines, there is a concern that we are forgetting how to deal with people on a personal level. This tends to pull societies apart instead of bringing individuals together to deal with common concerns.

■ Despite the many attempts at providing security, many people are still concerned that data placed in computers can be altered by mistake or illegally, causing grave consequences. Sensitive records and personal information, if changed, could threaten the people's privacy.

resource inputs. For example, a user at a terminal types in a program (command input). He or she then types in data (resource input).

The computer's processor acts on the resources in response to the command. The processor is the process in the system model. The output of the computer is the output of the system. Feedback may come from a person, or through hardware or software.

Computers are sometimes used as stand-alone systems. Very often, however, a computer is part of a larger system. In this case, the computer is only a subsystem. It may provide part of the larger system's feedback, process, or comparison. It may provide the command input to the larger system. Computers come in many sizes and can be programmed to do a wide variety of jobs. Because of this, they are one of the most useful and widespread tools of technology.

Computers can be used as systems or can be small parts of larger systems.

Computer Applications

Computers are often parts of other, larger systems. In a manufacturing system, a computer may supply command inputs to a production line. It might also compare the actual output to the desired output. Then it can order any necessary changes. In a system that addresses and mails thousands of letters a day, the computer is the process part of the system. Its output is addressed envelopes, letters, and bills.

A computer is often used to help people form a feedback loop. For example, a computer can be a central control center in a transportation system. It can help switch trains from track to track or bring planes in to land safely.

A computer can have several roles. In a microwave oven, a microcomputer sets the energy level based on the cook's input. It can turn the oven on. It can sense when food is done and shut the oven off or keep it on at lower power to keep the food warm.

A control computer for automated manufacturing and warehousing systems. (Courtesy of Gould, Inc.)

The control center for Buffalo, NY's light rail transit system. (Courtesy of General Signal Corporation, Stamford, CT)

SUMMARY

The use of electronics has changed technology greatly over the last hundred years. Electricity can help represent, store, change, and communicate information.

All materials are made from atoms. Atoms contain smaller particles called protons, neutrons, and electrons. Protons have a positive electric charge, while electrons have a negative electric charge. Particles with the same charge repel each other. Particles with opposite charges attract each other.

Materials whose atoms give up electrons easily are called conductors. Materials whose atoms hold tightly to their electrons are called insulators. Electric current flows when electrons move through a material. Electromotive force makes the electrons flow, forming a current. The measure of how strongly a conductor opposes current flow is called resistance.

Electronic components include resistors, diodes, and transistors. Components connected together to do different jobs are called circuits. Printed circuits and integrated circuits make it possible to put more circuits into smaller spaces.

Analog circuits work with signals that vary smoothly as the information they represent changes. Digital circuits work with information that has been converted into binary codes (1s and 0s). Digital circuits are best for sending information over long distances, or for ensuring accuracy.

Computers are general-purpose tools of the information age. They can be programmed, or instructed, to do many different jobs. They have processors, main memory, secondary storage, and input/output components. Computing power has greatly increased during the past fifty years, while the cost of computer systems has fallen.

Computers can be classed as microcomputers, minicomputers, mainframe computers, or supercomputers. These classes are constantly changing as computers become more powerful.

Computer software is divided into operating systems and applications programs, which can be purchased already written, or be written by the user.

The computer may be easily described by the general system model. Computer terms are much like system terms.

Computers come in many sizes. They can be programmed to do many different jobs. Because they are so useful, and because their cost has gone steadily down, they are widely used. They are used in appliances and automobiles, as well as in word processing, science research, manufacturing, and business.

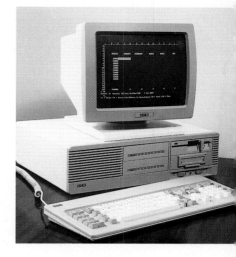

(Courtesy of NCR Corporation)

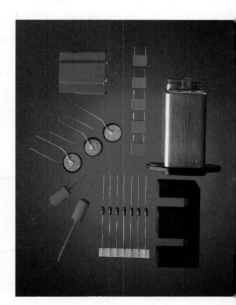

(Courtesy of Siemens Components, Inc.)

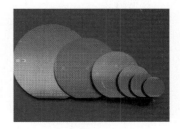

(Courtesy of National Semiconductor Corporation)

1. Name three particles found inside an atom.
2. In an electric circuit powered by a 9-volt battery, one ampere of current flows. If the 9-volt battery is replaced by a 20-volt battery, does more or less current flow? Why?
3. Name five electronic components that can be used to build a circuit.
4. Why was the invention of the integrated circuit important in the history of technology?
5. Based on your own experience and observation, give one example each of how the use of electronics has changed manufacturing, transportation, communication, and health care technologies.
6. Should a telephone be an analog or a digital instrument? Why?
7. Name four parts of a computer.
8. Describe the difference between operating system software and applications software.
9. A computer can be used to mail letters to thousands of people. List the major components that you would expect to find in such a computer system. Draw the system using a general system model.
10. Do you think it's a good idea to have a totally automated system, with no involvement by people? Why or why not? Give an example to support your answer.

(Courtesy of International Business Machines Corporation)

(Courtesy of International Business Machines Corporation)

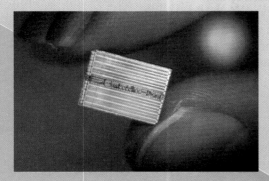

(Courtesy of International Business Machines Corporation)

(Courtesy of Matsushita/Panasonic)

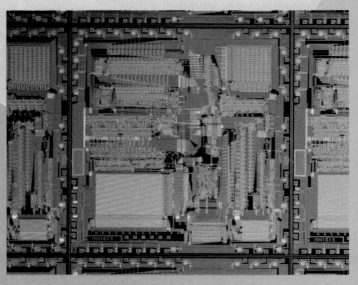

CHAPTER KEY TERMS

Across

1. Controls the computer memory, output devices, and use of software. DOS is an example.
3. Binary digit (abbreviation).
6. A material (like silicon) that is neither a good conductor nor a good insulator.
9. Information that is coded into a series of 0s and 1s.
10. The force that makes current flow through a conductor.
11. An electronic component that allows a small amount of current to control a larger current.
14. The heart of the computer system. It controls the flow, storage, and manipulation of data.
15. A material that permits electrons to flow through it.
16. A material that does not conduct electricity.

Down

2. A complete circuit on a semiconductor chip.
3. Eight bits.
4. A signal that is similar to the information it represents.
5. The fastest and largest kind of computer.
7. A flow of electrons through a conductor.
8. Opposition to electron flow, measured in ohms.
12. A set of instructions that tells a computer to perform a task.
13. A combination of electronic components.

Ampere
Analog
Bit
Byte
Circuit
Component
Compounds
Conductor
Current
Digital
Electron
I/O
Insulator
Integrated circuit
Memory
Operating system
Printed circuit board
Printer
Processor
Program
Random access memory (RAM)
Resistance
Semiconductor
Supercomputer
Transistor
Voltage

SEE YOUR TEACHER FOR THE CROSSTECH SHEET

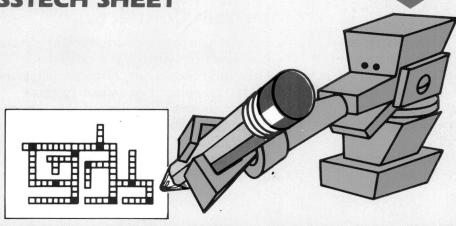

ACTIVITIES

BUTTONS R US

Problem Situation

Microchip technology has found another home in the toy and novelty industry. Computer-controlled video games, remote-control vehicles, and even talking dolls that are voice activated top the toy wish list this year. In order for manufacturers to be competitive, toy designers must be creative in adapting the newest technologies to their products.

Design Brief

You are a product designer for a novelty toy company and have been asked to develop a new product line using flashing light-emitting diodes. The company specializes in novelty buttons but is willing to expand its product line to any item you wish to incorporate the LED circuit into.

Design and build a product utilizing the flashing LED circuit.

Procedure

1. Decide on a product to incorporate the flashing diode circuit. A few ideas you might want to work on are holiday buttons, name buttons, hats, and sunglasses.
2. Lay out and etch a circuit board according to the diagram provided. Follow the safety procedures described by your teacher.
3. Solder the electronic components and wires to the board according to the schematic provided.
4. Attach the diodes to the product you have chosen.
5. Modify the plastic box as needed to hold the battery, circuit board, and switch.

Suggested Resources

Flashing LED (driver circuit) Radio Shack Part #276-036
LED's — assorted colors
Micro slide switch
Circuit board materials
#22 gauge stranded wire
9-volt battery and snap
Plastic box
Button machine and supplies
Soldering equipment and supplies

Classroom Connections

1. Electronic circuits are composed of components, each with its own function within the circuit.
2. Integrated circuits are complete circuits that have been miniaturized and placed on semiconductors.
3. Electric current is the flow of electrons through a conductor.
4. When we control the flow of electrons through a circuit, we can control the output of the circuit.
5. An insulator has very few free electrons and therefore does not allow electrons to flow easily through it.

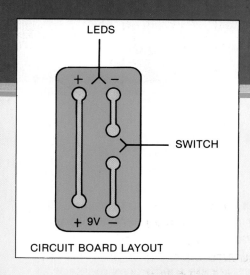

LEDS

+ / −

SWITCH

+ 9V −

CIRCUIT BOARD LAYOUT

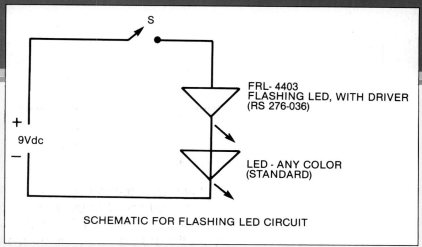

S

+
9Vdc
−

FRL- 4403
FLASHING LED, WITH DRIVER
(RS 276-036)

LED - ANY COLOR
(STANDARD)

SCHEMATIC FOR FLASHING LED CIRCUIT

6. The application of electricity and basic circuits made the telegraph and telephone the first successful rapid communication
devices. These two devices united people across our nation
and the world.

Technology in the Real World

1. Lazer Tag™, one of the newest in electronic games crazes, utilizes photonics (light-activated electronics).
2. As the size of electronic circuits becomes smaller and smaller
so can the size of hand-held electronic games.

Summing Up

1. What electronic components do you think are incorporated
into the driver circuit of the Radio Shack Flashing LED?
2. What are the advantages of using circuit boards?
3. How does a switch open and close a circuit?
4. Define the following words:

- electron
- solder
- voltage
- ampere
- resistance
- current

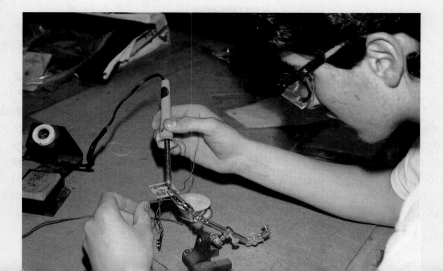

ACTIVITIES

DESKTOP PRESENTATION

Problem Situation

Desktop presentation is a new and exciting way to prepare audio visual programs using a computer instead of a camera. A program such as the Slide Shop™ can be used to produce slide shows that include color, graphics, sound, and special effects to get the attention of an audience and help communicate the desired message.

Desktop presentation slide shows can be used for many different purposes. Businesses use them to provide information to customers and to train staff members. In school they can be used for teaching, to announce events such as school dances and to send computerized messages and greeting cards to others.

Design Brief

In this activity you will use a computer and the Slide Shop™ to prepare a presentation that will be used to announce an upcoming school event. Your presentation should include at least five slides. Use sound, color and graphics to add interest. Add at least two different special effects as transitions between slides.

Procedure

1. Your teacher will demonstrate how to use the computer system and some of the important features of the Slide Shop™.
2. Review the Slide Shop™ manual to become familiar with the graphics, sound and special effects that are available.
3. Practice using the Slide Shop™ by working with another student for one class period. Take notes to help you remember key details.
4. Select an upcoming school event such as a game, drama production, back-to-school night, or pep rally to announce with your slide show.
5. Gather important facts about the event.
6. Organize the facts in a logical order.
7. Construct a storyboard using the format provided by your teacher. The storyboard will describe the text, graphics, sound and special effects for each frame.
8. Load the Slide Shop™ master disk. Use the storyboard to prepare each frame. Experiment with the many features offered by the program. Be sure to save each frame when it is complete.

Suggested Resources

Apple II or IBM compatible computer system
Slide Shop™
Mouse or joystick
Blank disk

9. When all the frames are complete, create the script for your presentation. The script will determine the order of the slides and the special effects used for transition between slides.
10. Have several students preview your show to make sure that it is successful in communicating the desired message. Make necessary changes.

Classroom Connections

1. Computers are general purpose tools of technology. They can be programmed to perform a wide variety of tasks.
2. Computers have inputs, processors, outputs and memories. The mouse is an input device used with desktop presentation programs.
3. When preparing a desktop presentation you need to consider the purpose of the presentation and the audience. The information to be presented should be organized in a logical way.
4. A storyboard is a tool that can be used to organize information for any kind of audiovisual presentation.

Technology in the Real World

1. Desktop presentation is a new way of preparing audiovisual presentations.
2. In addition to slide shows, desktop presentation programs can be used to make videos, overhead transparencies, and 35mm slides.

Summing Up

1. Describe three different ways that desktop presentations can be used.
2. Use the storyboard format to prepare a message to be sent to a student in another school.
3. Was your presentation successful? How do you know?

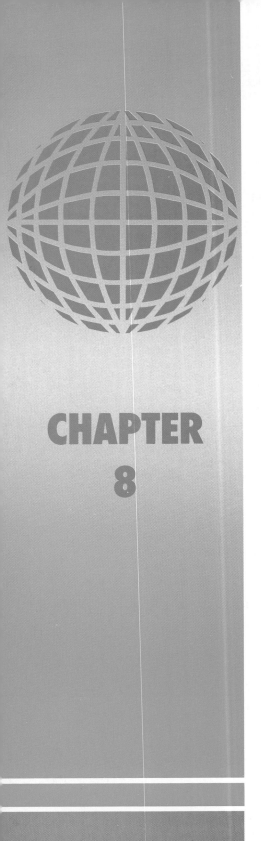

COMMUNICATION SYSTEMS

MAJOR CONCEPTS

■ Communication includes sending a message and having it received and understood.

■ Humans, animals, and machines can all communicate.

■ Communications systems use sound and speech, symbols and codes, printed words, and drawings and pictures.

■ A code is a set of signals or symbols that have some specific meaning to both the sender and the receiver of the message.

■ All communications systems have a transmitter, a receiver, and a channel over which the message travels.

■ Electricity and electronics have revolutionized communication technology.

■ Computing and communication technologies are merging to produce powerful tools for information processing.

WHAT IS COMMUNICATION?

As long as humans have lived on the earth, they have needed to communicate their needs and wants to each other. All societies use some sort of spoken language to communicate. Hundreds of different languages are spoken around the world. Each has words that are used to express what people need to communicate.

Communication can take many forms. You may talk face to face with a friend or write a letter to a relative in another country. You may read a book or a newspaper or listen to radio or watch television. Pictures can be a form of communication. Road signs use simple pictures to communicate. They give useful information or warn of hazards ahead.

To have good communication with someone else, we must be sure that what we are saying is clearly understood. We must know that our words have the same meaning to the other person as they do to us. Just speaking to someone else doesn't mean that we have communicated.

Technology has had an enormous impact on communication. Because of technological advances, communication has become more rapid and accurate. People are able to communicate more effectively, and across greater distances, than ever before.

Communication includes sending a message and having it received and understood.

People come from many different backgrounds, but they all need to communicate.
(Courtesy of Southern California Edison)

Problem Situation 1:   The Morse Code is one of the most successful means of communicating over long distances.

Design Brief 1:   Using simple circuitry, design and construct a device for flashing Morse Code signals. The circuit should use a battery for a current source. The apparatus must be neatly housed and self-contained.

Problem Situation 2:   Your school is familiar to you now, but do you remember how hard it was to find your way around when you first attended?

Design Brief 2:   Design and construct a display board and signpost system which will enable visitors and newer students to find their way around school.

Problem Situation 3:   Fiber-optic cable uses glass fibers to transmit information from one location to another with digital pulses of light.

Design Brief 3:   Design a system using fiber-optic cable to transmit a coded message from one place to another.

TYPES OF COMMUNICATION

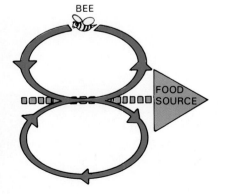

Animals communicate with each other in many ways. This bee is showing the direction and distance of a food source.

People communicate with each other. Most often, they use words, but body language and facial expression are also ways to communicate. You squirm in your seat when you're bored, smile when you're happy, or wave at a friend across the room to say hello.

People also communicate with animals. A dog can follow commands such as sit, come, and heel when taught by a patient owner. Animals can communicate with people, too. When a dog or cat is hungry, it lets you know in a hurry!

Animals can communicate with other animals. Bees dance in a figure-eight pattern when they return to the hive to tell other bees how to find a source of nectar. In an interesting case of animal-to-human communication, a chimpanzee was taught sign language. She, in turn, taught it to her own chimpanzee babies.

People can also communicate with machines. A computer user communicates by typing a command on a keyboard. When you play a video game, the motion of the joystick as you move it is changed into electrical signals that position images on the screen. The machine communicates with you, too, showing you the points you have earned or the time you have left until the game ends.

Machines can also communicate with each other. In a factory, computers are connected to other machines to control them. For example, a computer controls the operation of a robot that spray-paints automobiles. The computer tells the robot where to spray, how long to spray, and when to shut down. The computer and the robot communicate with each other.

Machines also communicate in a home heating system. A furnace heats the room to a preset temperature. When the room reaches this temperature, the thermostat sends an electrical signal to the furnace. The signal turns the furnace off.

Humans, animals, and machines can all communicate.

As automation becomes more widespread, machine-to-machine and machine-to-human communications become more important. (Courtesy of Allen-Bradley, a Rockwell International Company)

COMMUNICATION SYSTEMS

A **system** has been defined in this book as a method of achieving the results that we want. When people speak to each other, they form a **communication system**. Like all systems, a communication system has an **input**, a **process**, and an **output**.

When two people talk to each other, the input (desired result) is the message you want to communicate. The process is how you communicate the message. In this case, it's through the process of speaking. The output (actual result) is the message that is actually received by the other person. **Feedback**, in the form of speech or action, tells you whether the other person has understood your message.

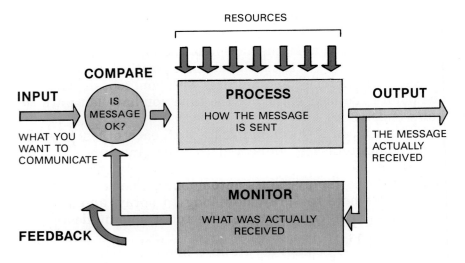

The general system diagram for a communication system.

Sometimes, the actual result is not the same as the desired result. Your mother may say, "Your room is messy." Her real message is, "Go and clean up your room." But you may not receive that message clearly. Your mother receives feedback when she looks in your room!

CATEGORIES OF COMMUNICATION SYSTEMS

Communications systems use sound and speech; symbols and codes; printed words; and drawings and pictures.

Many different communication methods have been used by people now and in the past. These methods include systems that use

- sound and speech;
- symbols and codes;
- printed words; and
- drawings and pictures.

Communication Systems Using Sound and Speech

Most person-to-person communication comes about through the spoken word. Our language makes it possible to express almost every shade of thought and feeling. Tone of voice, gestures, facial expressions, and body language help us when we communicate through speech.

Sounds are also used for communicating. Examples are grunts and moans, a church bell ringing, a railroad crossing bell, a car horn, and a dog barking.

Communication Systems Using Symbols and Codes

Codes have been used as a means of communication for thousands of years. Early examples of codes are signals beaten on a drum and rock piles (cairns) used as markers along trails.

In ancient Rome, army leaders set up signal posts to communicate with soldiers. In the sixth century A.D., the Chinese used oil lamps attached to kites to signal the advance of enemy armies. In 1588, fires warned English soldiers that the Spanish Armada was sailing up the English Channel.

A method for rapid communication was developed in France during the French Revolution (1789-1799). Claude Chappe, along with his brothers, developed a system that let the government contact its forces quickly. Chappe's system was a **visual telegraph**.

Chappe's telegraph stations were mounted on towers, five to six miles apart. They were made from wooden beams that could be swung into different positions using ropes. The positions of the wooden arms stood for words and phrases. They were read through a telescope by a person in the next tower, then passed on. By 1852, there were 556 visual telegraphs in France, stretched over 3,000 miles.

Chappe's idea is still in use today. On railroads, signals using pivoting mechanical arms transmit information to train engineers. At sea, sailors use flags held in different positions to communicate with other ships. This signaling system is called **semaphore**.

Morse code is a very effective method of communication. Over long distances, it may even be more effective than voice transmission. In Morse code, dots and dashes stand for letters. If, because of noise or static, one dot or dash is missed, it is still possible to figure out the word from the other letters received. In voice transmission, entire words or sentences may be lost. This is one reason why Morse code is used when very reliable communication is needed.

A code is a set of signals or symbols that have some specific meaning to both the sender and the receiver of the message.

Chappe's visual telegraph

Communication Systems Using Printed Words

The earliest known writing is **cuneiform writing**. Cuneiform writing was first used by people who lived in the Middle East about 6,000 years ago. This kind of writing used symbols made of wedge-shaped marks. The symbols were pressed into tablets of soft clay or cut into rock. Later, the Egyptians developed a form of writing called **hieroglyphics**. At first, hieroglypics were chiseled into stone. Later they were written on papyrus, paper made of a reed plant.

The first alphabet was also developed in the Middle East. The Hebrew alphabet started out with the letters **aleph** and **bet**. Later, the first two letters of the Greek alphabet were **alpha** and **beta**. These two letters together give us the word **alphabet**.

The earliest form of paper was made from the papyrus plant around 2500 B.C. The fibers of the plant were soaked in water. They were then mashed together and matted to form thin sheets. Paper similar to the paper we use today was developed by the Chinese about 2,000 years ago. Paper was not really of good quality, however, until about A.D. 1400. Cheap, lightweight paper made it easier for people to record information and share it with others. Technological progress depended in large part on this sharing of information. People could read what others had learned and had done, and could build on this information.

An important development in communication technology took place in Germany in 1450. Johannes Gutenberg perfected a way of making letters to print with by pouring hot metal into molds. These molded letters are called **type**. Before Gutenberg's invention, type was made out of wood. Each letter was hand-carved, taking a great deal of time. Gutenberg also designed a method to hold type in place and used a screw press (like the kind used in a home workshop vise) to do the printing.

Gutenberg had created a printing system. The system used a press, good paper, ink, and metal type. The printing system made it possible to print books in large numbers. More people could afford to buy books.

Gutenberg's printing method used raised surfaces. This is called **relief printing**. In relief printing, only the type's raised surfaces are inked. Lower surfaces are not. When inked type is pressed against a piece of paper, only the inked type surfaces print.

Relief printing is also known as **letterpress printing**. The method is still used, though with some changes. Newspapers and greeting cards are printed using relief printing. Typewriters print using a relief-printing process.

Printing can also be done using a lowered surface. Lines are scratched into the surface of a piece of metal. Ink is spread on the surface and then wiped off. The ink remains in the scratch. When a piece of paper is pressed against the metal surface, the paper is forced into the scratch and pulls the ink out. This type of printing is called **gravure**, or **intaglio**, printing. Gravure is used to print some magazines, such as the Sunday newspaper magazine section.

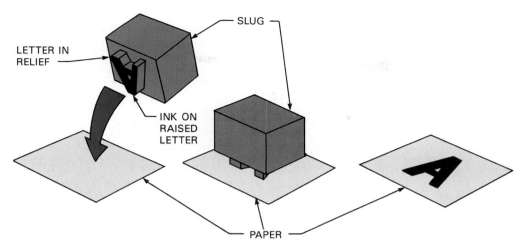

Relief printing

Another kind of printing is **screen printing**. Screen printing is often used to print words and pictures on T-shirts and posters. In screen printing, ink is pressed through holes in a stencil, or mask, onto the surface to be printed. The holes are made in the shape of the object to be printed. The mask is attached to a fine-mesh silk screen that gives an even, controlled flow of ink onto the printed surface.

Communication Systems Using Drawings and Pictures

Prehistoric people drew on cave walls. Often, these cave drawings were of hunting scenes. The pictures included people, animals, and weapons such as spears and arrows.

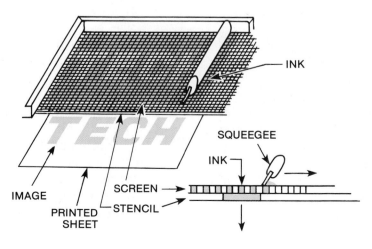

In screen printing, ink is pressed through a screen and stencil onto the surface to be printed. (Reprinted from COMMUNICATION TECHNOLOGY by Barden and Hacker, copyright 1990 by Delmar Publishers Inc.)

Offset Lithography

Today, most printing is done by **offset lithography**. **Lithography** means writing (**graphy**) on stone (**litho**). It is based on the principle that oil and water do not mix. The process was first used by a German artist, Alois Senefelder. Senefelder drew a line with a waxy crayon on a smooth, flat piece of limestone. He then wet the entire surface. He found that an oil-base ink applied to the limestone stuck only to the crayon lines, and not to the wet part of the limestone. When the surface of the limestone was pressed against a piece of paper, only the inked part (the crayon lines) printed.

Modern offset printing uses metal sheets rather than pieces of limestone, but the process is much the same. A photographic process is used to put a greasy image on a sheet of aluminum. (The aluminum sheet has been treated to make it sensitive to light.) This sheet, called an **offset plate**, is wrapped around a metal cylinder. The cylinder is wetted with a water solution.

Ink is applied to the plate. It sticks only to the greasy image, not to the wetted areas. As the plate turns on the cylinder, it presses against another cylinder, which is covered with a thin rubber blanket. The image is transferred (or offset) to the rubber blanket. The offset image is in reverse, as if seen in a mirror. The rubber blanket rotates, and paper is fed through the press. The image is again offset from the rubber blanket to the paper. The image is no longer reversed.

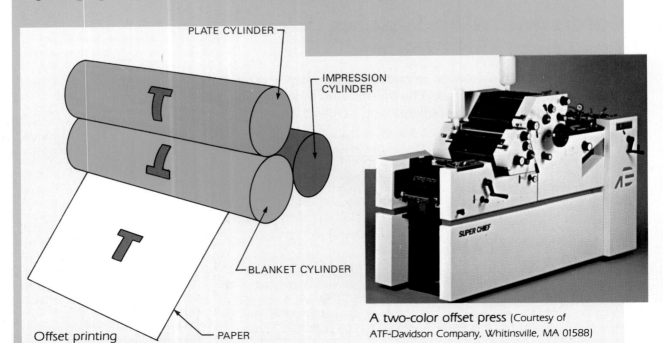

PLATE CYLINDER

IMPRESSION CYLINDER

BLANKET CYLINDER

Offset printing

PAPER

A two-color offset press (Courtesy of ATF-Davidson Company, Whitinsville, MA 01588)

An early cave drawing

In technical drawing, a relatively small number of tools is used to produce complex drawings. (Courtesy of Simpson Industries, Inc.)

Computers are changing technical drawing. With computer-aided drafting (CAD) systems, drafters can create, change, and document more complex drawings than ever before.
(Courtesy of International Business Machines Corporation)

Three views are enough to completely describe an object.

TOP

FRONT   SIDE

Early cultures such as the Greeks made art an important part of daily life. Later, during a period called the Renaissance (about A.D. 1350-1500), art went through great changes.

The most important artistic invention of the Renaissance was **perspective drawing**. Perspective drawing is a way of making things look real by making distances look correct to the viewer. This technique became more and more important as people needed to communicate the real shapes and sizes of objects. Craftspeople, for example, had to work from accurate drawings.

Technical drawing was developed to provide even more accuracy in drawings. Technical drawing is done using special instruments and tools (see Chapter 3). These include T-squares, triangles, drawing boards, and CAD systems. Technical drawing is used by artists, engineers, architects, designers, and drafters.

Technical drawings can show objects in two ways. One is to draw the object in three dimensions. This is called an **isometric**

A nineteenth-century daguerreotype of feminist Lucy Stone. *(Courtesy of National Portrait Gallery, Smithsonian Institution, Washington, D.C.)*

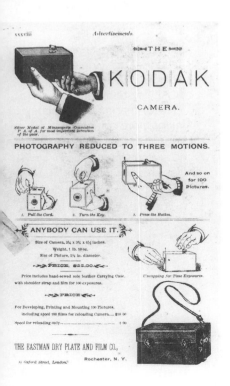

Advertisement for an early Kodak camera. *(Courtesy of Eastman Kodak Company)*

drawing. A second is to draw three views of the object: top, front, and right side. This is called **an orthographic** drawing.

Photography

Another important way of communicating through pictures is **photography**. The word photography means "to write with light." The device used to make the pictures is called a **camera**.

All cameras have three common elements. They have a dark chamber, a lens that admits light and controls the amount of time in which light can enter, and a medium that records the image of the light. The earliest camera-like device was the **camera obscura**, invented in Italy in the 1500s. The term is Italian for "darkened room." It consisted of a darkened room with a lens in the wall facing the street. Light came through the lens, falling on the opposite wall. An artist would put a canvas in the way of the image from the street, then trace and paint over it.

In 1839, a Frenchman named Louis Daguerre invented the first camera. He used a light-sensitive film made of silver-coated metal with an iodine solution on it. The film was not very sensitive, so a person whose picture was being taken had to sit still for about half an hour. The pictures taken were called **daguerreotypes**.

In the late 1800s, George Eastman introduced the Kodak camera. Until then, most photography was done by professionals in studios. The Kodak camera made it so easy to take a picture that anyone could do it. The company advertised, "You press the button, we do the rest." The first Kodak included a roll of film that could take 100 pictures. After the roll was used, the film could be sent to an Eastman processing plant to be developed. Soon, almost every family owned a camera and a collection of snapshots.

Five elements are necessary for photography. They are

1. **light** (the sun, a light bulb, or a flash);
2. **film** (color or black and white);
3. a **camera** (large or small, with a lens);
4. **chemicals** (for developing film and printing pictures); and
5. a **darkroom** to process the pictures in.

Light is reflected by an object and permanently recorded on light-sensitive film. The first photographs were taken in indoor studios. A bright source of light was needed. Photographers burned magnesium powder, which gave off an intense light. The powder, however, was smoky and dangerous.

Today, **film** is made out of acetate (a plastic). Films are available for all kinds of purposes. Some films are sensitive to all colors except red. They can be developed in a room that

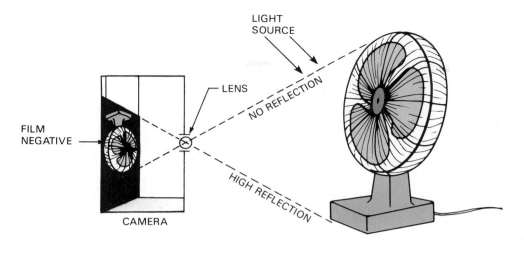

Formation of a negative image on film.

has red lights. Some films are sensitive to ultraviolet light. Some are sensitive to heat. Some can be used with very low light levels.

Photographic film is coated with tiny grains of silver. These particles are so small that they are hardly visible. When the lens focuses an image on the film and the camera's shutter is opened, the silver grains are exposed to light. When these exposed grains are developed in chemicals, they turn black. The unexposed grains are washed away during the developing process. This leaves clear areas on the film. In this way, a **negative** is formed.

There are four major types of cameras. The **view camera** is the simplest. It has a lens on one end and a place for film and focusing on the other. Behind the lens is a shutter and behind that is a dark chamber (called a bellows) made of flexible material. The camera is usually very large. It must be supported by a tripod.

In a view camera you look through the back of the camera to focus. In a **viewfinder camera**, you look through a separate viewfinder to compose the picture. "Instant" cameras and disk cameras are viewfinder cameras.

The **twin-lens reflex camera** uses two lenses. One is for viewing and one is for focusing light on the film. This camera has a mirror that reflects the light from the viewing lens to the eyepiece at the top, for easy focusing.

A **single-lens reflex camera** uses one lens for both viewing and focusing. The mirror, which reflects the light, is movable. In its normal position, the mirror reflects the light from the lens to the viewing eyepiece. A person can see what he or she is taking a picture of. When the picture is taken, the mirror swings up out of the way. The light that enters the lens can then reach the film.

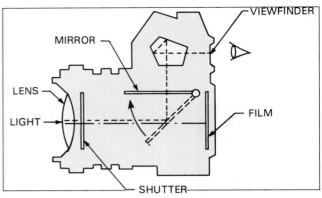

A single lens reflex (SLR) camera. (Courtesy of Minolta Corp.)

Diagram of a single lens reflex camera. (From Dennis, APPLIED PHOTOGRAPHY, copyright 1985 by Delmar Publishers Inc. Used with permission)

A darkroom is needed to develop photographs. (Photo by Michael Hacker)

Chemicals develop the film and print the photograph. Three kinds of chemicals are used. They are **developer**, **stop bath**, and **fixer**. Developer turns the exposed silver particles black. Stop bath stops the developing action. Fixer washes away all the unexposed silver particles and clears the parts of the film that received no light.

Photographic film and paper are very sensitive to light. The chemical processing, therefore, must be done in a **darkroom**. This darkroom can be very large, with expensive equipment. Or it can be a closet in one's home. Amateur photographers often have a home darkroom that uses simple, inexpensive equipment.

SIMILARITIES AMONG COMMUNICATION SYSTEMS

Communication systems are alike in many ways. Since they are systems, they all have inputs, processes, and outputs. We can model each of the systems with our basic systems model.

The Process

All communications systems have a transmitter, a receiver, and a channel over which the message travels.

All communication systems have a three-part process. There is a **transmitter**, or means of sending the message. There is a **channel**, the route the message takes. There is a **receiver**, which accepts the message.

The communication channel is never perfect. Often, when we try to send a message, some imperfection in the channel affects

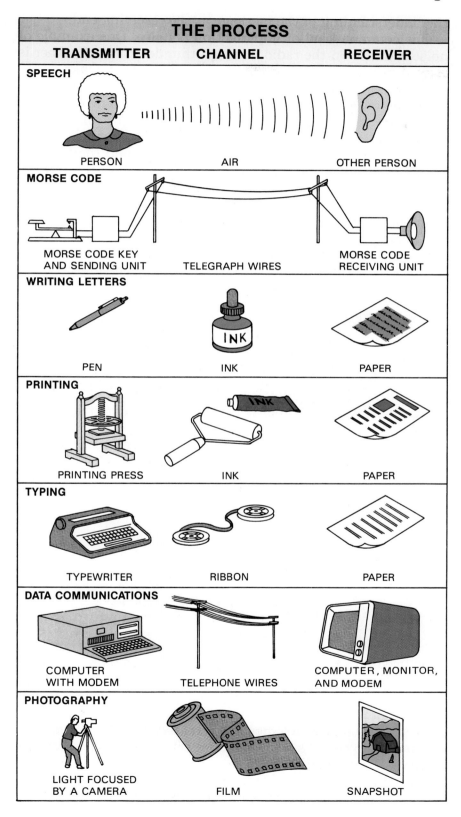

THE PROCESS

TRANSMITTER	CHANNEL	RECEIVER

SPEECH

PERSON — AIR — OTHER PERSON

MORSE CODE

MORSE CODE KEY AND SENDING UNIT — TELEGRAPH WIRES — MORSE CODE RECEIVING UNIT

WRITING LETTERS

PEN — INK — PAPER

PRINTING

PRINTING PRESS — INK — PAPER

TYPING

TYPEWRITER — RIBBON — PAPER

DATA COMMUNICATIONS

COMPUTER WITH MODEM — TELEPHONE WIRES — COMPUTER, MONITOR, AND MODEM

PHOTOGRAPHY

LIGHT FOCUSED BY A CAMERA — FILM — SNAPSHOT

The communication process consists of a transmitter, a channel, and a receiver.

the understanding of the message at the receiving end.

One common kind of imperfection is **noise**. One example of noise is static on the radio during a thunderstorm. Another is caused by airplanes when they fly overhead; they make a television picture flutter. Dust on a record can cause noise, as can smudges on a drawing. A system must be designed with as little noise as possible to make it as reliable as it can be.

Bandwidth is one measure of the ability of a channel to carry information. It is measured in cycles per second (hertz). The larger the bandwidth, the more information a channel can carry. Some signals, such as motion pictures, contain a large amount of information. These signals require channels with bandwidths much larger than those needed by signals with a smaller information content, such as voice. TV requires about a thousand times the bandwidth of a telephone voice signal. That is why normal TV signals are not carried over telephone lines.

Resources

As do all systems, communication systems make use of the seven types of resources. **People** design the systems, provide the message to be sent, receive the message, and take part in delivering it. **Energy** moves the message from one location to another. Moving the message from the transmitter to the receiver takes **time**.

The **tools and machines** used are radio transmitters, printing presses, cameras, and other devices. **Information** is needed to design and operate the system. **Materials** (such as paper, film, and tape) are often used to transport the message. **Capital** is needed to set up and operate the system.

EFFECTS OF MODERN TECHNOLOGY ON COMMUNICATION SYSTEMS

So far, we have discussed communication systems that have been used for many years. They are still in use today, but they have become much more useful because of modern technology. Electronic technology especially has improved communication systems.

Electricity and electronics have revolutionized communication technology.

Electricity and electronics have created the information age, just as steam engines brought about the industrial age. We can watch TV programs from other continents. We can speak on the telephone with friends and relatives thousands of miles away. We use automated bank teller machines to bank after closing hours. These are a few of the many ways electricity and electronics have changed the way we communicate.

MODERN COMMUNICATION SYSTEMS USING SOUND AND SPEECH

Many modern communication devices are electronic systems that reproduce **audio** (sound, like speech and music). These include stereos, radios, telephones, and compact disk players. During the past century, electronic communication has changed the world.

The Telephone

Telephones have changed over the years, from the crank-box phones of the late 1800s to the sleek push-button phones of

---

CONNECTIONS: Technology and Mathematics

■ For a telephone communication system to be effective it must be possible to connect any telephone in the system to any other telephone. The number of connections needed poses an interesting mathematical problem.

Suppose a town had just 10 telephones. How many connections would be needed so that any person could speak to any other person in town? We can make a chart and look for a pattern.

Number of telephones	Number of connections
2	1
3	3
4	6
5	10
:	:
10	45

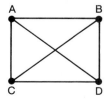

Two telephones, A and B, need only 1 connection, The third telephone, C, needs to be connected to the other two. The fourth phone needs to be connected to the other three, etc. The fifth telephone connects to four others, 5×4 connections. Since two phones share a connection, we divide by 2. For 10 telephones, multiply 10 × 9 and divide by 2.

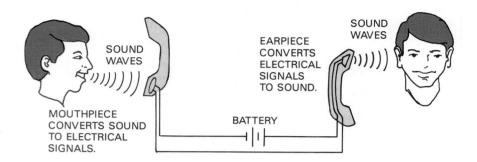

SOUND WAVES

MOUTHPIECE CONVERTS SOUND TO ELECTRICAL SIGNALS.

EARPIECE CONVERTS ELECTRICAL SIGNALS TO SOUND.

SOUND WAVES

BATTERY

A telephone converts sound to electrical signals and electrical signals back to sound.

today. Today, there are nearly 150 million telephones in the United States.

Alexander Graham Bell, the inventor of the telephone, was a teacher of the deaf. He knew how the human ear works, and used his knowledge in his invention. Today's telephone is based on his ideas.

A telephone mouthpiece contains tiny carbon particles. When you speak into it, the sound of your voice creates pressure on the particles in the telephone microphone. When the pressure increases (you speak more loudly), the particles become more tightly packed. This lets more electricity flow through the telephone circuit. When you speak softly, the particles are loosely packed. Less electricity can flow. The electric current that flows through the telephone line changes in response to your voice.

As this changing current flows through the telephone earpiece on the receiving end, the current makes a thin piece of metal vibrate. These vibrations reproduce the sound made by the person's speech on the other telephone.

Sometimes people speak on telephones a very long distance apart. In this case, amplifiers or repeaters boost the electric current. Many telephone circuits change electric currents to digital pulses. This makes it easier to send the signals over long distances.

The connection of your telephone to another telephone is called **telephone switching**. Telephone switching is an important job. It must be possible to connect any one of the 150 million telephones in the United States to any other telephone, not only in this country but in the world.

In today's telephone network, a smoothly varying electric current is produced by the mouthpiece microphone. This current, an analog signal, is changed into digital pulses like computer data. The pulses are then sent to the central office telephone switch. The switch is a computer, designed especially for telephone switching. At the receiving end, the digital pulses are changed into analog electric current. They are sent to the earpiece of the receiving telephone.

A modern telephone switch is a special-purpose computer. (Photo by Robert Barden)

Digital technology makes special telephone services possible. Digital switches help provide services like call forwarding and call waiting. Voice synthesizers are used to provide telephone numbers and time of day.

New technology makes it possible to talk on the telephone almost anywhere. The cordless telephone contains a small radio transmitter in the part that you carry around. The transmitter sends a signal to the telephone in the house, which sends the signal on as usual. Automobile telephones work in much the same way. However, they use a new technology called **cellular radio**. Cordless phone signals can travel only a short distance. Cellular radio lets people talk on mobile telephones over a much wider area, such as a city.

In cellular radio, radio transmitters placed around a city send signals to car phones within an area, or cell, of the city. Each

Combining telephone and radio technologies has produced cordless phones and mobile (cellular) phones.
(Photo courtesy of Southwestern Bell Corporation)

The Automatic Telephone Switch: An Invention Born of Necessity

Inventions are often made by people who need them. In the late 1800s, Almon Strowger was a funeral home director in Kansas City, Missouri. The town's telephone operator was the wife of the owner of another funeral home. When people called the operator and asked for a funeral home, she always connected them with her husband's.

To save his business, Strowger designed a device that made telephone connections automatically. An operator was no longer needed. Strowger switches provided the first automatic system to be used in a public exchange. The first Strowger switch went into use in 1892. They are still in use today in some telephone offices.

(Courtesy of Science Museum Library, London)

Almon Strowger and his automatic telephone switch.

(Courtesy of AT&T Bell Laboratories)

219

transmitter is connected to the public telephone system. A person using a car phone can speak with anyone else in the world.

In the future, the telephone will be an even more useful device than it is today. Most home telephones will be connected to a computer. You will be able to find a telephone number without looking it up. The telephone will call for help automatically when the police or medical aid are needed.

Sound Recording

Sound recording has come a long way since Thomas Edison invented the phonograph in 1877. His device could both record and play back. To record, a metal needle scratched a groove into a rotating cylinder covered with a thin sheet of tin. The depth of the groove depended on the loudness of the voice or music being recorded. The metal needle could also play back the recording.

Today, recording is usually done on plastic disks or audio tape. A 12-inch long-playing record (LP) is pressed from a master disk. The metal master disk is made in much the same way as Edison's cylinder. A needle scratches a spiral groove into a thin coating of lacquer on the metal disk. The needle vibrates in response to the voice or music that drives it. After the master is recorded, it is used to make a secondary master, called a **stamper**. The stamper stamps its grooves into blank plastic disks. These plastic disks become the LPs.

Grooves can pick up dust and dirt. A record can therefore be noisy when it is played. The newest type of record, the **laser disk**, has no grooves at all. Laser disks are often called **compact disks (CDs)** because they are smaller than the 12-inch LPs.

The spiral tracks on CDs are made by laser beam. The laser beam burns millions of tiny pits into the disk's surface. The pits are so shallow that you can't tell they are below the surface. A protective coating is then placed over the entire surface.

When a CD is played, a laser beam shines light on it. A light-sensing device "reads" the reflection from the pits. A deep pit reflects only a little bit of light. One that is shallow reflects more light. These reflections are changed into digital pulses of electricity. The pulses are then turned into sound.

Radio

During the mid-1800s, James Maxwell, a Scottish physicist, showed mathematically that signals could be sent through the air. Many years passed before his idea was turned into a device for sending and receiving radio waves.

On December 12, 1901, in a deserted old hospital in New-

Edison's phonograph was able to record and play back sound. (Courtesy of RCA)

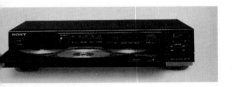

Compact disks combine optical and digital technologies to produce extremely high-quality audio and video recordings. (Courtesy of Sony Corporation of America)

When the *Titanic* sank in 1912, radio was used to call for help from other ships. (Courtesy of the Trustees of the Science Museum, London)

foundland, Canada, the Morse code for the letter "S"—three dots—was heard by Guglielmo Marconi. The code had been sent across the Atlantic Ocean by wireless. It was an exciting breakthrough. Now people could communicate over great distances without wires.

Sailors were the first to use radio communication. They could communicate with land and other ships, and send calls for help. New inventions allowed voice and music to be carried on radio waves. Regular radio broadcasts began in the 1920s at KDKA in Pittsburgh. Today, almost any area of the country has dozens of radio stations.

Radios allow us to listen to music, news, sporting events, and other programming. Radios let us communicate between our homes and cars, boats, or airplanes. We can even communicate with people on the space shuttle.

Most radios use the same basic principles. Radio transmitters send messages into the air at a specific frequency in the spectrum. All sound waves have **frequency**. In sound, the frequency refers to the number of times the object making the sound vibrates in one second. For example, when you beat a drum, the drum skin vibrates. When you pluck a guitar string, it vibrates, producing a tone. When a note is said to be high or low (pitch), it is high or low in frequency. The higher the note, the more vibrations there are per second. High notes have high frequencies, while low notes have low frequencies. Frequency is measured in vibrations per second or cycles per second (hertz).

Radio waves are not mechanical vibrations like sound waves. They are a combination of electric and magnetic fields that change very rapidly. They have frequency in the same way that sound waves do, but it is so high that it is measured in millions of cycles per second (Megahertz, or MHz).

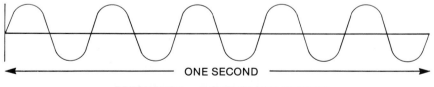

ONE SECOND

FREQUENCY = 5 CYCLES PER SECOND
FREQUENCY = 5 Hz

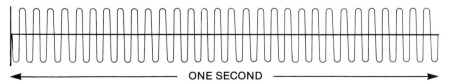

ONE SECOND

FREQUENCY = 30 CYCLES PER SECOND
FREQUENCY = 30 Hz

The **wavelength** of a radio wave is the distance from the beginning of one cycle of the wave to the end of the same cycle as the radio wave travels through the air. Radio waves exist at many different frequencies, from about 0.1 MHz (100,000 Hz) to over 100,000 MHz (100 GHz). The total range of available frequencies and wavelengths is called the radio **spectrum**.

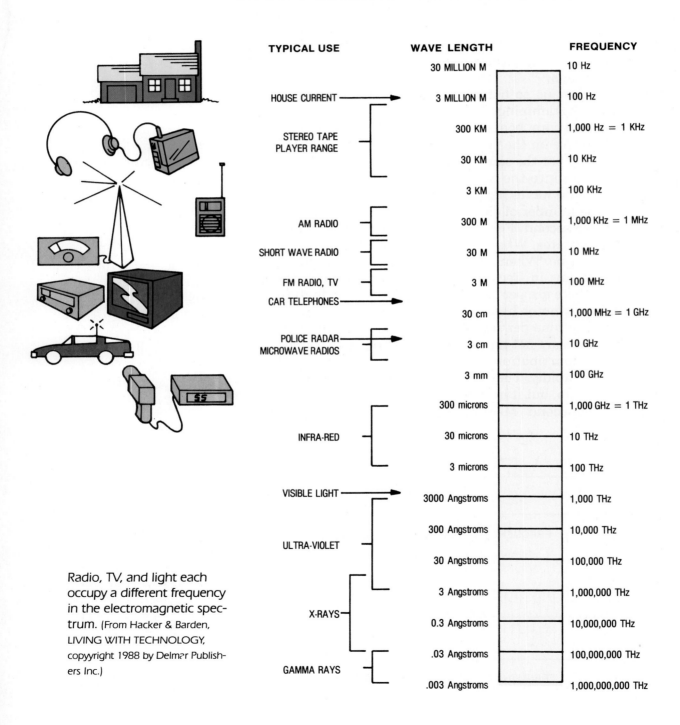

TYPICAL USE	WAVE LENGTH	FREQUENCY
	30 MILLION M	10 Hz
HOUSE CURRENT	3 MILLION M	100 Hz
STEREO TAPE PLAYER RANGE	300 KM	1,000 Hz = 1 KHz
	30 KM	10 KHz
	3 KM	100 KHz
AM RADIO	300 M	1,000 KHz = 1 MHz
SHORT WAVE RADIO	30 M	10 MHz
FM RADIO, TV	3 M	100 MHz
CAR TELEPHONES	30 cm	1,000 MHz = 1 GHz
POLICE RADAR / MICROWAVE RADIOS	3 cm	10 GHz
	3 mm	100 GHz
	300 microns	1,000 GHz = 1 THz
INFRA-RED	30 microns	10 THz
	3 microns	100 THz
VISIBLE LIGHT	3000 Angstroms	1,000 THz
	300 Angstroms	10,000 THz
ULTRA-VIOLET	30 Angstroms	100,000 THz
	3 Angstroms	1,000,000 THz
X-RAYS	0.3 Angstroms	10,000,000 THz
	.03 Angstroms	100,000,000 THz
GAMMA RAYS	.003 Angstroms	1,000,000,000 THz

Radio, TV, and light each occupy a different frequency in the electromagnetic spectrum. (From Hacker & Barden, LIVING WITH TECHNOLOGY, copyyright 1988 by Delmer Publishers Inc.)

Radio signals at low frequencies (LF) have long wavelengths. Radio signals at high frequencies (HF) have shorter wavelengths and are sometimes called **shortwave**. Radio signals at super-high frequencies have even shorter wavelengths, called **microwaves**.

Radio waves of different wavelengths travel in different ways, so they are used in different ways. High-frequency radio waves bounce off the atmosphere's upper layer, the **ionosphere**. They come back to earth far away from the sending point. This makes HF useful for communicating over long distances. Super-high-frequency (SHF) waves travel right through the ionosphere. They can be used to communicate with satellites.

Wavelength and Frequency

All electromagnetic waves travel through space at the same speed. Perhaps you know what this speed is, because light is a form of electromagnetic wave. Radio waves and light waves travel at about 186,000 miles per second. This is about 300 million meters per second. A wave sent by a radio transmitter (or a light source) is 186,000 miles (or about 300 million meters) away one second later. This is true no matter what the frequency of the wave is.

During that second, the radio transmitter sends the number of wave cycles equal to its frequency. Thus, a 10 MHz (megahertz) transmitter will generate 10 million cycles in one second. These 10 million cycles cover 300 million meters in one second. Therefore, its **wavelength**, or distance covered by one cycle, is:

$$\text{Wavelength} = \frac{300 \text{ million meters per second}}{10 \text{ million cycles per second}}$$

$$\text{Wavelength} = 30 \text{ meters}$$

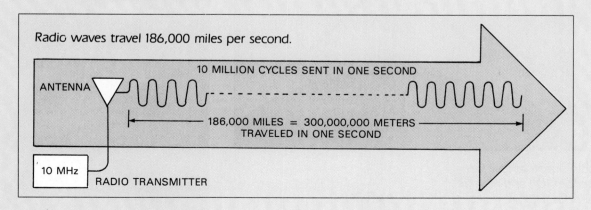

Radio waves travel 186,000 miles per second.

10 MILLION CYCLES SENT IN ONE SECOND

ANTENNA

186,000 MILES = 300,000,000 METERS TRAVELED IN ONE SECOND

10 MHz

RADIO TRANSMITTER

In radio communication, information is sent from a transmitting antenna to a receiving antenna at some distance away. The channel is the part of the electromagnetic spectrum used in the signal. Because radio stations each transmit on a different frequency, many transmitters can send at the same time. The radio receiver can select the frequency of the desired TV or radio station, ignoring the others. That's why you can choose from many TV or radio stations.

Radio communication can be used in different ways. A transmitter can send out a signal to many receivers. This is called **broadcasting**. AM and FM radio broadcasters often have thousands of listeners. When a transmitter sends a signal to only

A radio communication system contains a transmitter, a receiver, an antenna for each, and a channel.

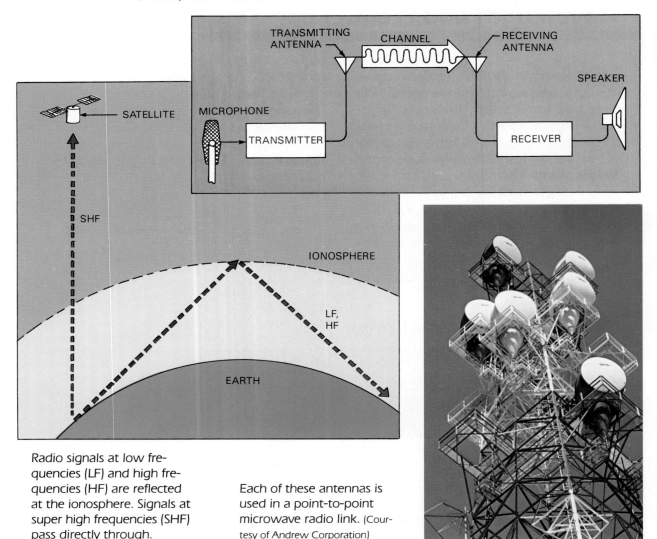

Radio signals at low frequencies (LF) and high frequencies (HF) are reflected at the ionosphere. Signals at super high frequencies (SHF) pass directly through.

Each of these antennas is used in a point-to-point microwave radio link. (Courtesy of Andrew Corporation)

one receiver, it is called **point-to-point** transmission. Such a system can provide communication for remote places, like fire-ranger towers, or mobile receivers, like those used in repair service trucks.

Satellite Communication

Satellite communication makes it possible to have instant communication with the most remote parts of the world. It has played an important part in "shrinking" the globe.

Communication satellites are radio relay stations, called **repeaters**. Because they are located so far above the earth, they can beam signals to a very large area. Satellite communication requires many different technologies. The satellites themselves are complex electronic systems. They are transported by space vehicles. They must be sent into space and remain in orbit around the earth.

Perhaps you have tried tying a rope to a bucket of water and swinging it in a circle. If you swing the bucket fast enough, you can swing it upside down without spilling a drop. The water seems to be stuck inside.

A satellite is like the bucket of water. Gravity is the "rope" that keeps the satellite from flying off into space. At the same time, circular motion keeps the satellite (and the water) from

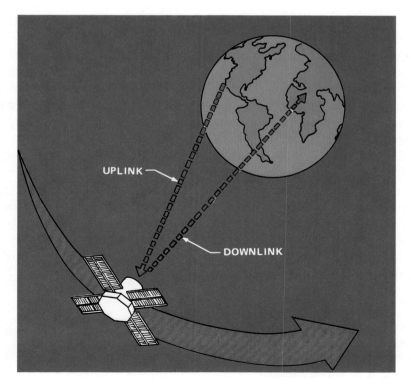

Both the satellite and the earth make a complete revolution in the same time (24 hours). Viewed from the earth, the satellite appears not to move.

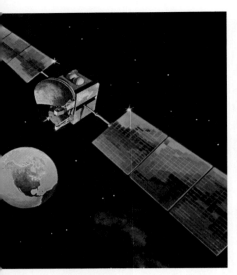

Spacenet I (Copyright 1984 GTE Spacenet Corporation. Spacenet is a registered service mark of GTE Spacenet Corporation)

falling back to earth. At low altitudes, a satellite must circle the earth very quickly to stay up because gravity is very strong. At an altitude of about 22,500 miles, gravity is much less. The satellite must circle the earth only once every 24 hours to stay up.

Since the earth also turns once every twenty-four hours, the satellite moves along with it. To someone on earth, it appears not to move at all. This means that a satellite can be put into an orbit so it stays over a particular location. Such a stationary satellite is said to be in **geosynchronous orbit** (**geo** means "earth" and **synchronous** means "same time").

Since the satellite doesn't appear to move, an antenna pointed at it doesn't have to move. Because the satellite is so high, its signals can be received over a large area.

A transmitter on earth sends a transmission to the satellite on one frequency (the **uplink**). A receiver on the satellite accepts the signal and changes it to a second frequency (the **downlink**). A transmitter on the satellite then sends the changed signal toward the earth. Anyone with a satellite receiver in the coverage area can get the transmission. Millions of homes now receive TV broadcasts directly from satellites.

Fiber-Optic Communication

Telephone and data networks use light to send messages. A **fiber-optic cable** is a cable made of many very thin strands of coated glass fibers. It can be used to carry light long distances, even around corners.

A very bright light source, such as a **laser** or **light emitting diode (LED)** is the transmitter. Information is sent on the light wave by making the light bright and dim in a code. The fiber-optic cable is the channel. The receiver is a light-sensitive semiconductor device called a **photodiode**. The photodiode changes light into electricity.

Light has a very large bandwidth. It can therefore carry huge amounts of information. A single pair of fibers in a cable can carry as many as 20,000 telephone conversations at once. (A cable can have 144 or more fibers.) In most big cities fiber-optic cables play a large part in telephone communication. Fiber-optic cables are being used more and more between cities, as well. They are even installed under the sea and used to connect cities on different continents.

Optical fibers used in communications are very small. Each point of light in this picture is the end of an optical fiber. (Courtesy of United Telecom)

MODERN COMMUNICATION SYSTEMS USING SYMBOLS AND CODES

During the 1800s, telegraph became the first way of communicating instantaneously over long distances. In 1843, Samuel F. B. Morse constructed the first telegraph line between Washington, D.C., and Baltimore. Morse's first system used a pen that made marks on a piece of paper. It soon gave way to the telegraph sounder, a machine in which clacking sounds represented Morse code. Operators listened to the clacks and wrote down the letters and numbers they stood for. The Morse code, though somewhat changed, is still used today.

This fiber-optic cable is being laid in the Atlantic. (Courtesy of AT&T Bell Laboratories)

Telegraph wires soon crisscrossed Europe and the United States. In 1858, the first transatlantic telegraph cable was laid. People in Europe and the United States could communicate instantaneously for the first time.

Codes are also used to help people communicate with machines. One such code is the Universal Product Code, described in Chapter 7. Two codes used between computers and their terminals are the American Standard Code for Information Interchange, or **ASCII** (pronounced as-key), and the Extended Binary Coded Decimal Interchange Code, or **EBCDIC** (pronounced ebb-sa-dick).

MODERN COMMUNICATION SYSTEMS USING PRINTED WORDS

Newspapers

Since the days of Gutenberg, the printing industry has spread far and wide. Millions of books and magazines are printed every year. People can share their knowledge with others in even the most remote places in the world.

Newspapers were the first means of mass communication. The first regularly published newspaper in the United States was printed in Boston in 1704. This early newspaper, like those that followed, was printed on a hand-operated press. About 150 years passed before the Industrial Revolution brought machine-driven presses.

In 1840, the *New York Sun* printed 40,000 newspapers a day. This circulation was the highest in the country. Today, the *New*

Time, Newsweek, U.S. News and World Report, and other magazines are transmitted via satellite from New York to regional printing centers. (Courtesy of U.S. News and World Report)

York Times prints 1.5 million Sunday papers. News stories are composed on computers instead of typewriters. The pages are optically scanned by lasers and sent from one plant to another by microwave or satellite transmission. Today's presses are huge, feeding large spools of paper through their rollers.

Some newspapers, like the *Wall Street Journal* and *USA Today*, print their papers in several different locations. This cuts transportation costs. The information is transmitted to satellites, which send it on to receivers in many cities. The papers are printed and distributed in each of the cities.

Out of Gutenberg's system, in which letters were cast one at a time, came **linotype** (line of type). Newspapers used linotype to cast entire lines of type on a single piece of metal. Today, some newspapers use a technology called **pagination**. Pagination means that entire pages, including words and pictures, are composed on a computer terminal. The whole page can be viewed. Or the operator can zoom in on a particular column or line. Once the page is composed, it is sent electronically to a **phototypesetter**. The phototypesetter is a computerized printer. It prints out a newspaper page on a single piece of paper. An offset plate is made of the printout, and the presses roll.

Typewriters

Mechanical typewriters were replaced by electric machines in the 1940s. In 1961, IBM developed a machine that used a ball with the type on it. Balls were available with different type styles. By changing the ball, the typist could change the type style.

Today's newest typewriters are electronic. Many mechanical parts have been replaced by integrated circuits. These typewriters have many useful features. Some have a memory, and can store pages of text. A letter can be stored and, with slight changes, sent to many people. Electronic typewriters also do things like centering titles, underlining words, and checking spelling.

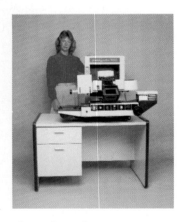

A modern phototypesetter (Photo Typositor is a registered trademark of Visual Graphics Corp.)

Computer Printers and Word Processors

The number of computer printers is growing as the use of computers grows. Printer types include dot matrix, daisy wheel, laser, and ink jet. Printers are available that can print one character, or one line, or one entire page at a time.

Word processors combine the typewriter with the computer. What you type on the keyboard is stored in computer memory.

Many office functions are now automated through a combination of computer and communication technologies. (Courtesy of NCR Corp.)

The text is displayed on a video monitor instead of on a piece of paper. If you make a mistake, you can go back to it and correct it. Only after the text is exactly as you want it do you print the text on paper.

Documents are saved on floppy disks instead of on paper. One floppy disk can hold more than 500 pages of text. Most offices are replacing their typewriters with word processors to save typists time.

Computing and communication technologies are merging to produce powerful tools for information processing.

Copying Machines

In 1937, a law student from New York City named Chester Carlson developed a way of making copies of documents. He called his invention **xerography**, which means "dry copying." In 1959, the Xerox Company developed Model 914, the first copier that could be used in an office. It made copies without using messy inks or other fluids.

CONNECTIONS: Technology and Social Studies

- Each time a new system has been introduced to convey ideas, more people have become better informed, not only about past history but also about current happenings.
- As mass communication systems have grown they have placed enormous power in the hands of the people who control them. The media are able to influence governmental, corporate, and individual decisions by the manner in which they present news and information.
- Many technological advances are very costly, and today's "wonder machine" may be outdated tomorrow. This poses a problem for people trying to use their money as efficiently as possible.

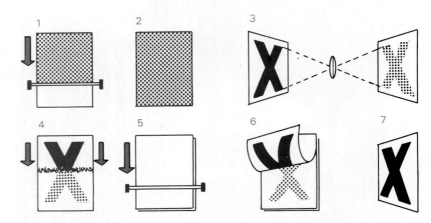

Diagram of the xerographic process. (Courtesy of Xerox Corp.)

Copying machines use static electricity. Static electricity is what is created when you walk across a rug on a cold, dry day, then touch a doorknob. You build up an electrical charge on your body. The charge is discharged when you touch an object with the opposite charge. (Remember from Chapter 7 that like charges repel and unlike charges attract.)

A copying machine has a metal plate coated with a light-sensitive material. (See part 1 of the diagram above.) The metal plate gets a positive charge as it passes under a wire in the copying machine (2). The paper to be copied is exposed to a very bright light (3). The light is reflected by the white areas on the page to be copied. It is not reflected by the areas dark with printing or writing. The light that is reflected by the white areas removes the positive charge from the metal plate. The rest of the positive charge (where the black image is) remains. A black powder called **toner**, which has a negative charge, is dusted over the plate. Toner is attracted to the positive charge of the dark image area (4). Another sheet of paper (5) is charged positively. It attracts the negative toner to it, and pulls it off the dark image area of the metal plate (6). Finally, the paper is heated by a roller. The heat bonds the toner to the paper and makes a finished copy.

Office Automation

Office automation means using computers and communications in an office setting. Many of today's typewriters, copiers, and other information machines contain microprocessors. Microprocessors enable the machines to do more jobs. Some of these machines have data communication connections. These connections let them work together, making them more useful. Many offices now have a data communication network, called a **local area network (LAN)**. The computerized equipment can communicate over the network.

MODERN COMMUNICATION SYSTEMS USING DRAWINGS AND PICTURES

Facsimile

Facsimile (FAX) is a method for sending information electronically. Newspapers use FAX to send news photos from one place to another. Weather maps have been sent this way since the 1920s. Businesses use FAX to send memos and graphs to each other. Recently, FAX has become a very widespread communication method. FAX machines have become so inexpensive that some homes, as well as most businesses, now have them.

In FAX, an optical scanning device moves across a page. The scanner converts the black-and-white patterns it picks up into electrical impulses. These impulses are sent over wires to another location. A FAX machine there turns the impulses back into a perfect copy of the original page.

Television

In **television**, visual information is changed into electrical signals. The signals are sent across a distance. Finally, the signals are changed back to visual images on a screen.

In a television system, the camera is part of the transmitter.

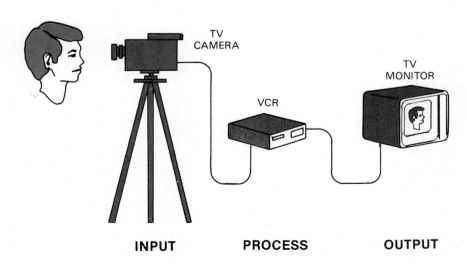

TV CAMERA

TV MONITOR

VCR

INPUT   **PROCESS**   **OUTPUT**

A TV system converts a visual image into an electrical video signal. The signal can be stored in a video recorder or broadcast live.

The camera changes what is seen to electrical signals **(video)**. A monitor changes the signals into images on a screen. This is the receiver—the part that is watched by you, the viewer. Video signals may be stored in video recorders (for example, a video cassette recorder, or VCR). Video may also be sent by a wire (as in closed-circuit TV), sent by cable TV, or broadcast using TV transmitters. The videotape or the cable is the channel over which the video signal is sent.

The signals TV cameras produce can be either black-and-white or color. Color signals break down light into red, blue, and green for transmission to the monitor.

In closed-circuit TV, the camera is connected directly to the monitor by a cable. This TV system is often used to serve over-flow audiences at special events. Closed-circuit TV is also used for security. Guards can watch several areas at the same time.

Broadcast TV is the most common way of sending video signals. In broadcast TV, each station has its own frequency, or channel. TV sets are video monitors that can be tuned to a specific frequency. TV transmitters send sound at the same time as the picture.

Television transmissions have a limited range. Many people live beyond that range. To provide good reception in these areas, cable TV systems use very good antennas or relay links or both. The system then provides all the channels to subscriber homes by way of a cable.

Cable can also provide programming other than that provided by broadcast stations. Movies, sports, weather, and

Using a closed circuit TV system, security guards can watch many areas at once. (Courtesy of ADT, Inc.)

This newscast is being broadcast live. (Courtesy of WRGB—Newscenter 6)

community news are some of these. Because of these extra programs, cable TV systems have been built in most big cities, even where broadcast TV reception is good.

MODERN TELECOMMUNICATION SERVICES

Video conferences carried over microwave or satellite circuits allow people at many locations to conduct face-to-face meetings without traveling. (Courtesy of Compression Labs, Inc.)

Teleconferencing

A **teleconference** is a meeting held by people at different locations with the help of electronic communications. With travel costs rising, this saves money.

Several kinds of teleconferences are possible. In the simplest kind, only voice lines are used. People hear each other as if they were sitting in a group. This is sometimes called a conference call. It can be set up using telephone lines, with each person on a telephone. If a telephone with a loudspeaker is used, a small group of people in one room can take part in the conference.

In **videoconferencing**, television is used. People not only hear each other but see each other too. It is costly, and is most often used over long distances, when travel costs for a face-to-face meeting would be high.

Telecommuting

Using personal computers, some people are able to work at home. Programmers, word processor operators, and data-entry clerks, for example, can work at home. The completed work can be brought to the office on a floppy disk or sent to the

Effective data communication allows some information workers to work at home rather than in an office. (Courtesy of AT&T Bell Laboratories)

company's computer by telephone lines and modems.

Some workers like **telecommuting** because they can choose their own work hours. They spend less time traveling to and from the office. It is useful for parents who want to stay home with their children. Employers like telecommuting, too. They pay less for office space and utilities.

There are some problems with telecommuting. Employers worry that the quality of work will suffer because workers are not in everyday contact with each other. Also, workers may find they are constantly interrupted at home by visitors and telephone calls.

Automated teller machines allow customers to bank at any time of the day or night. (Photo taken by Jay Fries with permission, Tandem Computers, Inc.)

Electronic Banking

Computing and communications have combined to change the way banking is done. In most parts of the country, automated teller machines (ATMs) are available. They even outnumber bank branches in some places. ATMs let customers deposit money, withdraw money, get their bank balance, and pay bills any time of the day or night.

ATMs are computer terminals connected by telephone lines to the bank's main computer. In some places, banks share ATMs.

Two-Way Cable TV

At first, cable TV systems could only send signals from the control center (the **head end**) to the subscribers. Two-way cable systems, however, can carry signals from the subscriber back to the head end, as well.

Two-way cable TV has made several new services possible. With **pay-per-view**, customers are charged only for the programs they watch rather than paying a monthly fee. Some-

Communications are changing the way we shop. (Courtesy of AT&T Bell Laboratories)

CONNECTIONS: Technology and Science

■ Waves are regular disturbances moving outward from a source. They transfer energy. Waves are classified as **transverse** and **longitudinal**. In transverse waves, the wave vibrates at right angles to the direction in which the wave is moving. Electromagnetic waves are transverse waves.

In longitudinal waves, the wave vibrates in the same direction as the wave moves. Sound travels in longitudinal waves.

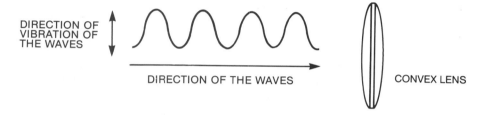

DIRECTION OF VIBRATION OF THE WAVES

DIRECTION OF THE WAVES

CONVEX LENS

■ A lens is a piece of transparent material such as glass that can bend (refract) rays of light entering it. It causes these parallel rays to meet at a point known as the focal point. Convex lenses are thicker in the center than at the ends. They can be used to magnify objects. Cameras contain convex lenses, which focus images onto the photographic film that contains light-sensitive chemicals.

times the billing is done automatically when the pay channel is selected. In other systems, the viewer must ask for programming from the cable operator. This system is now used at many hotels to provide movies.

Shop at home allows viewers to get information on all kinds of products. A keyboard near the TV is used to ask for information. When a request is received, the products are demonstrated on a TV channel. The viewer can then order the product.

Using **data base inquiry**, subscribers can look through information contained in data bases. The data bases are maintained by major newspapers, organizations, and financial institutions. A viewer can use the service to look through the latest copy of the *Washington Post* or the *Wall Street Journal*.

SUMMARY

Humans, animals, and machines can communicate among themselves and with each other. Communication includes sending a message and having it received and understood. In a communication system, feedback shows whether the message was received and understood.

Four types of communication systems are: sound and speech, symbols and codes, printed words, and drawings and pictures.

The invention of the printing press was an important advance not only in communication technology but in human history. The printing press made information available to many more people than ever before. Gutenberg's press used a form of printing called relief printing. In relief printing, raised letters are inked and then pressed against paper. Other methods of printing include gravure, screen, and offset printing. Today, offset printing is the most commonly used of these methods.

Technical drawing is an important method of communication. Many drawings are still made by hand, but computer-aided design (CAD) is being used more and more.

The five elements required for photography are light, film, a camera, chemicals, and a darkroom. Four types of cameras are the view camera, the viewfinder camera, the twin-lens reflex camera, and the single-lens reflex camera.

All communication systems have a transmitter, a channel, and a receiver. The channel often adds imperfections to the transmitted signal, affecting our ability to receive the signal clearly. One imperfection is noise. Bandwidth is the measure of a channel's information-carrying capacity.

Electricity and electronics have revolutionized communication. Old communication systems have been improved, and new ones have been invented using electronic technology.

Modern telephones use both analog (smoothly varying voltages) and digital (1s and 0s) technologies. Many telephone calls are converted into digital (computer-like) signals before being routed by special-purpose computers. Digital technology has given us telephone services such as call waiting and call forwarding. Combining telephone and radio technologies has brought about cordless telephones and mobile telephones.

Radio sends information from one location to another without the use of wires. Different kinds of radios are used at different frequencies within the electromagnetic spectrum. The spectrum ranges from very low frequencies through very high frequencies to light frequencies.

Communication satellites are radio and TV relay stations located 22,500 miles above the earth's equator. They revolve around the earth once every 24 hours, the same time that the earth takes to make one rotation. The satellites thus remain over one location.

Fiber-optic cable has extremely wide bandwidth. It can carry a very large number of telephone conversations or data signals at the same time. Light from a laser or a light-emitting diode (LED) is transmitted down a fiber-optic cable. The light is received by a photodiode at the receiver.

Modern newspapers are composed by computer systems. They are often sent by satellite to regional printing presses. Other newspapers use semiautomatic machinery called linotype. Linotype casts entire lines of type at a time.

Microcomputers are integrated into typewriters, word-processing machines, copying machines, and other office equipment. Many times, these machines can communicate with each other. Such communication makes each a more powerful and useful tool.

Facsimile (FAX) transmits drawings and pictures from one location to another. Telephone lines are often used for FAX. Television transmits audio and moving visual images from one place to another.

Teleconferencing, home shopping, and various other services are provided by two-way cable TV.

(Courtesy of NASA)

1. What three elements make up the process part of a communications system?
2. In a communication system, what is noise? Give three examples.
3. Define bandwidth. Does a color TV signal need more or less bandwidth than a telephone conversation?
4. What kind of printing process did Gutenberg's press use? Why was Gutenberg's press important?
5. Name a camera and describe which of the four kinds of cameras it is.
6. Describe the five elements needed for photography.
7. Are telephone circuits analog or digital? Explain your answer.
8. What new products resulted from combining radio technology with telephone technology?
9. In what way is fiber-optic cable better than copper wire for telephone circuits?
10. Would you like to be a telecommuter? Why or why not?

(Courtesy of AT&T Bell Laboratories)

(Courtesy of Sperry Corporation)

(Courtesy of NASA)

CHAPTER KEY TERMS

Across

3. A popular computer applications program used for writing.
6. A record with no grooves. It uses lasers to sense reflections from pits on the surface.
8. A conference held with the participants at a distance from each other.
9. The frequency used by an earth station to send a transmission to a satellite.
11. This printing process uses a stencil and silk screen to print designs on T-shirts and posters.
12. This is measured in cycles per second, or Hertz.
16. This includes several elements: a dark chamber, a lens and shutter, and film.
17. This can interfere with communication.
18. A measure of the ability of a channel to carry information.

Down

1. To transmit a radio or TV program to many receivers.
2. A system of signals or symbols that have some specific meaning.
4. Printing by transferring an image from a metal plate to a rubber blanket to a piece of paper.
5. Drawing using instruments like T-squares, triangles, drawing board, and CAD systems.
7. A binary code that represents letters, numbers, and punctuation marks.
10. Communication that uses a thin glass fiber cable to guide a light source.
13. This component of the communication process serves as the route the message takes.
14. This component of the communication process accepts the message.
15. A plastic material covered with light-sensitive silver particles.

ASCII
Bandwidth
Broadcast
Camera
Channel
Code
Compact disk
Downlink
EBCDIC
Fiber optic
Film
Frequency
Noise
Offset printing
Receiver
Satellite communications
Screen printing
Technical drawing
Telecommute
Teleconference
Telephone
Television
Transmitter
Uplink
Wavelength
Word processing

SEE YOUR TEACHER FOR THE CROSSTECH SHEET

ACTIVITIES

DESKTOP PUBLISHING

Problem Situation

Desktop publishing systems require hardware and software. Computers, printers, monitors, and scanners are hardware. Software programs tell the computer how to process the information you have entered into the computer.

An important advantage of desktop publishing is that text, headings, and graphics are combined to produce an entire page. Desktop publishing programs include a word processor and allow the user to select different sizes and styles of type. Graphics are selected from "clip art" disks or artwork can be scanned. Scanners change artwork images into a form that can be processed by the computer.

Design Brief

In this activity your class will use computers and desktop publishing software to publish a newsmagazine. The publication should include articles of interest to students, teachers, and parents. You will write at least one article, learn to use desktop publishing software, and help to produce the newsmagazine.

Procedure

1. Choose the kind of story you want to write for the newsmagazine. Here are several ideas.
   EDITORIAL—Express your opinion about something in school or in the community.
   REVIEW—Describe a book, film, television show, restaurant, or product and explain why you liked or disliked it.
   FEATURE—Write about an activity or event that is newsworthy.
2. Your teacher will provide guidelines for writing the story. Use the standard reporter's questions: who?, what?, when?, where?, and why?.
3. Your teacher will demonstrate how to use the computer and software. After a short time you will be able to work on your own. Note: Be sure to follow your teacher's directions.
4. Use the newsletter format without a heading to write your first draft.
5. Format a storage disk and save your story.
6. Print one copy. Proofread and correct any errors you find.
7. After your teacher reviews the story, make any additional changes that are necessary.

Suggested Resources

Apple II or IBM compatible computer
Dot matrix printer
The Children's Writing and Publishing Center™
Access to a photocopier

8. Add a headline to your story using large type. Add a graphic from one of the available picture disks.
9. Save the complete article on your storage disk. Print a copy. Turn in both the hard copy and the storage disk.
10. Your teacher will demonstrate how the stories written by each student will be combined to produce the newsmagazine. You will also be given a production assignment.

Classroom Connections

1. Communication includes having a message sent, received, and understood.
2. Word processors combine typewriter and computer technologies.
3. Articles written for newspapers and newsmagazines should address the 5 Ws: who?, what?, when?, where? and why?
4. The kinds of articles written for newsmagazines include feature stories, reviews, and editorials.
5. Newspapers were the first medium of communication for the masses.
6. Newspapers bring people closer together because of better communication. Satellite communication systems allow newspapers such as *USA Today* and the *Wall Street Journal* to be published worldwide.
7. Most computers use digital (1s and 0s) technologies.

Technology in the Real World

1. Desktop publishing systems enable individuals to produce quality printed materials.
2. Scanners are used so that photographs and other artwork can be included in computerized page layout.
3. ASCII is a code that allows communication between computers.

Summing Up

1. Describe an important advantage of desktop publishing systems.
2. How did you use the Children's Writing and Publishing Center™ to revise your article?
3. Explain the difference between hardware and software.

COMMUNICATING WITH FIBER OPTICS

Problem Situation

One of the most promising new communication systems involves converting information into impulses of light and channeling them to their destination inside fiber-optic cables. Fiber-optic cables are fine glass fibers that allow light to pass through them. Fiber-optic technology is rapidly replacing electrical carrier lines for phone transmission. The advantages of fiber optics are many. Among them are greater clarity in transmission, and the fact that a fiber-optic carrier can carry a higher volume of information than copper wire.

Design Brief

Design and build a fiber-optic communication system that can be used for human-to-human and human-to-machine communication.

Procedure

1. Set up the lamp, push-button switch, and battery circuit to make a light transmitter. Punch a small hole in the paper towel tube so a section of fiber-optic cable can be fit into it. Place the tube over the lamp so it acts as a cowling.
2. Activate the light with the switch and observe how the fiber acts as a carrier for the beam of light.
3. Set up a light receiver circuit with a piezo buzzer, 9-volt battery, and the phototransistor.
4. Place the loose end of the fiber-optic cable securely in front of the phototransistor. You should now be able to operate the buzzer by transmitting impulses of light with the light transmitter.
5. Develop a code and transmit a message.
6. Replace the buzzer and 9-volt battery with the 1.5-volt battery and the DC motor. Operate the motor with the transmitter.

Suggested Resources

Fiber-optic cable
Phototransistor
Lamp and ceramic socket
    (1.5-volt)
Push-button switch
22-gauge wire
1.25-volt DC motor
1.5-volt batteries
Stiff cardboard
Paper towel center core
9-volt piezo buzzer
9-volt battery and battery snap.

Classroom Connections

1. All communication systems have a transmitter, a receiver, and a channel through which the information travels.
2. Fiber-optic communication systems use light as the information carrier.
3. Light travels at a speed of 186,000 miles per second.

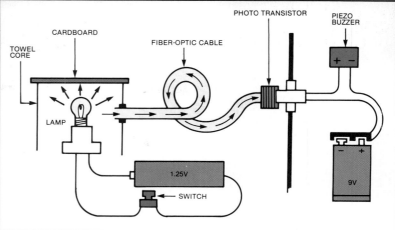

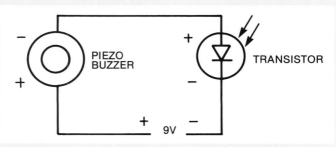

4. Light rays are refracted, or bent, when they pass through materials with different densities.
5. For communication to take place, the information must be understood. Information can be transmitted in the form of codes that use symbols to represent specific meanings.
6. Early communication systems such as the telegraph and telephone helped the countries of the world become interdependent.

Technology in the Real World

1. When adapted to medical technologies, fiber-optic technology can be used to observe the inside of the body without surgery. An instrument called an endoscope can be fed into the stomach and other parts of the body. Images of these areas travel up the fiber-optic carrier and into the doctor's eyepiece.

Summing Up

1. What is the purpose of the sleeving surrounding the glass fibers in a fiber-optic cable?
2. How does a phototransistor operate?
3. Why can fiber-optic cables still carry light impulses after being tied in a knot?
4. List three advantages of fiber-optic communications.

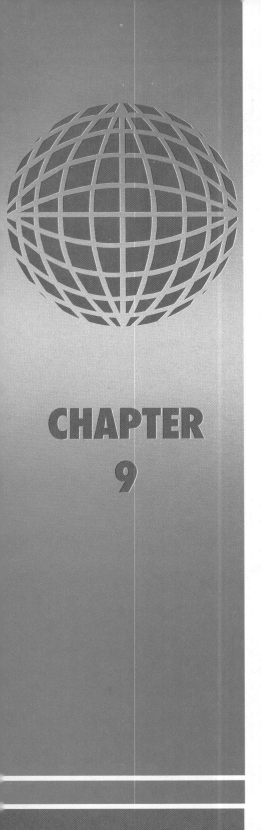

CHAPTER 9

PRODUCTION SYSTEMS: MANUFACTURING

MAJOR CONCEPTS

After reading this chapter, you will know that:

- Production technologies fill many of people's needs and wants by means of manufacturing and construction systems.
- Manufacturing is making goods in a workshop or factory. Construction is building a structure on a site.
- Mass production and the factory system brought prices down. People were able to improve their standard of living.
- Manufacturing systems make use of the seven types of technological resources.
- There are two subsystems within the manufacturing system: the material processing system, and the business and management system.
- Automation has greatly increased productivity in manufacturing.
- Computers and robots have improved product quality while bringing manufacturing costs down.

PRODUCTION SYSTEMS

Our environment has been created by people. We live, play, learn, and work in structures built by people. Our food, furniture, clothing, and automobiles are made by people.

Production systems are the means by which our needs are met. There are two kinds of production systems. **Manufacturing** is making goods in a workshop or **factory**. **Construction** is building a structure on a site.

Construction systems are covered in Chapter 10. In this chapter you will learn about manufacturing systems.

Production technologies fill many of people's needs by means of manufacturing and construction systems.

(Courtesy of Maytag)

(Photo by Jonathan Plant)

(Courtesy of Sony Corp. of America)

(Courtesy of Saab)

(Courtesy of Campbell's)

People use two kinds of production systems (manufacturing and construction) to satisfy many of their needs and wants.

245

Problem Situation 1: Toys that consist of a basic unit that can be assembled in a variety of forms are particularly entertaining.

Design Brief 1: Design and construct a shape so that several pieces of the same shape can be connected together in a number of ways to make a vertical "sculpture."

Problem Situation 2: Many products are used in health care and personal hygiene. These include cosmetics, skin creams, perfumes, soaps, deodorants, and lotions.

Design Brief 2: Design and produce a product from one of the above categories and test market it using other students in the class. Design a label and package for the product you produce.

Problem Situation 3: Some manufacturing processes require counting devices.

Design Brief 3: Design and construct a device for counting the rotations of a bicycle wheel.

MANUFACTURING SYSTEMS

Manufacturing is making goods in a workshop or factory. Construction is building a structure on a site.

People have used manufacturing technology for many thousands of years. Prehistoric people made use of natural materials to make weapons, food, and clothing. Stone, bone, wood, and clay were the materials most often used. Tools were made by chopping or scraping a piece of material to the desired shape and size.

Evidence shows that people were making pottery from clay over 30,000 years ago. The firing process was probably discovered by accident, when people found hardened clay under the ashes of their cooking fires. They then began to fire clay items on purpose. This is one of the first examples we have of manufacturing.

Natural metals, like copper, gold, and silver, were made into ornaments and tools. During the Bronze Age (starting about 3500 B.C.), people began casting metal.

Glass may have been the first synthetic (human-made) material. Among some early people, glass beads were considered to be very valuable. They were used as money in some places.

The Craft Approach

For centuries, objects were manufactured by people only for their and their families' use. Then people began to specialize. Shoemakers made shoes not only for family and personal use but for other people as well. In return, they received goods from other craftspeople. There were many kinds of craftspeople: candlemakers, weavers, spinners, glassblowers, silversmiths (who made tableware), coopers (barrel makers), gunsmiths, and tailors.

These craftspeople made their products one at a time, working alone from start to finish. Most often, they worked at home or in small workshops. Their tools were costly. Not everyone who wanted to become a craftsperson could afford to do so. Often, the master craftsperson hired young people who wanted to learn the trade. These young workers, called apprentices, were trained on the job.

People today still want one-of-a-kind products. We still have craft production. Of course, we pay more for products made by craftspeople. A hand-knit sweater, for example, is much more expensive than one made by machine in a factory.

Some products must be designed and made to meet particular needs. Handicapped people often buy **custom-made** items that make daily activities easier. Craftspeople today make pottery, jewelry, clothing, furniture, and many other items to suit individual needs and tastes.

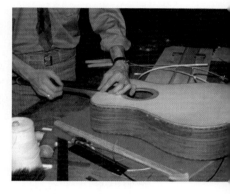

Many musicians have their instruments custom-made.
(Photo by Michael Hacker)

The Factory System

As America grew, railroads, canals, and highways opened up new markets. More goods were needed.

During the Industrial Revolution, which started in the late 1700s, there were many new inventions. Machines such as the steam engine, the cotton gin, and the sewing machine helped with many tasks. Businesspeople used the new machinery to improve production. Soon goods were being made in factories by large numbers of workers. Over the years, the factory system replaced the craft system in the manufacture of most goods.

MASS PRODUCTION AND THE ASSEMBLY LINE

Mass production is the production of goods in large quantities by groups of workers in factories. In mass production, the pro-

Computers and Cowboy Boots

A modern-day industry that still employs craftspeople is the shoemaking industry. What is happening to the shoemaking craft in this modern technological age? It is an industry that is in the sunset of its life.

Because shoemaking requires skilled workers, labor costs are high. U.S. shoe manufacturers are finding it hard to compete with foreign competition. Today, most shoes that are purchased by people in the United States are made overseas. Between 1968 and 1986, the shoemaking industry in the United States lost about 100,000 jobs. Many companies have gone out of business.

Cowboy boots require fancy stitching.

It would be very expensive for companies to hire U.S. workers to do this stitching by hand. The handwork is now done by workers in Spain and Portugal, who provide labor at a lower cost.

Some years ago, the Shoe Machinery Group of Emhart Corporation developed computer-controlled stitching machines. Computerized stitchers are bringing the manufacture of cowboy boots back to the United States. Computer-controlled systems are being used for many other operations, including design. While the days of the cowboys and Indians may be over, it still looks good for cowboy boots in the United States.

This computerized stitcher is sewing a pattern on a cowboy boot. (Courtesy of Shoe Machinery Group—Emhart Corp.)

duction process is divided into steps. Each worker does one step, passing the item on for the next step. Work is carried out on an **assembly line**, a system by which the item is moved quickly from one workstation to the next. Through the use of mass-production methods and the assembly line, more goods can be produced in a given period of time.

Mass Production and the Automobile Industry

One of the first moving assembly lines was at the Ford Motor Company in 1913. Auto parts were pushed from one worker to the next. This reduced production time by about one-half. By applying the same principle to the assembly of a total car, Henry Ford speeded up car production. A finished Model T Ford came off the assembly line every ten seconds.

The interchangeability of parts is one of the most important characteristics of a mass-production system. Modern automobiles, for example, are manufactured with interchangeable parts. Door latches for the Honda are alike. If one part breaks down, we can purchase another just like it from an automobile dealer.

A 1913 auto parts assembly line
(Courtesy of Ford Motor Company)

Workers making door latches for Honda automobiles
(Courtesy of Rockwell International)

Eli Whitney started one of the earliest assembly lines in 1789. Whitney signed a contract with the U.S. Army to make 10,000 rifles (a huge number in those days) in two years. He succeeded by making a large number of each kind of rifle part at once and keeping the parts in separate bins. The parts were **standardized**, exactly alike. This made them **interchangeable**—any of them could be used in assembly. Before this, rifles had been made one at a time, so the parts of a rifle would fit only that rifle.

This student is using a jig. The jig holds a plastic strip in place so that the ends of the strip can be rounded evenly by a belt sander. (Photo by Michael Hacker)

Jigs and Fixtures

One way manufacturers make standardized parts exactly alike is to use **jigs** and **fixtures** on the machinery. A jig holds and guides the item being processed. It also guides the tool that does the processing. A fixture is used to keep the item being processed in the proper position.

IMPACTS OF THE FACTORY SYSTEM

Mass production and the factory system brought prices down. People were able to improve their standard of living.

The factory system produced a larger supply of goods. Luxury items became less expensive. More people could afford to buy the products of mass production.

The factory system also brought about other changes. New industries sprang up. The steel, automobile, and clothing industries provided millions of new jobs. Craftspeople and farmers became assembly-line workers.

Factories created wealth by adding value to the resources that were processed into goods. As a result, the standard of living improved for people who lived in industrialized nations.

Craft production and mass production were quite different. Craftspeople could stop working to take care of household chores. Factory workers, however, had to work without stopping. Time was very important. Manufacturing operations were **synchronized**. That is, one process had to be followed immediately by the next. If one step were delayed, all production would slow down.

Workers were often paid by piecework. They were paid for the number of items they completed each day. During the early 1900s, it was not unusual for people to work twelve hours a day. **Unions** were formed to protect the rights of workers. **Child labor laws** were passed to ensure that children were treated fairly.

During the craft era, families often worked together at home. Someone was always around to care for the children. When people began to work in factories, they had to leave home.

Children were often used as factory workers in the late nineteenth and early twentieth centuries. They were often forced to work under harsh conditions and during evening hours. (Courtesy of the Bettman Archive)

Finding a place for the children became a problem. The number and importance of schools increased partly because of the factory system.

THE BUSINESS SIDE OF MANUFACTURING

Until the 1930s, most businesspeople wanted only to improve the production process to make a product cheaper and faster. They wanted to be able to offer a standard product at a low price. A suggestion was once made to Henry Ford that he paint his Model Ts different colors. He replied, "Give it to them in any color, so long as it is black."

This attitude changed with competition. In the 1930s, General Motors started producing a new model of automobile every year. The way automobiles were advertised and sold was suddenly as big a concern as how they were produced. Marketing and business management became important systems in manufacturing.

Differences Between Craft Manufacture and Mass Production

CRAFT MANUFACTURE	MASS PRODUCTION
1. Workers are very skilled.	1. Workers need limited skill.
2. Workers make a product from start to finish by themselves.	2. Workers work on only one part of the product.
3. Work is varied and interesting.	3. Work is routine and often dull.
4. Craftspeople get satisfaction by seeing the finished product (like a completed chair).	4. Factory workers see only the one part that they produce (like the chair leg).
5. Each part is hand crafted so no two are exactly alike.	5. Parts are machine made and are interchangeable.
6. Only one item is produced at a time.	6. Many items are produced during the production run.
7. It takes a long time to produce each item.	7. The average time it takes to produce each item is reduced.
8. The cost of each item is high.	8. The cost of each item is lowered.
9. Quality depends mainly upon the skill of the craftsperson.	9. Quality depends mainly upon the accuracy of the machines and how well they have been set up by people.

Fire Escape Parachute
Patent No. 221,855
March 26, 1879

In March 1879, Benjamin Oppenheimer patented a fire escape parachute, complete with headpiece and sponge-bottom shoes.

Manufacturing systems make use of the seven types of technological resources.

Employees at Pittsburgh Plate Glass developed a better method for attaching the "button" that secures the rearview mirror to an automobile windshield. (Courtesy of PPG Industries)

THE ENTREPRENEURS

An **entrepreneur** is a person who comes up with a good idea and uses that idea to make money. An entrepreneur might improve a product, or improve the way a product is made, or even come up with an idea for a new product.

Some entrepreneurs are **inventors**. An inventor comes up with a totally new idea. The safety razor, the laser, and the contact lens are all inventions. Inventions can be protected by a patent. When a device is patented, no one but the patent holder can make and sell it for seventeen years. Many inventions are now in everyday use. Others, such as air-conditioned suits and hats with fans, did not catch on with the public.

Some entrepreneurs are **innovators**. An innovation is an improvement in an invention. Innovations lead to new uses. They can start new industries or change existing ones. The electric guitar is an innovation. A new music industry sprang up around the music played on this instrument. Other innovations are the diesel engine and power steering.

RESOURCES FOR MANUFACTURING SYSTEMS

Manufacturing systems use the seven technological resources to make a product. Value is added to these resources along the way. The finished product is worth more than the cost of the resources.

People

People design the products. They decide how to produce them. They choose the materials and the best tools and machines. People organize the production lines. People obtain the necessary capital. They advertise, distribute, and sell the products.

In the past, people provided more of the labor in manufacturing plants than they do today. In the 1950s, about 30 percent of workers in the United States worked in manufacturing jobs. Today, machines do many factory jobs. About 16 percent of the workforce works in manufacturing.

The roles of people in the manufacturing system have been changing. As workers have become more educated, they have asked to take part in decisions that affect the company. **Quality circles** are groups of workers and managers who get together during the work day. Workers discuss problems to be solved and ways to improve production. Management can explain problems to workers about costs, profits, and competition. Quality circles let workers and management discuss issues openly. The result is a better work environment and improved production.

American industries have had a hard time competing with foreign companies whose workers are paid much less. These industries have been forced to improve their production methods

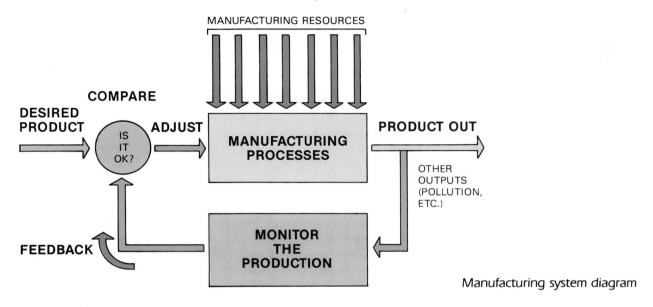

Manufacturing system diagram

Entrepreneurship and McDonald's Restaurants

One famous entrepreneur is Ray Kroc. Ray was a salesman who sold electric mixers, called multimixers, to restaurants. One of the restaurants he serviced was owned by two brothers. In 1954, Ray made an agreement with the two brothers to franchise the concept of the restaurant they were operating. The menu included 15¢ hamburgers, 10¢ french fries, and 20¢ shakes. The brothers were Mac and Dick McDonald. The restaurant was McDonald's. There are now close to 10,000 McDonald's restaurants worldwide.

Ray A. Kroc (Courtesy of McDonald's)

Here, information is being collected about how accurately the machine tool is performing. (Courtesy of Deere & Co.)

to remain in business. Some industries have been unable to do so. For example, U.S. steelworkers are paid six times as much as steelworkers in Brazil and eight times as much as Korean steelworkers. Between 1975 and 1990, about 300,000 American steelworkers lost their jobs. Workers who lose manufacturing jobs often have a hard time. They find that they can get work only in lower-paying service industries. They must be trained to do other kinds of work.

In years to come, many of the manufacturing jobs in the United States will center on the design and engineering of products. The assembly-line work will be done by automated machines or by workers overseas.

Information

Factories are often built in places where there are educated workers like engineers and computer programmers. Some companies build their plants near universities. Professors can then

---

CONNECTIONS: Technology and Science

- Copper, gold, and silver are relatively unreactive metals. Therefore, they are sometimes found in nature as elements rather than in compound form. The following are some physical properties of these metals.
  a) Copper is malleable, ductile, and a very good conductor of heat and electricity. As a result, copper is used in wires.
  b) Silver is a soft metal with a very high luster. It is the best conductor of heat and electricity. Because silver is expensive, it is not used in most wiring.
  c) Gold is a soft metal, and is a very good conductor of heat and electricity. It is the most malleable and ductile of all metals.
- Efficiency is an important scientific concept. Machines are never 100% efficient, although 100% efficiency is a desirable goal.

$$\% \text{ Efficiency} = \frac{\text{Work Output}}{\text{Work Input}} \times 100$$

Work input is always greater than the work output, mainly because of friction. By reducing friction, machines can become more efficient.

help company planners. Companies can also use university research to learn about new materials and production methods.

Companies must find out what people will buy and how their tastes are changing. They also gather information about the costs of materials. This information helps in choosing which materials to buy. Companies must keep track of production to help them adjust the manufacturing process.

Materials

Raw materials are made into basic industrial materials, which are made into finished products. For example, the steel industry uses coal, limestone, and iron ore to make iron ingots. Iron ingots are made into steel sheets. These steel sheets are made into finished products like automobile frames and bodies. (See Chapter 5.)

The cost of raw materials is important. The cost is not only the cost of the material itself, but other costs such as shipping. If a steel company in Indiana needs iron ore, should it buy ore from a mine in Wisconsin, where it is only 20 percent pure? Or should it buy the ore from Brazil, where it is 65 percent pure, and pay the extra shipping charges? This is the kind of decision manufacturers must make in choosing materials.

Tools and Machines

The tools used in modern factories are very advanced and precise. Most of them are automatic. Computer programs and sensing devices provide feedback and guide the machine operation.

Some machines, called **numerical control** machines, are controlled by punched tapes. The holes punched in the tapes direct the machines. A **control unit** receives and stores all the directions.

Manufacturing systems sometimes require specialized tools. This mustard dispenser used by McDonald's employees puts the same amount of mustard on each hamburger. (Courtesy of McDonald's)

This machine is controlled by a punched tape; the machine's control unit reads the tape and directs the machine's actions. (Courtesy of MHP Machines, Inc., Buffalo, New York)

Energy

About 40 percent of the energy used in the United States is used in manufacturing. Some of the biggest energy users are factories that make metals, chemicals, ceramics, paper, food, and equipment. Most manufacturers use electricity from fossil fuels (coal, oil, and natural gas). Other sources of electricity are hydroelectric and nuclear energy plants.

Factories are often built in places where energy costs are low. The glass industry grew up in West Virginia because the state had plenty of natural gas. Small steel mills using electric furnaces are built in places where electricity is cheap.

Some industries use the heat given off during manufacturing. The paper-making industry, for example, uses this energy to heat water and make steam. The steam turns steam turbines and produces electricity. This process of energy re-use is called **cogeneration**.

Capital

Companies must have capital to finance their operations. They must buy land, build factories, purchase equipment, pay

CONNECTIONS: Technology and Mathematics

■ Manufacturers raise money to finance their operations by selling shares of stock. Investors buy and sell stock to make a profit on their investments. The listing below shows that 80 shares of stock in the NEW DESIGN greeting card company were purchased at 12½ ($12.50 per share) and sold for 13¾ ($13.75 per share). The profit on each share was $1.25, which amounted to a 10% return on the investment.

Company	Sales in 100's	Purchase Price	Selling Price	Amount of Change	Percent of Change
New Design	80	12½	13¾	+1¼	$\frac{\$1.25}{\$12.50} = 0.10 = 10\%$

■ Manufacturing uses about 40% of all the energy consumed in the U.S. A recent figure for the amount of energy used yearly in the U.S. is the equivalent of 2,276.3 million metric tons of coal. This means that 40% of 2,276.3 (or .40 × 2,276.3) or the equivalent of about 910.5 million metric tons of coal are used each year by the manufacturing systems in the U.S.

workers, maintain machines, and advertise their products. Capital is often obtained by selling shares of stock to the public. Stockholders become partners in the corporation. If the company makes a profit, the value of its stock may go up. Stockholders then make a profit, too.

Private companies may raise money from investors who contribute venture capital. A venture, like an adventure, is a trip into the unknown. **Venture capital** is money used to finance the costs of starting a new company. Investors take big risks. They expect to make big profits once a company starts production.

Time

In manufacturing, time is money. The faster products are made, the more profitable the company will be. **Productivity** is how quickly and cheaply a product is made. Since a large part of the cost of a product is the cost of paying workers, if workers do their jobs faster, productivity will increase.

Much thought has gone into improving productivity. In 1910, Frederick W. Taylor developed an idea called **scientific management**. His idea was to study every movement that a worker made. Then the worker's routine was changed to cut out any wasted movements. Such changes resulted in increased output without making people do any more work.

Today, companies try to make sure that products are constantly moving from one process to the next. The longer an item sits waiting for the next step in assembly, the lower productivity will be.

HOW MANUFACTURING IS DONE

Before a company starts making a product, it carries out **market research**. Market research helps the company find out what customers want in a product. Companies survey a sample group of people. They try to pick a sample that represents the people who might buy their product. Market researchers ask the sample group about the product. The feedback they receive helps the company decide whether to make the new product, or how to change it to make it better.

Using what has been learned from market research, product designers and engineers prepare drawings, sometimes using computers, and develop finished design ideas.

Research and Development (R & D) is done by product designers and engineers. R & D is used to come up with ideas

Expert craftspeople make a clay automobile model.
(Courtesy of Ford Motor Company)

There are two subsystems within the manufacturing system: the material processing system, and the business and management system.

OBTAIN MATERIALS

HEAT AND FORGE BLADE

ANNEAL (SOFTEN) BLADE BY HEATING TO CHERRY-RED COLOR AND COOLING SLOWLY

FILE TWO FACES OF BLADE

GRIND TIP TO PROPER ANGLE

REMOVE ALL SCRATCHES WITH EMERY CLOTH

HARDEN BLADE BY HEATING TO CHERRY-RED COLOR AND COOLING QUICKLY

CLEAN BLADE WITH EMERY CLOTH

TEMPER BLADE BY HEATING TO A STRAW COLOR AND COOLING QUICKLY

CLEAN BLADE WITH FINE EMERY CLOTH

BUFF BLADE

INJECTION-MOLD PLASTIC HANDLE

This flowchart shows the steps in making a metal screwdriver with a plastic handle.

for new products and improve ways of making old ones. R & D often leads to inventions or innovations.

Once designs are ready, management takes a close look at them. The company must decide whether the product can be made at a cost and sold at a price that will bring in a profit. If costs are too high, the product may be changed to use cheaper or different materials. Or a new production idea might help to lower costs.

When the engineers and businesspeople agree on a design, a model called a **prototype** is made. Craftspeople build these prototypes. Prototypes help in solving design and engineering problems. They are tested for a length of time before machinery is bought to begin factory production.

The last step in manufacturing is setting up the production line. Forming, separating, combining, and conditioning tools must be chosen. Operations must be organized to manufacture the product. A **flowchart** shows the operations of a production line in diagram form.

People who work in advertising must think of ways to sell the product. They create advertisements and commercials that will interest the public.

Salespeople sell and distribute the finished product. They are a very important part of the business side of manufacturing.

The company's job does not end when the product is sold. Products may need servicing or repair. Defective products may have to be replaced. Customer service departments help with these very important jobs.

ENSURING QUALITY IN MANUFACTURING

Companies want their products to be of the highest quality possible. Making sure that quality stays high during manufacture is the job of **quality control**. Quality control workers inspect

CONNECTIONS: Technology and Social Studies

- With the benefits of mass production also came problems. As workers engaged in repetitive tasks, boredom and carelessness tended to develop and the quality of the product was often affected negatively.
- As the same general products were mass produced, manufacturers needed to highlight the virtues of their specific model. Thus, advertising became important, and an industry developed that has grown in size over the years. Presently, billions of dollars are spent each year to promote every good and service imaginable.
- Robots and automation have helped increase efficiency and reduce production costs. However, in the process, they have also displaced or kept thousands of men and women out of the workforce. Some of these people can be retrained to do new jobs. What do we do about the others, with limited education, who cannot be retrained easily?
- A major concern of American industry is improving product quality. With increasing competition from foreign companies, U.S. manufacturers have found that consumers look at the price of goods and services but also are interested in getting the best value for their money.

products to make sure they conform to the desired result. Feedback from quality control allows workers to change manufacturing processes when necessary. It also helps make sure that products are all alike, or uniform. **Uniformity** is an important concern of quality control.

Without quality control, many products would have to be thrown away. Good-quality products keep customers happy. Fewer items are returned or need repair.

AUTOMATED MANUFACTURING

Automation is the process of controlling machines automatically. For example, a robot can be programmed to pick up a part, move it a certain distance, and drop it into a bin. This kind of control is called **program control**.

Feedback control uses feedback to adjust the way a machine is working. Feedback control depends on the **sensor**, a device that gathers information about its environment. For example, a sensor "senses" when a drill has drilled deep enough. It then sends a signal to switch off the drill. A machine operator is no

In some factories, people inspect the parts by eye at various stages of the manufacturing line. (Courtesy of Sperry Corporation)

Automation has greatly increased productivity in manufacturing.

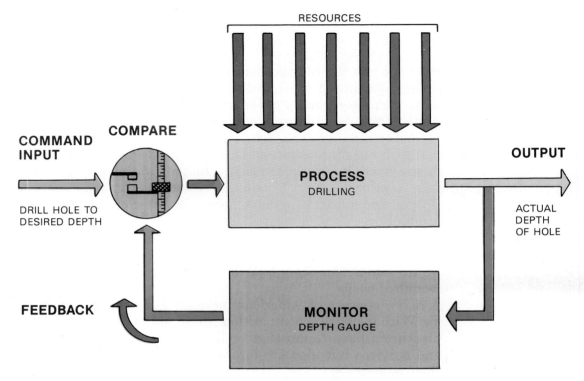

System diagram for an automatically controlled drill press. A switch activated by a depth gauge controls the drill press.

Automated machines attach heavy wheels to tractors. In the past it took three people to do this work. (Courtesy of Deere & Co.)

longer needed to turn the drill on and off. In automated factories, quality control is provided by sensors on machines. These sensors make sure the machines do their jobs exactly right.

Automation saves money on labor costs. To compete with lower-cost foreign products, more and more factories are being automated.

Robotics

Robots are automated machines that are controlled by computers. Many have sensors. Some of these robots can "feel." They can hold parts with just the right amount of pressure so they won't break. Other robots "see" with television eyes. They are able to tell the difference between parts of different shapes.

Robots are taking the place of people in many factories. Robots do almost all the work in one factory in Japan. People work there, too, making sure everything is operating the way it should. People are needed to install, service, and program robots.

Robots have allowed manufacturers to make products of higher quality at lower price. Robots don't take coffee breaks.

The Robot Revolution

Often we hear of the "robot revolution." Why is the use of robots a "revolution"? The Industrial Revolution replaced human energy (muscle power) with mechanical energy (machine power). Despite the new technology, human beings were still needed to operate the machines. Robots can actually replace human beings in a manufacturing plant. Although robots don't look human, they can act very much like people. They can perform movements just like a human arm, wrist, and hand. They can use sensors to "see" and "feel." Robots used in industry are called **industrial robots.**

This robotic hand is light and compact. It has three human-like fingers and fourteen joints (Courtesy of Hitachi, Ltd.)

They can work twenty-four hours a day. Most of the jobs they do relieve people from dangerous, boring, heavy, or unpleasant work. Robots are used for welding, spray-painting, picking up parts and placing them into machines. They are used for loading objects onto platforms and conveyors.

Robots are better than ordinary machine tools because they can be **reprogrammed**. They can be programmed to weld a fender on one kind of car. Later, they can be programmed to do the same job on a different model.

CAD/CAM

CAD is **computer-aided design**. With CAD, a computer is used by designers, drafters, and engineers to do drawings and designs. The drawings are then stored in the computer. (See Chapter 6 for more on CAD.) CAD can also be used to test a design. In a **simulation**, a design is run through a series of tests, all on computer. The part itself is not needed.

CAM is **computer-aided manufacturing**. In CAM, computers are used to control factory machines. A computer might

Robots welding automobile parts. (Courtesy of Ford Motor Company)

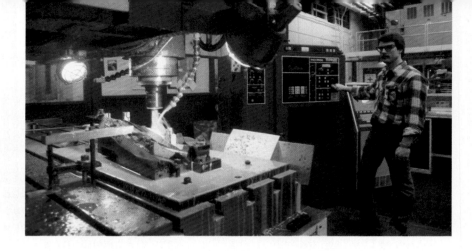

In a CAD/CAM operation, a designer stores a drawing in a computer and sends it to a computer-controlled machine. There a prototype is produced. (Courtesy of Grumman Aerospace)

direct machines to drill holes or spray-paint a part.

CAD/CAM is a new technology that joins CAD and CAM. CAD/CAM lets a person create a design on the computer screen, then send it directly to a machine tool, which makes the part.

Computer-Integrated Manufacturing (CIM)

CIM is **computer-integrated manufacturing**. In CIM, computers are used not only for CAD/CAM, but for business needs as well. They are used to store information about raw materials and parts. They set times for the purchase and delivery of materials, report on finished goods, and do the billing and accounting.

Managers involved in purchasing, shipping, accounting, and manufacturing can get a total picture of all the factory conditions by looking at the computer screen. (Courtesy of International Business Machines Corp.)

Computers and robots have improved product quality while bringing manufacturing costs down.

CIM combines the manufacturing, design, and business functions of a company under the control of a computer system.

Flexible Manufacturing at Deere & Company

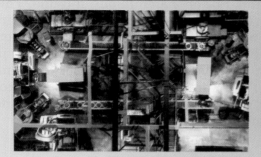

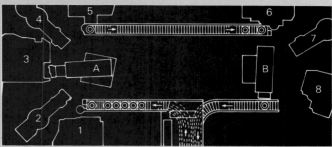

This plant makes construction equipment called backhoe loaders. Once a part enters the manufacturing system, it is tracked and controlled by a computer. It enters one end of the system and comes out the other end completely finished in as little as five minutes. The machines within the system are all computer-controlled. Each machine can be programmed to do many different kinds of machining jobs. Thus, the machines can turn out parts with different dimensions. An employee monitors the entire process.

The manufacturing system is diagrammed here to show part flow and system components. Parts progress from the number 1 machining center to the last machining stop at number 8. Parts move on automated conveyors until picked up by one of two robots (A and B). These robots feed parts to appropriate cutting machines (1, 3, 5, 6, or 8) at appropriate times. Stations 2, 4, and 7 are parts storage areas and automated tool carousels.

The entire manufacturing process is computer controlled.

A finished backhoe loader doing its job
(All photos courtesy of Deere & Co.)

Flexible Manufacturing

Flexible manufacturing is the efficient production of small amounts of products. Today, many products are made in batches of a hundred or a thousand, rather than millions of items. This is because customers have special needs. For example, the General Electric Company in New Hampshire makes 2,000 different versions of its basic electric meter. A John Deere factory in Iowa can make 5,000 different versions of its tractor for farmers with different needs. The same production line is used for all versions of a product. The machines are programmed to do different operations to make the different versions.

Just-in-Time Manufacturing

Raw materials come to factories by truck, train, or ship. They are most often delivered in large amounts that take up a lot of storage space. Companies must pay to rent, heat, and light warehouse space. They must pay people to move the materials and deliver them to the factory where they are needed.

Just-in-Time (JIT) manufacturing is manufacturing in which materials and parts are ordered so that they arrive at the factory when they are needed. Also, the product is immediately shipped to the customer. With JIT manufacturing, there is no need for storage space and workers.

SUMMARY

The two production systems are manufacturing and construction. Manufacturing is making goods in a workshop or factory. Construction is building a structure on a site.

For thousands of years, people made goods by hand. Craftspeople made items for their own and others' use. As the nation grew, people needed more goods. Factories began to mass-produce goods using the assembly line.

Assembly lines divide production into steps. Each worker does one job. The assembly line makes and uses standardized parts that are interchangeable.

The factory system improved the way Americans lived. It brought prices down. With the factory system, time became important since operations were synchronized.

The business side of manufacturing is important. Entrepreneurs start companies, invent new products, and change old ones. Many of today's businesses were started by entrepreneurs.

Manufacturing uses the seven technological resources. They are: people, information, materials, tools and machines, energy, capital, and time.

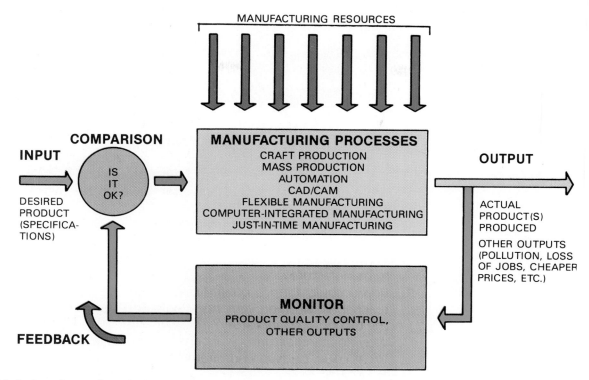

MANUFACTURING RESOURCES

COMPARISON

INPUT

IS
IT
OK?

DESIRED
PRODUCT
(SPECIFICA-
TIONS)

MANUFACTURING PROCESSES
CRAFT PRODUCTION
MASS PRODUCTION
AUTOMATION
CAD/CAM
FLEXIBLE MANUFACTURING
COMPUTER-INTEGRATED MANUFACTURING
JUST-IN-TIME MANUFACTURING

OUTPUT

ACTUAL
PRODUCT(S)
PRODUCED

OTHER OUTPUTS
(POLLUTION, LOSS
OF JOBS, CHEAPER
PRICES, ETC.)

MONITOR
PRODUCT QUALITY CONTROL,
OTHER OUTPUTS

FEEDBACK

Well-designed manufacturing systems use modern processes to make products of good quality. The products must meet the desired specifications without harmful effects on people or the environment.

Automation has made factories much more efficient. More goods can be made in a shorter time with less labor cost. Automated machines are programmed to do a series of steps. They often make use of feedback from sensors. The sensors keep track of the manufacturing processes.

Making sure products are uniform and of high quality is an important goal for manufacturers. Quality control lowers costs and means fewer rejects and repairs.

Robots are being used in modern manufacturing plants. They do jobs that are dangerous or unpleasant for humans.

Computer-aided design and computer-aided manufacturing (CAD/CAM) link engineering and design with the factory floor. Designs for parts can be drawn using a computer. They can then be changed or stored for later use. The data can be sent to a machine that makes the part.

Computer-integrated manufacturing (CIM) uses the computer in engineering, production, and business. With flexible manufacturing, it is possible to produce special products for customers' special needs. Materials come in and products go out at exactly the right time in Just-in-Time manufacturing. These new methods lower costs and result in a better product.

(Courtesy of Ford Motor Company)

1. Give five examples of how manufacturing technology has helped satisfy people's needs and wants.
2. Explain how craft production differs from mass production.
3. How did mass production improve people's standard of living?
4. Draw a flowchart of an assembly line for making a greeting card.
5. What are two disadvantages of mass production?
6. What are the two subsystems that make up the manufacturing system?
7. Suggest an invention that would help you with your homework.
8. How might you innovate a soda can?
9. What would be a suitable product for your technology class to manufacture?
10. If you could form a quality circle with your classmates and teacher, what improvements would you suggest be made in the technology laboratory?
11. Why do manufacturers want their products to be uniform?
12. What could be two undesirable outcomes from a system that manufactures computers?
13. How is feedback used to ensure good quality in manufactured products?
14. Do you think that robots should be used instead of people on assembly lines? Explain why or why not.
15. How have computers affected the manufacturing industry?
16. Draw a system diagram for a system that manufactures chewing gum. Label the input command, resources, process, output, monitor, and comparison.

(Courtesy of General Motors Design)

(Courtesy of International Business Machines Corp.)

(Photo by Michael Hacker)

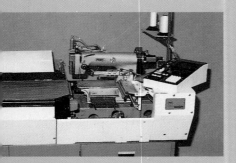

(Courtesy of Shoe Machinery Group — Emhart Corp.)

CHAPTER KEY TERMS

Across

3. Until the middle 1900s, many people were workers on these.
5. Parts that are alike are _____.
7. A model of a product that will later be produced in quantity.
8. A place where products are produced in quantity.
9. A computerized system that links design and manufacturing (abbreviation).
12. A system of producing products in quantity using assembly-line techniques.
13. A reprogrammable device that can replace a human in a factory.
14. A type of manufacturing where machine tools can be reprogrammed to do different jobs.
16. A system of producing products in a factory or workshop.
17. This enables us to automatically control a machine tool.
18. A person who comes up with a brand-new idea for a product.

Down

1. A system for ensuring that the quality of a product is what is desired.
2. Computer-integrated manufacturing (abbreviation).
4. Manufacturing where parts arrive just when they are needed.
6. A system of using feedback to control the operation of machines.
10. Products that are made to meet an individual's specific needs are _____.
11. A system that includes manufacturing and construction.
15. An organized group of workers.

Assembly line
Automation
CAD/CAM
CIM
Custom-made
Entrepreneur
Factory
Feedback
   control
Flexible
   manufacturing
Interchangeable
Inventor
Just-In-Time
   manufacturing
Manufacturing
Mass production
Numerical
   control
Production
Prototype
Quality control
Robot
Uniformity
Union

SEE YOUR TEACHER FOR THE CROSSTECH SHEET

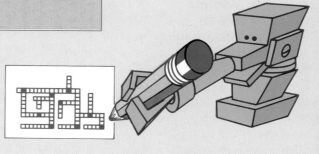

267

ACTIVITIES

FRAGILE: HANDLE WITH CARE

Problem Situation

Packaging serves several purposes. Many items are packaged so that they are attractive to consumers. Tamper-proof packaging, originally used for medicines, is now being used on many food items to protect consumers.

Packages designed for shipping must protect the product. Small, valuable items, including medical specimens and electronic components, must be carefully packaged for shipping. The ideal package is inexpensive, lightweight, and effective in protecting the item it contains. It is also desirable to have a container that can be either recycled or disposed of without creating environmental problems.

Design Brief

Package and mail a raw egg so that it will arrive at its destination intact. Use corrugated cardboard and other recycled materials. The volume of the package should not exceed 64 cubic inches and must conform to all U.S. Post Office regulations.

Suggested Resources

One egg
Corrugated cardboard
Hot glue
Glue gun
Postal scale
Utility knife
Kraft paper
Computer
Mailing label program
Mailing labels
Color markers
Additional packaging
  materials to be
  brought in by each
  student

Procedure

Note: Follow the safety procedures taught by your teacher for using the utility knife and hot glue gun.

1. Use the brainstorming process to generate a list of possible packaging materials. Bring in an assortment of suitable materials for the next day of class.
2. Sketch several possible designs for a container to be made from corrugated cardboard. Remember that it should have a volume of 64 cubic inches or less.
3. Use drawing tools to do the package layout. Cut the cardboard using a utility knife.
4. Use hot glue to assemble the container.
5. Experiment with the packaging materials that you brought in until you are certain that the egg will be adequately protected.
6. Place the egg inside the container along with a packing slip that includes your name, date, and class period.
7. Seal the container and neatly cover it with kraft paper. Use Post Office-approved packing tape.
8. Use a computer and software provided by your teacher to prepare mailing and return address labels.

9. Make a series of idea sketches for two additional labels that will communicate the need for careful handling of your package. Use color and be creative.
10. Attach the labels and give the package to your teacher for mailing.

Classroom Connections

1. Innovators improve on existing ideas and products. Manufacturers create innovative package designs to increase sales.
2. Manufacturers need to evaluate cost-benefit trade-offs when choosing materials.
3. Aseptic (sterile) packages have a long shelf life and can be used to protect products like milk, which are usually refrigerated.
4. Many packaging materials like plastic wrap and foam create environmental problems because they are not biodegradable.
5. To calculate the volume of a rectangular container, multiply length $\times$ width $\times$ height. For example, a $3" \times 4" \times 4"$ box has a volume of 48 cubic inches.

Technology in the Real World

1. Market researchers survey groups of people to determine if new package designs will help increase sales of a product.
2. To reduce disposal problems, government regulations in some communities are restricting the kinds of packaging materials that can be used.

Summing Up

1. Name several purposes of packaging.
2. Explain why you chose the materials you used for your package.
3. Did your egg arrive intact? Why do you think it did or didn't?

ACTIVITIES

MANUFACTURING WITH RECYCLED MATERIALS

Problem Situation

Cooperative efforts among environmentalists, municipalities, and industry are turning mountains of garbage into usable products. Communities nationwide have begun recycling programs to help reduce the huge quantities of garbage we produce every day. Recycling begins when each family separates materials that can be recycled from its weekly garbage. These materials may be metals, plastics, glass, and other items. Town sanitation services then pick up the separated materials and bring them to a central storage area.

Over the years, research has come up with many uses for these materials, making them useful to the manufacturing and production industries. Using processes such as crushing, shearing, melting, and mixing, these techniques put many disposable products back into use as new products or as part of other new products: For example, crushed glass from jars and bottles has become filler in concrete roads; the ash from town incinerators has been used to produce concrete blocks; and old car tires have been used as a fuel to fire furnaces.

The use of recycled materials in manufacturing provides many solutions to our waste-management problems.

Suggested Resources

Waste products (plastic, glass, paper, sawdust, and so on)
Dry-mix brick mortar
Plaster of Paris
Dixie cups
Gram scale
Mortar and pestle
Graduated cylinder
File
Machinist's vise

Design Brief

You are a research material scientist for a concrete block and gypsum manufacturer. You have been asked to conduct research into the possibility of using recycled materials as filler in concrete and plaster products. You are to collect sample disposable products, process them, add them to batches of concrete and plaster, and test the resulting new products.

Procedure

1. Form teams of three students each to conduct research.
2. Develop a data sheet to keep a record of the quantity of each material in the test batches you process.
3. Using the safety procedures described by your teacher, process your waste materials by cutting, grinding, shredding, or burning.
4. Mix up five small batches of dry-mix brick mortar (enough

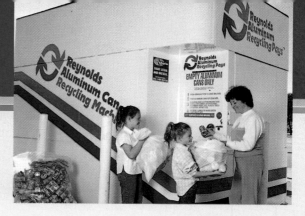

(Courtesy of Reynolds Metals)

to fill the Dixie cup). Add enough water to obtain a "thick shake" consistency.

5. Add different percentages of the processed recycled materials to each batch of mortar. Mix one batch without the recycled additive to remain as a basis for comparison. Allow to cure a few days.
6. Test each batch for hardness by scratching it with a file. Test compression strength by placing the sample in a vise and crushing it.
7. Record your observations. Does the additive affect the mix? Does the percentage affect the hardness or strength of the mix?
8. Try the same procedures, this time using plaster instead of cement.

Classroom Connections

1. When involved in experimental research, it is necessary to follow scientific laboratory techniques and keep accurate records.
2. When batching materials it is important to know the properties of solutions, suspensions, supersaturated solutions, and colloids, since materials may be batched in any of these forms.
3. In order to vary the percentage of material by weight, it will be necessary to calculate mathematically what percentage of each material has been added to the mixture.

Technology in the Real World

1. Research, now being conducted at the Waste Management Institute, centers on manufacturing concrete blocks from the ash produced by garbage incinerators. The ash is used as a substitute for rock and gravel in the concrete mix.

Summing Up

1. What are the ingredients found in brick mortar mix?
2. What value would the ability to use recycled waste material in the concrete mix be to a concrete-block manufacturer?
3. Describe how concrete blocks are manufactured.
4. What does the word "cure" mean?

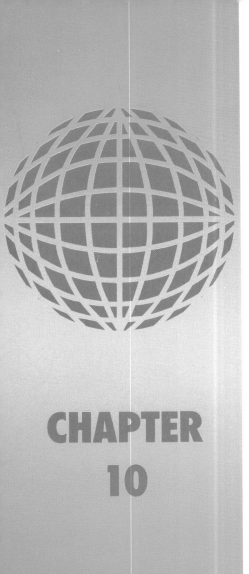

CHAPTER 10

PRODUCTION SYSTEMS: CONSTRUCTION

MAJOR CONCEPTS

After reading this chapter, you will know that:

- Construction refers to producing a structure on a site.
- A construction system combines resources to provide a structure as an output.
- Three subsystems within the construction system are designing, managing, and building.
- Construction sites must be chosen to fit in with the needs of people and the environment.
- A foundation is built to support a structure.
- The usable part of a structure is called the superstructure.
- Structures include bridges, buildings, dams, harbors, roads, towers, and tunnels.

The Romans used the arch to support their bridges, which were called viaducts. (Courtesy of Harvey Binder)

CONSTRUCTION SYSTEMS

Early people lived in caves or in the brush. They did not have the technology to build structures that would shelter them from the weather or dangerous animals.

The first buildings people constructed were probably simple shelters. Teepees made from animal hides stretched over a wooden frame were used as long ago as 20,000 B.C. These homes were movable. People could take them along as they searched for new sources of food. As people began to settle in villages, they needed more permanent houses. Materials like wood, stone, and mud were used. Later, bricks made from straw and mud were used in construction.

The early Egyptians, Greeks, and Romans were very good builders. The great pyramids in Egypt were built from blocks of limestone 5,000 years ago. The largest pyramid is almost 500 feet tall and is made of over two million blocks of limestone, each weighing more than a ton.

The Romans were skilled engineers. They built cities, roads, and bridges. About 300 B.C. they started building systems called aqueducts for supplying water to their cities. These pipelines, built from stone and cement, brought water from nearby rivers. They also built sewer systems to carry waste from the city of Rome into the Tiber River. One of their greatest contributions to construction technology was the use of the **arch** to hold up buildings.

Construction means building a structure at the place where it will be used. The location is called a **construction site**. Today, about six million people work in the construction industry in the United States. The industry produces over a hundred billion dollars in goods and services in the United States each year.

Even today, people in some parts of the world use natural materials and anything else they can find to construct housing. (Photo by Michael Hacker)

An example of modern construction technology (Courtesy of NY Convention & Visitors Bureau)

Problem Situation 1:   Patio paving can be very attractive.
Design Brief 1:   Design a shape for a concrete paving stone 2″ thick that can be interlocking and form a regular pattern.

Problem Situation 2:   Unless properly designed, bridges will fail under load.
Design Brief 2:   Design and construct a bridge to span 12″. Make it as strong as possible using two pieces of ½″ × ⅛″ × 14″ pine, and 35 or fewer 6″ craft sticks. Load your bridge until it fails and compare it to bridges built by your classmates.

Problem Situation 3:   Construction projects must be designed to fit in with the environment in a community.
Design Brief 3:   Design and sketch a new department store or industrial building for your community. Consider how the project would affect people and the environment. List the factors that must be considered when choosing a construction site for the building.

RESOURCES FOR CONSTRUCTION TECHNOLOGY

A construction system combines resources to provide a structure as an output.

A construction system uses the seven technological resources. The failure of any one resource could result in poor work or an unsafe structure.

People

Three subsystems within the construction system are designing, managing, and building.

People are needed to design and engineer structures. They are needed to manage the business of construction. Many workers are needed to do the actual building.

The **landowner** decides that there is a need for a structure.

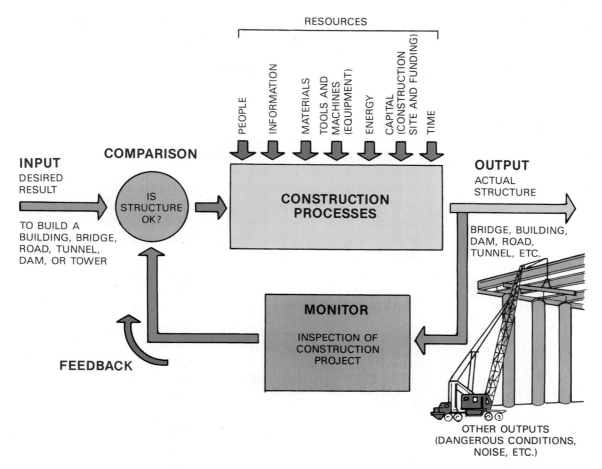

A construction system combines the technological resources to produce a structure on a site.

This could be a home, a housing development, or a shopping center.

Architects design buildings. Their plans show how a structure will be built and where it will be placed on a site. Architects are trained for their work in college. They take courses in mathematics, architectural design, technical and architectural drawing, and art.

Civil engineers work with architects. They help the architects decide if the building can be built as desired. Civil engineers prepare exact drawings and plans for the building framework and foundation. The drawings and plans give information about the size and strength of materials used. They also give details about **utilities** (plumbing, heating, air conditioning, and electrical wiring).

Civil engineers plan bridges, roads, tunnels, dams, and towers. Engineers must go to college. They study mathematics, science, engineering design, technology of materials, and technical drawing. The **structural engineer** plans the building's structure. The structure must be strong enough to support the load it must carry.

General contractors own their own construction companies. They hire workers and oversee part or all of a project. General contractors work closely with everyone on the project.

Estimators work for the contractor. They make an estimate of the project's cost. They use this estimate to prepare a proposal called a **bid**. If the customer accepts the bid, he or she agrees with the cost and **specifications** in the bid.

Project managers make sure construction is carried out properly and meets building codes. They hire and supervise the workers. They try to keep the cost of the project as planned. Some colleges have programs to train construction project managers. Courses are taken in materials and techniques, planning and scheduling, engineering, and accounting.

Tradespeople work on projects from houses to skyscrapers. They may build a swimming pool or a dam. Tradespeople must

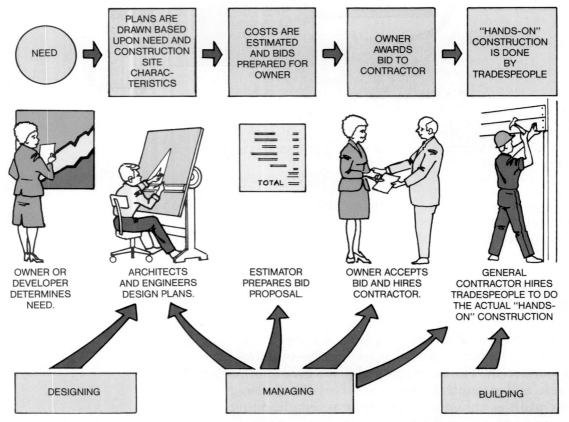

THE THREE SUBSYSTEMS OF CONSTRUCTION (DESIGNING, MANAGING, AND BUILDING) INVOLVE CONTINUOUS COOPERATION AMONG THE OWNER, THE ARCHITECTS AND ENGINEERS, THE CONTRACTOR, AND THE TRADESPEOPLE.

People in the construction industry

Project managers must review construction plans with the architects and engineers before actual construction is begun. (Courtesy of The Turner Corporation, photograph by Michael Sporzarsky)

know how to use the tools and materials of the trade. Tradespeople include carpenters, electricians, plumbers, and masons. Some run heavy equipment such as bulldozers.

Tradespeople may study their trade in high school or at a technical or vocational school. Or they can learn on the job, as apprentices to experienced tradespeople.

Information

In construction, information comes from the people who want the construction done and those who will do it. Information takes the form of plans, bids, and specifications. Specifications include data like the materials to be used, the way the foundation will be built, and even the kind of trees and bushes to be planted around the structure. These are the command inputs to the construction system.

People in construction must be able to make and read mechanical and architectural drawings and blueprints. Tradespeople must have information about building techniques. Engineers must have information about building materials and the loads the structure will support.

Materials

Building materials include concrete, lumber (wood), steel, glass, and brick.

Concrete is made from stone, sand, water, and **cement** (a mixture of limestone and clay). Wet concrete looks and feels like mud. It can be poured into molds to make any shape needed. These molds are called **forms** and are often made out of lumber. Once mixed and poured, concrete sets (gets hard).

Concrete is brittle. To give it more strength, steel rods called reinforcing rods are placed in the forms before the concrete is poured. Sometimes wire screens, called reinforcing mesh, are used. Concrete is used in dams, roadways, tunnels, and build-

This six-building complex uses energy-efficient glass as a primary building material. (Courtesy of PPG Industries)

ings—just about every kind of structure.

Lumber is a building material used to provide the framework of homes. Wood is easy to work with and not too expensive. Some composite materials used in building (plywood and particle board) are made from wood. These are strong and their cost is reasonable.

Steel is used for the framework of skyscrapers, bridges, and towers. It is very strong and can be made into cables, beams, and columns.

Glass not only lets the light in but adds great beauty to a structure. Glass can be installed as a single pane. Thermal glass (two or three thicknesses) is used to conserve energy.

Brick is made from clay. The clay is fired (heated) in an oven called a kiln. It becomes very hard. Brick houses are expensive. Labor costs are high because each brick must be set in place by a mason. Brick houses last longer and are more fire-resistant than wood houses.

Tools and Machines

Carpenters, plumbers, and other tradespeople use small hand and electric tools. Construction equipment can be huge, too;

Construction equipment has come a long way since this steam shovel from the early 1900s. (Courtesy of Perini Corporation)

A wide variety of power tools and equipment is used to do construction projects. (Courtesy of The Stanley Works)

examples of this are cranes and bulldozers. The largest pieces of equipment are used in earth-moving jobs. Special equipment is used for excavating (digging), lifting heavy materials, and building bridges, dams, roads, pipelines, and tunnels.

Robots are being used more and more often in construction. They are useful in dangerous jobs such as working high in the air, below ground, and deep under water.

Energy

Construction systems use energy to operate machines and tools. However, even more energy is used in industries that support the construction industry. Producing materials like concrete, bricks, and steel takes vast amounts of energy. Transporting these materials to building sites also uses energy.

Capital

Construction is expensive. Machines and tools, building materials, and labor are all costly. People who want to build must find money to finance their project. A bank will lend money only if it is fairly certain it will be paid back. A loan for a home is called a **mortgage**. Big projects often use money from both private and government sources.

A loan is paid back over a period of years with interest added to each payment. **Interest** is the fee the bank charges for the loan. If a person borrows $100,000 to build a house, the amount paid back is much more. However, repayment is spread out over many years so the borrower can afford the monthly payments. Home mortgages are usually paid back within fifteen to thirty years.

Land is often the most costly part of construction. In the middle of a city like New York or San Francisco, land could cost as much as $1,000 per square foot. That comes to about $40 million per acre. People who build in big cities must charge high rents for offices and apartments because land costs are so high. Skyscrapers make the best use of expensive land. People who own valuable land usually have an easy time getting a construction loan from a bank. If the loan is not repaid, the bank takes possession of the land.

Time

Construction takes a lot of time. A bridge or tunnel may take years to complete. Houses can be built in a few months.

Modern tools and equipment save time. For example, air-driven staple guns are replacing hammers and nails for fastening shingles to roofs.

New building techniques also reduce construction time. For example, parts of a structure can be **prefabricated** in a factory. Walls, for example, can be built in a factory, then moved to the construction site. Time is saved because the walls can be mass produced.

Modular construction is a technique that is being used more and more often. A module is a basic unit like a room. Modules can be combined to form structures of different shapes and sizes.

SELECTING THE CONSTRUCTION SITE

An important first step in construction planning is choosing a site. This is a decision made by management. Land costs and taxes can vary greatly from one place to another. Often, a few miles make the difference between high and low land costs. The site must be suitable for construction. If it is too hilly, much time and money must be spent on **grading** (leveling) the soil. If the ground is too rocky, it will be hard to dig a foundation.

There are other things to consider, as well. A site for a factory must provide for transportation needs, so goods can be delivered and shipped. Schools should be near the residential area they serve. Banks should be near the city's business district. A housing development should be located near schools and shopping. Bridges must be placed where the conditions are best for a strong foundation.

At the same time, the environment must be preserved. Community needs must be considered. For example, an airport should not be located in the middle of a quiet residential community. If the site is in a historical area, the construction should

The immense construction site for the Miho Dam dwarfs the huge pieces of machinery and equipment below.
(Courtesy of Kajima Corporation)

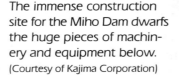

fit in with surrounding buildings. Sometimes, roads or tunnels must be routed around or under historical landmarks to preserve them.

A construction site must be near roads, railroads, or ports so materials and equipment can be delivered. Water and electricity must be available, as well as a way to dispose of wastes. Construction workers should be able to reach the site easily. The best construction site is one that meets all the specifications at the lowest cost.

PREPARING THE CONSTRUCTION SITE

The construction site must be prepared for building. Heavy equipment is used to clear the ground. If there are unwanted buildings on the site, they must be removed. Wrecking balls or explosives are used to demolish buildings. Bulldozers clear trees or brush from an area. Unwanted materials are hauled away by dump trucks.

Before construction starts, the structure must be laid out. A **surveyor** is a person who marks the site to show where the structure will be built. Surveyors use a **transit** to measure and lay out angles. They use an engineer's level to set the elevation (height above ground) of different points.

BUILDING THE FOUNDATION

A **foundation** supports the weight of a **structure**. If you were standing on soft earth, you might sink down. If you stepped on a wide board, your weight would be spread out over a larger area. You wouldn't sink down as far. A foundation spreads the weight of a structure over a larger area of ground so it is well supported. In cold climates, foundations also prevent frost damage to the structure. Foundations are sometimes called **substructures**.

A foundation has three parts. One is the earth on which it rests. The second is the footing, which transfers the structure's weight to the earth. The third is vertical supports which rest on the footing.

If the ground is hard, a **spread** footing is used. It spreads out the weight of the structure, just as your weight is spread over an area as wide as your foot. If the ground is soft, or the site is marshy or under water, **piles** are used. Piles are like stilts. They are driven into the earth until they reach hard ground or rock.

Construction sites must be chosen to fit in with the needs of people and the environment.

Giant shovels remove earth to prepare a site for construction. (Courtesy of RCA)

A foundation is built to support a structure.

Piles support the Columbia River Bridge in Portland, Oregon. (Courtesy of Sverdrup Corporation)

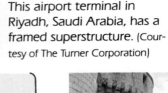

This airport terminal in Riyadh, Saudi Arabia, has a framed superstructure. (Courtesy of The Turner Corporation)

A dam is an example of a mass superstructure. (Courtesy of U.S. Department of the Interior, Bureau of Reclamation)

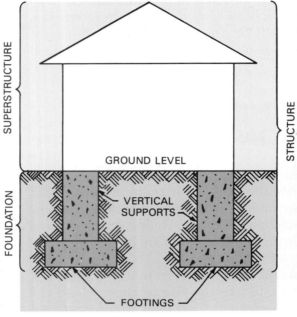

A structure includes a foundation and a superstructure.

This castle is an example of a bearing wall superstructure. (Photo by Michael Hacker)

BUILDING THE SUPERSTRUCTURE

The usable part of a structure is called the superstructure.

The **superstructure** is the part of a structure that is above the ground (unless the structure is a tunnel or pipeline).

Mass superstructures are made from large masses of materials. They have little or no space inside. Dams and monuments are mass superstructures. They are built from brick, concrete, earth, or stone.

Bearing wall superstructures enclose a space with walls. They are built from brick, concrete, or stone. The castles built in the Middle Ages are bearing wall superstructures. The walls of some of these castles were 20 feet thick or more at the base.

Framed superstructures use a framework to support the building. Today, most buildings are framed superstructures. Lumber is used for framing in most houses. In office and apartment buildings, reinforced concrete and steel are used for framing. Reinforced concrete and steel are stronger, longer-lasting, and more fire-resistant than wood.

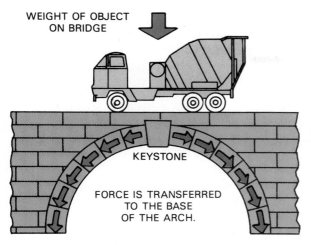

WEIGHT OF OBJECT
ON BRIDGE

KEYSTONE

FORCE IS TRANSFERRED
TO THE BASE
OF THE ARCH.

A stone arch bridge. The arch transfers weight to the supports at the base of the arch.

Cantilever bridges extend outwards and are secured at the ends.

TYPES OF STRUCTURES

Bridges, buildings, dams, harbors, roads, towers, and tunnels are different types of structures. Each requires special construction techniques.

Bridges

People have been building bridges since prehistoric times. Bridges that use one beam to cross a distance are called **beam bridges**. These are the simplest kinds of bridges. The strength of a beam bridge depends on how strong the beam is.

The Romans used the arch to support viaducts (bridges) and aqueducts (raised channels that carried water from one place to another). The first **arch bridges** were made of stone. They are now made from concrete or steel because of their strength.

A **cantilever** bridge works like two diving boards facing each other. The sections are firmly attached at their ends. If they do not meet, another section may be added to link them. A cantilever bridge requires a huge force to support each end. **Double cantilever** bridges have been designed to overcome this problem. The two sides of the cantilever section balance each other.

Suspension bridges are used to bridge wide spans. A suspension bridge uses steel cables to hang the deck (roadbed) from towers. Some of the world's most famous bridges are suspension bridges. The Golden Gate Bridge in San Francisco is 4,200 feet long. The George Washington Bridge, which links New Jersey and New York, is 3,500 feet long.

The Severn Bridge in Avon, England, is a suspension bridge, finished in 1966, that took twenty-one years to build. The deck

An early beam bridge

The Severn Bridge spans 3,250 feet. (Courtesy of British Tourist Authority)

was designed so it would not sway in the wind. Deck sections were built in a factory and moved to the bridge site. This kept costs down, since the deck was easier to build in a factory than a hundred feet in the air.

Buildings

There are four kinds of building construction:

1. residential (hotels and housing);
2. commercial (banks, stores, and offices);
3. institutional (schools and hospitals); and
4. industrial (factories).

In the first three kinds of buildings, aesthetic design and beauty are important. In industrial buildings, however, beauty is less important than usefulness. A factory must house many kinds of equipment as cheaply and efficiently as possible.

Land costs for industrial buildings are fairly low compared to land costs for other kinds of buildings. Houses, schools, offices, and hospitals must be located near residential areas. Factories can be built in the country, where land is cheap and noise and pollution do not affect so many people.

The lumber used in building houses is called **dimensional lumber** because it comes in many thicknesses and lengths. Most walls in residences are framed with 8-foot lengths of $2'' \times 4''$

CONNECTIONS: Technology and Mathematics

■ In the past rectangular shapes were often the only kind seen in buildings. Recently triangular shapes have appeared in construction because triangles make the structure rigid. Geodesic domes are spherically shaped structures composed of triangular faces.

   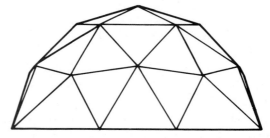

EQUILATERAL   ISOSCELES

These triangles are either *equilateral* triangles (all sides the same) or *isosceles* triangles (2 sides the same). Because of its shape the dome uses less material on its surface than a rectangular structure to enclose the same amount of space.

lumber. These lengths are spaced so the center of each is 16 inches from the center of the next. Insulation, electrical wiring, and plumbing pipes are placed in the framework before the walls are finished. The inside walls are usually made of **plasterboard.** Plasterboard (also known as **Sheetrock**) is sheets of plaster covered with heavy paper. Plasterboard sheets are nailed or glued to the framing.

Outside walls are made from composite or plywood panels. They may be covered with brick veneer or with wood, aluminum, or vinyl **siding.** Siding is attractive and sturdy. Wood siding needs to be painted or stained every few years, but aluminum and vinyl siding need little care. **Insulation** made of fiberglass or plastic foam is installed between the inside and outside walls to conserve energy. Insulation helps maintain inside temperatures and saves on heating and cooling costs.

Most housing construction is done on the site. Sometimes, materials are cut to size in a factory, then assembled on the site. A recent trend is **panelized construction**, in which parts of houses are built in a factory. The walls and roof trusses (framing for the roof) are prefabricated. Complete walls are mass-produced as finished sections that have wiring, plumbing, windows, and doors. This saves time and money.

Apartment houses and commercial and institutional buildings are framed with steel or reinforced concrete. Beams are the horizontal pieces that support the weight of the floors and walls. Columns are the vertical pieces. Columns transfer the weight from the beams to the foundation. A steel framework is often built elsewhere and moved to the construction site. There it is lifted into place by cranes. Reinforced concrete beams and columns are cast in place at the site. Floors may be made of concrete or steel. The outside walls are usually made of brick or panels of concrete, glass, metal, or plastic.

Electrical wiring and utilities are installed before the walls are completed. (Photo by Michael Hacker)

Tunnels

People might have got the idea for tunnels from watching animals burrow. Tunnels under water and through mountains have shortened travel routes since ancient times. The earliest tunnels were dug by the Babylonians about 2000 B.C.

Early tunnels were dug by hand because little machinery was available. The removed earth was hauled away in carts. Workers always feared that the tunnel would cave in around them. In 1818, the **tunneling shield** was invented. This device held the earth up while the tunnel was being dug. The shield was pushed along as the work progressed.

Tunnels dug into rock are drilled and blasted by explosives. The opening is supported by steel arches. A new technique

This building has a steel framework. (Photo by Michael Hacker)

CONNECTIONS: Technology and Science

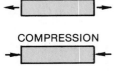

- Pressure (P) is a force (F) that is exerted over an area (A): $P = F/A$. According to the equation, as the area over which a force is exerted increases, the pressure that is produced decreases.
- The ability of water to drain through soil depends on factors such as porosity and permeability. Porosity refers to the amount of open space between soil particles and depends on the shape of the soil particles, how tightly they are packed together, and how well-sorted the particle sizes are. Rounded, well-sorted, lightly packed soils tend to be more porous.

  Permeability is a measure of the rate at which water passes through soil. It depends on the size of the pores and whether the pores are interconnected. The greater the pore size and the more the connections between pores, the more permeable the soil is.

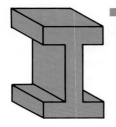

- Steel is a strong elastic material, which means that it can be stretched and compressed with little deformation. These properties make it valuable in construction. When a steel beam carries weight while in a horizontal position, it bends only a small amount because of the depth of the beam. I-shaped steel beams are used in construction because they use less steel than, but are similar in strength to, a solid beam.

The Tacoma Dome

An example of commercial construction is the Tacoma Dome in Tacoma, Washington. The dome is part of a $44 million sports and convention center. The 100,000-square-foot arena seats 26,000 people. It is covered by a 530-foot-diameter wooden dome, the largest in the world.

A dome is a lightweight structure that is very strong. It is like an arch, transferring weight from the top outward and downward to the base. The Tacoma Dome is built of triangular sections. A triangle is a rigid, strong shape.

(Courtesy of Sverdrup Corporation)

uses a concrete mixture called **shotcrete** to spray the walls. Shotcrete prevents water from seeping through the rock.

Tunnels drilled in soft ground may now be dug with machines. These machines have huge rotating cutters as large as 15 feet across. The cutters rotate about five times per minute and push against the earth with a force of nearly a million pounds. As the earth is cut, precast concrete or steel rings are put into place to support the opening.

One of the longest automobile tunnels in the world is the tunnel that connects France and Italy under Mont Blanc in the Alps. The nearly eight-mile-long tunnel saves about a hundred miles of driving. Teams started digging from opposite sides of the mountain and met in 1962, after two years of construction.

In the early 1900s, tunnels were dug by hand. Workers erected wooden arches and supports as the construction progressed. (Courtesy of Perini Corporation)

Roads

Before ancient Rome, roads were narrow paths used by two-wheeled carts. Roman engineers built over 40,000 miles of roads for the empire's armies to travel. Tunnels and bridges allowed the roads to go in a straight line from one place to another. The Romans even paved their roads, using stones.

Modern road building began in the late 1700s with the ideas of a Scotsman named John Loudon McAdam. McAdam made roads that lasted longer because of good drainage. On a base of hard soil, a stone layer was laid, topped by a layer of tar.

CONNECTIONS: *Technology and Social Studies*

- In order to prevent haphazard construction from taking place, communities enforce zoning laws that dictate what can be built where. For example, a zoning board is given the authority to ensure that a factory is not constructed in the middle of a residential area. Thus, they are able to preserve the beauty of the community while looking out for the safety of the people.

- Governments are confronted with the dilemma of finding funds to replace or repair public structures that are deteriorating after many years of use. Roads, bridges, and sewage systems are often in need of repair. Finding the billions of dollars needed to complete the task is a major problem.

- Contractors must pay attention to the impact on the environment of anything they build. More and more, communities and government organizations carefully review building proposals. They want to ensure that the environment will not be damaged by any construction projects.

The Fort McHenry Tunnel

Fort McHenry in Baltimore Harbor is a historic landmark. During the War of 1812, the British bombarded the fort. U.S. soldiers were able to fight off the British force of fifty ships. From an American ship in the harbor, a young attorney, Francis Scott Key, could see that the American flag still waved over the fort. This inspired him to write "The Star Spangled Banner."

Recently, a tunnel was built near the fort to link parts of the interstate highway system. It is hidden, completely out of sight and earshot of the fort. The tunnel has eight lanes, carrying 66,000 vehicles a day. The tunnel took seven years and $750 million to build.

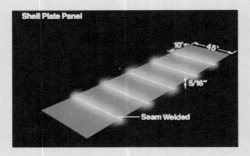

Tube making for the Fort McHenry Tunnel begins with steel panels. The panels are welded together to form a shell plate. Stiffeners are added for strength.

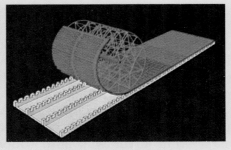

Modules are shaped by wrapping the shell plate around a specially designed reel. More structural pieces and various form plates are added.

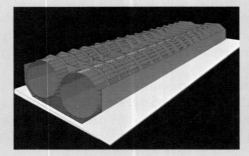

Sixteen modules (eight for each tube) are joined. They form one section of the double-barreled tube. Each tube holds two lanes of roadway.

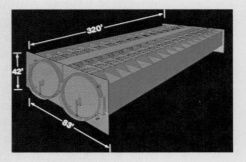

Dam plates seal each end of the tube. Keel concrete is added for strength and rigidity. The section is then launched for a 12-hour tow to the Fort McHenry Tunnel site.

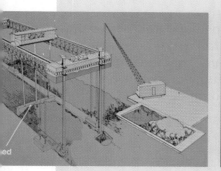

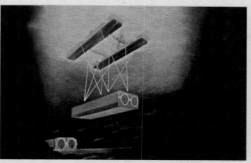

A heavy, plow-like beam from a screed barge dredges along the harbor bottom. This forms the trench that will hold the tunnel.

From a lay barge, the tube section is lowered into the trench. It is then attached to another tunnel segment. Thirty-two tube sections were carefully lowered and connected to form the Fort McHenry Tunnel.

The Fort McHenry Tunnel site

(All photos courtesy of Sverdrup Corporation)

McAdam's roads were higher in the center than at the edges, making water flow away from the road. McAdam's name was given to the material that is now used as a surface for many roads, **macadam**. Since the early 1900s, most roads have been built of concrete or macadam.

Before the twentieth century, few roads were paved. Most were made of crushed stone, which worked well for horses and buggies. But as people switched to automobiles, the rubber tires kicked out the stones, ruining the roads. A new type of surface was needed.

Today, roads are built to support heavy loads carried by high-speed vehicles. Safe, well-designed roads allow people to live miles from their work. They have linked our cities and opened up new markets for business.

Road construction begins with choosing the route. When possible, the route stays away from existing structures. Sometimes buildings must be removed to make way for a road, because it would cost too much to go around them.

Next, the ground is smoothed by bulldozers. The soil is pressed down by heavy rollers. It is then covered with stone, which spreads out the load and provides drainage. The pavement is made of concrete about one foot thick, or from blacktop materials like macadam or asphalt.

Center barriers, good lighting, and traffic control devices are also part of the road-building system.

Today's highways are safe for high-speed long distance travel. (Courtesy of New York State Department of Transportation —Clough, Harbour, & Associates)

Other Structures

Airports, canals, dams, harbors, pipelines, and towers are other structures that are built on a site. Large construction companies often have different divisions that build different structures. There might be a tunnel division, a pipeline division, and a building division. Smaller companies most often do only one or two kinds of construction.

Structures include bridges, buildings, dams, harbors, roads, towers, and tunnels.

RENOVATION

Renovation is the process of rebuilding an existing building. Sometimes this is done to change the style of a building. Sometimes it is done because a building needs repair. As the years pass, dust, wind, and water can cause damage. When a structure is renovated, old materials are replaced or renewed. Renovation is often less costly than demolishing a structure and building it all over again.

SUMMARY

Construction is the building of a structure on a site. Construction systems use the seven technological resources. People engineer, design, manage, and build structures. They use information to carry out these tasks. Building materials like concrete, steel, and wood are used, along with tools and heavy equipment. Energy is needed to run equipment. Capital costs are high because land, material, equipment, and labor are costly. Construction projects take a long time to complete, although new building techniques like prefabrication and modular construction reduce building time.

Construction involves choosing and preparing a site, building a foundation, building a superstructure, installing utilities, and finishing the inside and outside.

Site selection depends on the use of the structure. Commercial buildings are located near business districts. Schools are located in residential areas. The effect of a construction project on the environment and the community should be considered.

Before construction begins, the site must be cleared. Surveyors lay out the structure on the site. The structure must be supported by a foundation, which spreads the structure's weight over a larger area of ground. The superstructure is the usable part of the structure. Mass, bearing wall, and framed superstructures are three types.

The Statue of Liberty was recently renovated. After 100 years, salt water and air pollution had worn parts of the statue away. The renovated statue was rededicated on July 4, 1986.
(Courtesy of NASA)

Construction technology
provides us with many
benefits.

Building techniques have been learned through many years
of experience. Today, these techniques are used along with a
knowledge of materials to build airports, bridges, canals, dams,
harbors, pipelines, roads, towers, tunnels, and buildings.

(Courtesy of The Turner Corporation, photograph by Michael Sporzarsky)

(Courtesy of Mardian Construction Company)

1. What is the major difference between manufacturing and construction systems?
2. Draw a labeled systems diagram of the construction system.
3. Describe four career opportunities provided by the construction industry.
4. How could you finance the construction of a private home?
5. If you were to choose a site for a movie theater, what are five things you would have to consider?
6. What kinds of superstructures do the following structures have?
   a. the Washington Monument
   b. a tower that supports electrical wires
   c. a skyscraper
   d. a large dam
   e. a castle from the Middle Ages
7. Make a sketch of a suspension bridge.
8. Explain why an arch can support a great amount of weight.
9. Design and build a tower, using rolls of newspaper as your building material.
10. Name five different types of structures.

(Courtesy of British Rail)

(Courtesy of NY Convention & Visitors Bureau)

CHAPTER KEY TERMS

Across

3. Details given to a builder.
5. The usable part of a structure.
7. A strong curved support.
8. These are often made of wood and are used to hold concrete while it is hardening.
10. This is made from cement, sand, water, and stones.
12. A very strong material used as a building framework.
13. A structure rests on this.
14. A loan for a construction project.
15. To produce a part of a structure in a factory.

Down

1. A material used in road construction.
2. A person who takes charge of actual hands-on construction.
4. A person who draws designs for buildings.
5. A person who lays out a site.
6. A system of building a structure on a site.
9. An ingredient in concrete, made from limestone and clay.
11. A technical expert who is responsible for exact structural plans.

Arch
Architect
Cement
Concrete
Construction
Engineer
Estimator
Forms
Foundation
General contractor
Macadam
Mortgage
Prefabricate
Specifications
Steel
Structure
Superstructure
Surveyor

SEE YOUR TEACHER FOR THE CROSSTECH SHEET

293

ACTIVITIES

OFFSHORE OIL PLATFORM

Problem Situation

Crude oil is an important resource that is processed into gasoline, fuel oil, and dozens of other products. Most existing oil wells are on land, but drilling at sea is becoming more common because the best land areas have already been drilled.

Many offshore oil platforms rest on pilings in the seabed. In shallow water the platform can be built on site. For deeper water the platform will usually be towed to the desired location. Legs are then extended to anchor the platform.

Offshore platforms need to be designed to support the load of the drilling operations, the crew's quarters, and a helicopter landing pad that is included on most platforms. In addition, the structure must be designed to withstand strong winds and large waves that accompany storms.

Design Brief

An offshore oil platform owned by a major oil company was damaged in a recent storm and needs to be replaced as soon as possible. Design, construct, and test a model of a new structure that can be used at this site.

Simulated pilings made of 1½" lengths of 1" dowel must be part of the structure. The platform must be 10½" tall, free standing, and open at the top and bottom. Internal bracing may not be used.

Procedure

1. Use graph paper to design one side of the structure. A truss design of connected triangles will help distribute the load to be placed on the structures.
2. Place your plan on corrugated cardboard and cover with wax paper.
3. Cut the structural members for one side. Note: Follow your teacher's instructions for using the utility knife safely.
4. Assemble using white glue. Use straight pins to hold the structure together while the glue dries. Use the first side and your plans as a guide to complete the remaining sides.
5. Assemble your model platform on the simulated pilings. Masking tape and clothespins may be used temporarily.
6. Use the scale to weigh your structure. Record the weight in grams.
7. Test your platform to determine the load that it is able to sup-

Suggested Resources

Six lengths of ⅛" × ⅛" × 36" balsa
1" dowels, length as needed
White glue
¼" graph paper
Straight pins
Corrugated cardboard
Wax paper
Masking tape
Utility knife
Scale

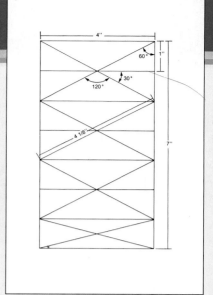

port. Your instructor will demonstrate how the structures will be tested.

8. Calculate the strength-to-weight ratio for your structure by using this formula:

$$\frac{\text{Failure weight (grams)}}{\text{Weight of structure (grams)}} = \text{Strength-to-weight ratio}$$

Classroom Connections

1. Construction sites must be chosen with the needs of people and the environment in mind.
2. The gram is a unit of metric measurement used to determine weight (mass). A kilogram is equal to 1,000 grams.
3. Ratios show the relationship between two numbers. A ratio of one number to another is the quotient of the first number divided by the second number.
4. Oil spills have occurred at many offshore oil platforms. Companies owning the platforms must take precautions to prevent spills. If they do occur, companies must accept responsibility and be prepared to contain and clean up spills.

Technology in the Real World

1. New techniques, including forcing water into wells to force oil up and out, are increasing the yield of existing oil wells.
2. Drilling for oil is expensive. New technologies enable geologists to find locations likely to produce oil without drilling.

Summing Up

1. Why is offshore drilling becoming more common?
2. Why is good design important for offshore oil platforms?
3. How much weight did your structure support before it failed? How could your design be improved?

ACTIVITIES

DEVELOPING A PLOT PLAN

Problem Situation

A plot plan is a drawing made by the architect that shows the location of the structure on the plot. When architects design a structure they must be sure that it fits properly onto the plot site, keeping within the property lines and setbacks. Setbacks are distances from the surrounding structures and property lines that your structure must maintain. Public utilities such as gas and electric companies often require that buildings maintain a setback from their wires and pipes.

The plot plan gives the overall dimensions of the structure and the plot. Plot plans may also include the following:

- compass locations
- sidewalks
- driveways
- setbacks
- noise areas
- swimming pool
- garage and other buildings
- property lines
- street names
- wind directions
- shrubs
- patio

Design Brief

Draw to scale a plot plan of your home, your school, or a structure in your community. Be sure to include the following as a minimum:

- plot dimension
- structure dimension and location
- setback distances
- adjoining street names
- sidewalk
- driveway
- compass location
- shrubs

Suggested Resources

Architect's scale
T-square and triangles
Drawing board
Compass
100' tape

Procedure

1. After selecting a structure, decide on an appropriate scale to use.
2. Using the 100' measuring tape, measure the structure, plot, setbacks, and so on. Record this information.
3. Using your drawing equipment, draw the plot plan as follows:
   a. Draw and dimension the property lines.
   b. Locate, draw, and dimension the structure within the property lines. You may wish to shade it in.
   c. Locate and draw the sidewalks, driveway, and adjoining streets.

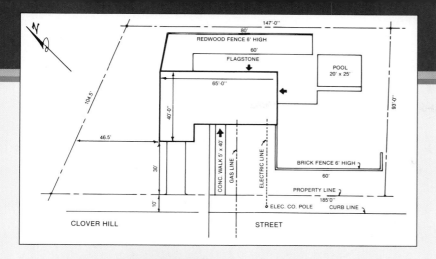

    d. Dimension all the setbacks.

    e. Locate and draw the shrubs, fences, and so on.

    f. Place an arrow on the drawing showing which side of the structure faces north.

4. Recheck your drawing at the actual site.

5. Check to be sure all your architectural symbols are correct.

Classroom Connections

1. Architects and engineers do the designing and planning for most structures.

2. The plans and specifications for a structure are the command inputs into the construction system.

3. Scale drawing is a system used to reduce large objects to sizes that will fit on a sheet of paper, while keeping the overall dimensions of the object in proportion.

4. Drafting uses symbols to communicate technical ideas.

5. Mapping boundary lines and property lines was a huge undertaking for early surveyors as this country spread westward.

Technology in the Real World

1. Modern surveyors and cartographers use aerial photography, satellite images, and lasers to calculate and draw maps, boundary lines, and property lines.

Summing Up

1. Why might the setback from a main road be greater than the setback from a neighbor's property?

2. What is an easement?

3. What is the value of showing north, south, east, and west directions on a plot plan?

4. Why is it helpful to show utility poles, water lines, and gas lines on a plot plan?

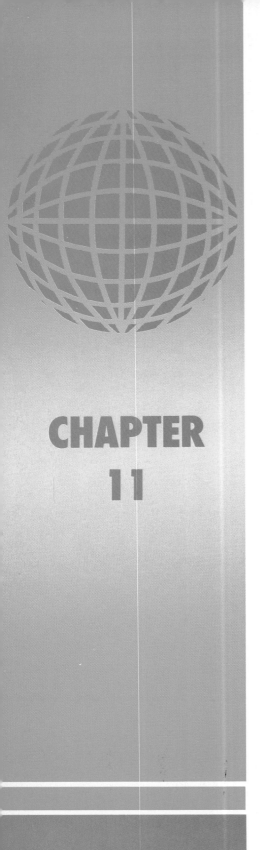

CHAPTER 11

TRANSPORTATION SYSTEMS

MAJOR CONCEPTS

After reading this chapter, you will know that:

- A transportation system is used to move people or goods from one location to another.
- Modern transportation systems have helped to make countries interdependent.
- The availability of rapid, efficient transportation systems has changed the way we live.
- Transportation systems convert energy into motion.
- Steam was the first important source of mechanical power for transportation systems.
- Modern transportation systems often use internal combustion engines or electric motors.
- Intermodal transportation systems make optimum use of each type of transportation used in the system.
- Most transportation systems use vehicles to carry people or goods, but some systems do not use any vehicles.

Today's traveler can travel around the world in eighty hours with plenty of time left over to sightsee along the way. (Courtesy of British Airways)

EXPLORING OUR WORLD

Humans have needed transportation since they first walked the earth. Prehistoric people traveled far in search of food and raw materials, which they had to carry home.

As farming developed, people found ways to move food from fields to storage places. Sometimes a sled was used, pulled by an ox or other animal or by people. Around 3500 B.C., the wheel was invented in Sumeria, in the Middle East. The wheel made it possible to move larger loads.

Transportation also became a means of communication. People on horses or horse-drawn wagons carried messages to and from distant places.

The early Phoenecians and Scandinavians traveled by boat, exploring new lands and trading with other people. A great age of exploration began in the late 1400s. European sailors explored and mapped much of the world.

The steam engine came into use during the Industrial Revolution of the 1700s and 1800s. Steam-powered boats and trains moved quickly across great distances. In his novel *Around the World in Eighty Days*, Jules Verne described a journey around the world in the 1800s. Flying in the *Concorde* supersonic transport (SST), today's traveler can make the same trip in much less than eighty hours.

Today we can travel faster and farther than ever before. We are exploring our solar system, as well as the ocean depths. With new transportation technology, we are at the start of an exciting new age of exploration.

A transportation system is used to move people or goods from one location to another.

The first fast forms of transportation were used to communicate with people in faraway places. (Courtesy of The Museum of Modern Art/Film Stills Archive, W. 53rd Street, New York City)

New types of vehicles enable us to explore the harsh environments of space and the deep sea.
(Courtesy of NASA)   (Courtesy of Lockheed Corp.)

Problem Situation 1:   How could you keep young brothers or sisters amused on long car rides?

Design Brief 1:   Design and make a board game suitable for children between the ages of 7-10 years that could be used in a car. Base the content of the game on a "car journey."

Problem Situation 2:   Water play is enjoyed by young and old alike.

Design Brief 2:   Design and construct an object, propelled by a rubber band and incorporating a propeller or paddle, that will travel for at least 10′ on water in a trough not more than 6″ wide.

Problem Situation 3:   When tire treads are worn down, the tires become unsafe to ride on.

Design Brief 3:   Design and construct a device to check the tire tread depth of an automobile tire.

CONNECTIONS: Technology and Social Studies

■ Since every government has limited financial resources, how to spend money becomes a major issue. With increasing demands for space and undersea exploration, officials are faced with deciding what should be given priority and what should be funded. The future for activities in these two environments may be determined more by the availability of funds than the technological advances achieved.

■ As people are able to move around the earth more rapidly, we have become a global society. Improved transportation systems have enabled us to send products to, and receive products from, all over the world. A fish caught in Massachusetts one day can be sold in Japan the next. With advanced air and land systems of transportation, countries have become truly interdependent. We can no longer ignore what occurs in another country, since it could have a direct effect on us.

THE WORLD IS A GLOBAL VILLAGE

Fast, cheap transportation has brought us closer to other countries. We can eat foods from other countries and buy goods made in other countries. We can sell our food and manufactured products to people in other countries. The economies of countries have come to depend on each other.

Transportation technology has made **tourism** possible. People are able to travel to distant places. Tourism promotes goodwill and understanding between people of different cultures and customs.

Modern transportation systems have helped to make countries interdependent.

CREATING A NEW STYLE OF LIVING

In 1900, the average U.S. citizen traveled 400 miles a year. In 1986, the average U.S. driver traveled 12,000 miles each year. Once people lived close to their work. Now, many must **commute**, or travel on a regular basis, to get to work.

Suburbs, or outlying areas around cities, have grown up. Many people commute to and from the suburbs each day. Some travel more than four hours a day, by train or car. Car pools, in which several people ride together and take turns driving, save money and wear-and-tear.

The availability of rapid, efficient transportation systems has changed the way we live.

RESOURCES FOR TRANSPORTATION SYSTEMS

The way resources are used makes one kind of technological system different from another. This is true in the case of transportation systems.

Energy

Early forms of transportation used wind, animal, water, and human power. Energy was changed to motion using devices such as sails and wheels.

Many of today's transportation systems are based on **vehicles**, containers that hold the people or goods being moved. Vehicles generally use stored energy. Automobiles, for example, use energy stored in chemical form such as gasoline. Subways

Transportation systems convert energy into motion.

and some trains use electricity. The part of the system that changes energy to motion is called the **engine** or motor.

Electric motors change electricity into rotary motion. The shaft of the motor is connected through gears or belts to the wheels of the vehicle. As the motor turns, the wheels turn and the vehicle moves. The gears or belts that connect the motor to the wheels are called the **transmission** or **drive** (sometimes called the drive train).

The advantages of electric motors are that they are small, easy to control, and don't pollute the area around the vehicle. A disadvantage is that getting electricity to the motor may be difficult. It must be supplied along the vehicle's path by wires or on a "third rail," or carried on the vehicle in the form of batteries. These batteries are heavy and must be recharged fairly often.

Gasoline is one of the most efficient ways to supply energy. It burns easily, producing a large amount of energy. It can be stored and moved in tanks. Because it burns easily, it must be used carefully. Gasoline engines are called **internal combustion engines** because the burning of the gasoline takes place within the engine.

Gasoline is made from oil that has been pumped from the ground. Diesel fuel, home heating oil, propane, and jet fuel are other oil-based fuels. Airplanes, trucks, and ships use oil-based fuels.

People

Transportation systems are used by people. They are also designed, built, and operated by people. Motormen drive trains and pilots fly jets. People handle the business of transportation, too. They sell the tickets, make the schedules, clean and maintain the vehicles, and buy supplies.

Many car manufacturers are now experimenting with cars powered by electricity. (Courtesy General Motors Corporation)

Many trains use electric power that comes from overhead wires. (Photo by Jeremy Plant)

Many kinds of vehicles must carry enough fuel with them to last the entire trip. (Courtesy of U.S. Navy)

Information

Designing and operating transportation systems requires information of many kinds. People who drive cars, fly planes, or sail ships need information to guide them. Information about location, course or route, speed, and vehicle operation is important for a safe and rapid trip. Many tools are used to provide such information. These include road signs, radar, two-way radios, and on-board computers.

Materials

Transportation systems use not only vehicles but sometimes roadways, as well. Tracks or canals are roadways. No roadways are used with airplanes or ships, but airports and seaports are needed for loading and unloading. The materials used for vehicles and roadways depend on the transportation system.

Airplanes must be made of strong, lightweight materials. Aluminum and titanium are often used. New **composite** (see Chapter 5) materials are also being used. Composites are fibers mixed with epoxies. They are stronger than metals. Cars and trucks are made of metal, usually steel. Other lightweight materials are used in cars and trucks to save on fuel. They include aluminum, plastics, fiberglass, and composites.

Materials such as concrete and asphalt are used to build roads. Roads must be strong enough to hold heavy weights, such as trailer trucks, without breaking. They must withstand heavy use and changes in temperature.

This advanced plane must use composites that are stronger than metal to withstand stresses during flight. (Courtesy of Grumman Corp.)

Tools and Machines

The tools and machines used in transportation systems are more than just the vehicles. They include support equipment such as automatic controllers and maintenance equipment. For example, test equipment is used to check systems on airplanes and automobiles.

Capital

A large amount of capital is needed for transportation systems. Building ports, roadways, and vehicles is costly. Maintenance is important, especially in systems that carry people, and it costs money, too.

Private companies most often pay for vehicles. Government or public agencies pay for roadways and ports, for two reasons.

Tools and machines include the support equipment for vehicles. (Courtesy of International Business Machines Corp.)

First, few companies can afford the huge costs of building a highway or an airport. Second, these facilities are shared by many. Roadways are built and maintained using money collected through taxes or tolls.

Time

In a transportation system, travel time depends on distance and technology. Time can be the few seconds it takes to move parts from one workstation to the next on an assembly line. It can be the weeks it takes for a ship to cross the ocean. It can be the years required to travel from our planet to Neptune.

Mass transit systems move thousands of people on a regular basis. Time is important to the smooth operation of these systems. Vehicles must move on a set schedule. Schedules must be kept so that people traveling one route arrive on time to transfer to a train, a plane, or a bus on another route.

TYPES OF TRANSPORTATION SYSTEMS

Transportation systems are alike in some ways and different in others. Most use vehicles to carry people or goods. But some, such as conveyor belts or pipelines that carry oil, do not. Most systems are made up of subsystems. For example, an automo-

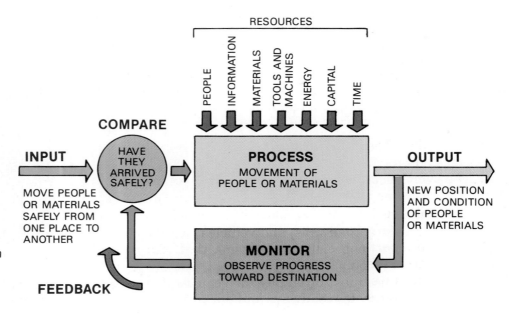

Like all other systems, transportation systems can be represented with system diagrams.

bile has a steering subsystem and a suspension subsystem, among others.

Subsystems can be put together in different ways to form different transportation systems. A diesel engine can be put in a floating hull to make a boat, a marine transportation system. It can also be placed in a vehicle with wheels to make a truck, a land transportation system. One way to classify systems is by the environment in which they move (land, sea, air, space).

In the Malagasy Republic, people-powered push-pushes provide transportation. (Photo by Michael Hacker)

LAND TRANSPORTATION

The earliest form of land transportation, after walking, was riding on animals. Next, animals were used to drag heavy loads on sleds. The wheel made it possible to pull heavier loads more quickly. Some cultures today still use animals and people for most transportation needs.

Steam-Powered Vehicles

When the **steam engine** was invented, people tried to use it to move vehicles. In 1769, Nicolas-Joseph Cugnot built the first steam-engine-powered vehicle in France. It was a tractor that moved at about 2 miles an hour. Its weight and steering system made it hard to control. It crashed into a wall. Cugnot's tractor idea failed to catch on. But steam engines were found useful in powering boats and trains.

The first **railroad** to use steam engines was opened in England in 1830. The engine could pull the train at up to 30 miles an hour. Both people and cargo (freight) were carried. Railroads were started in the U.S. about the same time. They quickly came into widespread use.

After the Civil War, several inventions made railroads much safer. One was the air brake, invented by George Westinghouse. All the cars on the train could be stopped at the same time using the air brake. In 1893 a law was passed that required its use on all trains. The air brake is still used on trains, as well as trucks and buses.

In 1862, construction was started on a railroad track between the Missouri River and Sacramento, California. One railroad company built west from the Missouri, the other east from Sacramento. In 1869, they met in Promontory, Utah. The eastern and western parts of the country were joined. This aided in the settlement of the American West.

Steam engines were improved. Better track was developed. The comfort of passengers and maintenance of systems im-

Many obstacles had to be overcome to build the railroad through the rugged west. (Courtesy of Oregon Historical Society)

Steam was the first important source of mechanical power for transportation systems.

A steam-powered logging engine. (Photo by Jeremy Plant)

Steam Engines

Steam engines are **external combustion engines**. Fuel, such as wood, coal, or oil, is burned in an open chamber. A boiler of water is heated, creating steam pressure, which is used to push a piston. The piston's reciprocating (back-and forth) motion is changed to rotary motion, which is transferred to wheels. Steam locomotives must stop often to take on supplies of water and fuel.

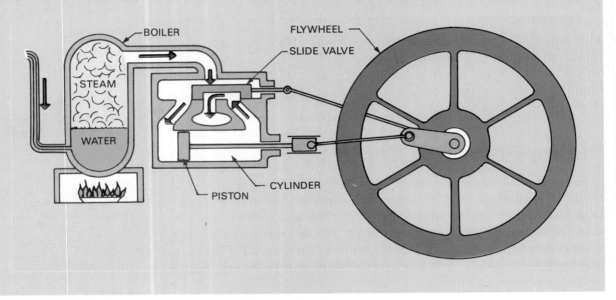

proved. By the late 1800s, trains could travel at speeds of 100 miles per hour. Special trains such as the Orient Express (London-Istanbul) let people travel in luxury.

Gasoline-Powered Vehicles

In 1876, the internal combustion engine was invented by Nickolas Otto of Germany. It was lightweight and could be used to power a carriage. Such a carriage could travel many miles on a small amount of gasoline. "Automobiles" quickly became popular. (See Chapter 5 for a description of how the internal combustion engine works.)

Because of its small size and greater fuel economy, this engine replaced steam engines in many vehicles. The Stanley Steamer was a steam-powered car that could go very fast. (One set a world's record of 122 miles per hour in 1906.) But it needed a fresh supply of water every fifty miles. It was soon replaced by

Cutaway view of
an automobile.
(Courtesy of Saab)

Henry Ford at the wheel of
his first car. (Courtesy of Ford
Motor Company)

A modern luxury car. (Courtesy of Ford Motor Company)

automobiles using internal combustion engines. By 1920, steamers were no longer being built.

The automobile was improved and improved again. Many companies were formed. They built many kinds of automobiles, from sporty "raceabouts" to family sedans. American families soon found it necessary to have an automobile. By the 1920s, millions had been sold. Since then, many companies have gone out of business. Automobiles built by some of them—the Pierce-Arrow, Duesenberg, Stutz, and Franklin—can now be seen only in museums. Only a few large companies build automobiles today. The automobile industry is America's largest industry.

A modern truck used to
deliver large amounts of
freight economically. (Courtesy
of Fruehauf)

Diesel-Powered Vehicles

The **diesel** engine also came into use during the early 1900s. A diesel engine is like a gasoline engine, except that it has no spark plugs. In a regular gasoline engine, spark plugs cause the fuel and air mixture to explode, providing power to the engine. In a diesel engine, this job is done by a piston that squeezes the mixture more tightly. When a gas is squeezed, or put under pressure, it gets hotter. With enough pressure, it will explode by itself.

Diesel engines are good for carrying heavy loads at a constant speed. They need less maintenance. A vehicle gets good mileage and lasts a long time. Diesel engines are used in cars, trucks, buses, locomotives, and construction machinery.

The railroads quickly changed over from steam engines to diesel engines after World War II. Railroad companies liked the better mileage, cleaner burning, and lower maintenance of diesel engines. In 1949, the last steam locomotive built for regular service was delivered. A few hundred steam locomotives remain in the U.S. Most are only on display, but a few are used as excursion trains and tourist attractions.

Modern diesel engines.
(Photo by Jeremy Plant)

This electric-powered subway runs both underground and above ground in San Fransisco, California. (Courtesy of Bay Area Rapid Transit District)

The electric-powered Japanese "Bullet Train" provides passenger service at 130 miles per hour. (Courtesy of Dave Bartruff)

Electric Vehicles

Electricity was used to power trains and cars almost as early as steam engines were. An electric car was first run in 1839. It used batteries to store energy and electric motors to turn the wheels. Electric cars ran very well, but their batteries had to be recharged before they could go far. The same is true today, although batteries have improved. With further advances in battery technology, it is likely that electric cars will come into general use.

Modern transportation systems often use internal combustion engines or electric motors.

Vehicles powered by electricity from overhead lines or a third rail are in wide use. Electric buses or trolleys running on overhead lines were common in cities. Much of the nation's railroad track is electrified, especially in cities, where clean electric engines are preferred. Most city subways are powered by electricity.

WATER TRANSPORTATION

People have always traveled on streams, rivers, lakes, and oceans. Large cities are often found on natural harbors because water makes travel and trade easier. Different forms of energy have been used to move boats throughout history.

Use of Natural Resources

At first, human muscle powered boats. With poles, paddles, and oars, people moved small boats on trips close to shore. They soon learned how to use the wind to move their boats. The ancient Egyptians, Phoenecians, and Romans had ships with dozens of rowers as well as sails.

With sails, people could travel farther. They began to explore. They found new trade routes, and people to trade with. The ancient Phoenecians, on the Mediterranean, traded with people of Great Britain. The Romans traded with people in the Far East. Early Scandinavian explorers may have been the first

Sailing ships explored and traded with all parts of the world. (Courtesy of the U.S. Navy)

Europeans to visit North America. From the 1400s through the 1800s, Portuguese, Spanish, English, and French sailors explored widely. Their two- and three-masted sailing ships carried enough supplies for long trips. North and South America, Africa, and many Pacific islands were explored, mapped, and

What Makes a Boat Float?

Early boats were made of wood. Today, boats are made of many different materials, including steel, fiberglass, and cement. A solid piece of most of these materials would sink, but boats made of them float. Why?

A boat floats for a reason stated as Archimedes' Principle. That is, an object placed in water (or any fluid) is pushed upward by a force equal to the weight of the water displaced (pushed aside) by the object. The object has **buoyancy**. Buoyancy makes things feel lighter under water than they are in air.

For example, a 32-pound piece of metal in the form of an open box displaces a cubic foot of water. Water weighs 64 pounds per cubic foot. The water pushes up on the box with 64 pounds of force. The box pushes down with only 32 pounds, so it floats.

A solid metal block weighing 32 pounds takes up only ⅛ of a cubic foot. That's 8 pounds of water. The water pushes up with 8 pounds of force, and the block pushes down with 32 pounds. The block sinks.

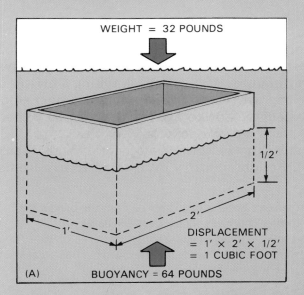

WEIGHT = 32 POUNDS

1/2′
2′
1′

DISPLACEMENT = 1′ × 2′ × 1/2′ = 1 CUBIC FOOT

(A)   BUOYANCY = 64 POUNDS

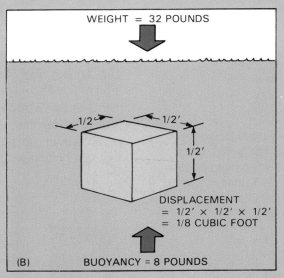

WEIGHT = 32 POUNDS

1/2′   1/2′
1/2′

DISPLACEMENT = 1/2′ × 1/2′ × 1/2′ = 1/8 CUBIC FOOT

(B)   BUOYANCY = 8 POUNDS

Buoyancy equals the weight of the water that is displaced by an object. (A) A box floats in water. (B) A solid piece of metal weighing the same as the box sinks in water.

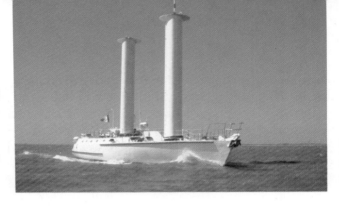

The use of wind power in a modern ship is demonstrated by the *Alcyone*, built by famed ocean researcher Jacques Cousteau. (Photo courtesy of The Cousteau Society, a member-supported environmental organization)

settled. Throughout this time, water travel was powered by natural resources (human, animal, and wind).

Steam-Powered Ships

In 1807, Robert Fulton built a ship powered by steam. The *Clermont* carried people and cargo between New York City and Albany, N.Y. It was the first steam-powered boat to be used successfully. The engine pushed a paddle wheel that pushed against the water, moving the boat.

More steamship designs followed. Some used paddle wheels on the sides, some on the back, and some used screw propellers. Some used sails, as well. They could sail when the wind was blowing, and stay underway when it died. Sails were used less and less as ocean-going ships grew much larger, carrying many passengers.

Modern Ships

Better hull design and engines made for faster ships. In 1952, the S.S. *United States* crossed the Atlantic Ocean in three days, ten hours, and forty minutes. But air travel, taking a matter of hours for the same trip, was even faster. Fewer passenger ships were built. Almost all of the great ocean liners have disappeared, but smaller cruise ships are still popular.

However, ships still carry most intercontinental freight. Tankers carry crude oil. Freighters carry everything from automo-

While other passenger ship travel has declined, cruise ships still remain popular, such as Cunard's flagship, *Queen Elizabeth II.* (Courtesy of Cunard)

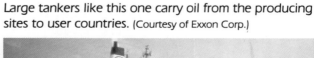

Large tankers like this one carry oil from the producing sites to user countries. (Courtesy of Exxon Corp.)

biles to bananas. Ships carry large, heavy cargoes more cheaply than airplanes. Because ships can carry such tremendous loads, extra care must be used to avoid spilling any of the load into the sea, where it might damage the environment.

Submersibles

Most ships travel over the water's surface. Some, however, also operate below the surface. They are called **submersibles** or **submarines**. They can operate either on or below the water's surface because they can change their weight without changing their buoyancy. Special tanks are filled with water, causing the ships to sink. Or the tanks can be filled with air, causing them to rise. These ships can float at any depth by changing the amount of air in the tanks.

Nuclear-powered submarines can stay submerged for months at a time. (Courtesy of the U.S. Navy)

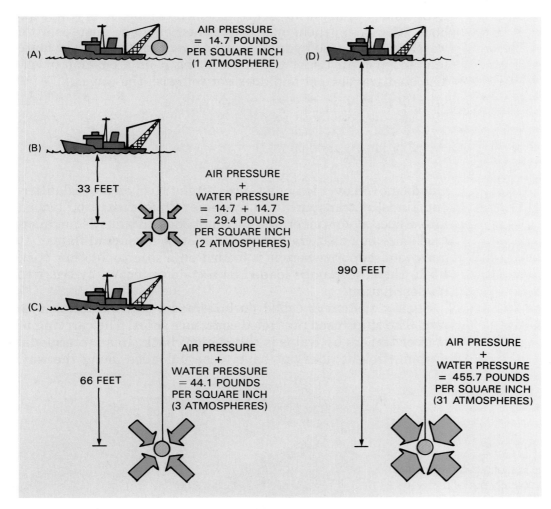

Pressure under water is a challenge to designers of underwater vehicles.

These underwater ships must withstand the huge pressures exerted by water at depth. We live at the bottom of an ocean of air. It has weight, and so it exerts pressure on us. Water is heavier than air, and exerts a much greater pressure. At a depth of 33 feet, water exerts 14.7 pounds of pressure per square inch. This is the same amount of pressure as the entire height of the atmosphere (more than 100,000 feet). At a depth of 66 feet, this pressure is doubled; at 99 feet, it is tripled. Spheres and cylinders are strong shapes for hollow containers. Most submersibles are made of one or both of these shapes.

Hydrofoils and Air Cushion Vehicles

Surface ships include hydrofoils and air cushion vehicles (ACVs). A boat with a fairly flat bottom rises up in the water as it goes faster. This idea is used in a hydrofoil. When small hydrofoils, or flat surfaces, are attached to a boat's bottom, the boat will ride on them once it goes fast enough. There is little water resistance, so hydrofoils can go very fast. ACVs use large fans to push air under the boat, lifting it on a cushion of air. The boat travels fast and does not roll with the waves.

INTERMODAL TRANSPORTATION

Intermodal transportation systems make optimum use of each type of transportation used in the system.

Goods moved over long distances often travel on several different kinds of transportation systems. Freight may be loaded into a special container at a factory. The container is a trailer-truck-size box that travels by truck or on a railroad flatcar. It goes to a seaport where it is loaded on a ship to another port. From there, it may be loaded on a train or moved by truck to its destination.

Such a system is called an **intermodal** transportation system. The ships used are called **container ships**. The carrying of tractor trailers by trains is called **piggyback**. In an intermodal system, freight does not have to be unloaded along the way.

The hull of a hydrofoil comes completely out of the water, reducing water resistance.
(Boeing photo)

Trucks are loaded piggyback onto a train.
(Courtesy of Santa Fe Railway)

There is less damage and loss. An intermodal system is so reliable that it is often used to supply parts for "just-in time" manufacturing. (See Chapter 9.)

AIR TRANSPORTATION

People have always dreamed of flying, but some of the earliest flyers were forced into it. Marco Polo was an Italian who traveled in China in the 1200s and later wrote about his adventures. He reported that sometimes Chinese sailors would tie a person to a kite and try to fly it. If the kite flew well, it meant a safe journey. If the kite crashed, it meant bad luck, and their ship remained in port for the rest of the year. Of course, it was also bad luck for the person tied to the kite!

Lighter-Than-Air (LTA) Vehicles

Flying really began in 1783 when two Frenchmen built a hot-air balloon that could carry people. Objects in water have buoyancy. Objects in air have **lift**, an upward force equal to the weight of the air displaced by the object. For an object to float in air, it must weigh less than the air that it has displaced. This occurs when a lightweight container is filled with a gas (hot

This container ship is used in an intermodal transportation system. (Courtesy of Sea-Land Service, Inc.)

Balloons float because they are filled with a gas that is lighter than air. (Courtesy of Albequerque Convention and Visitors Bureau)

The hydrogen gas in the *Hindenburg* caught fire and burned as the dirigible was docking in New Jersey after a trans-Atlantic crossing.
(Courtesy of New York Daily News)

air, hydrogen, or helium) that is lighter than air (LTA). The two together weigh less than the air that they displace.

In the early 1900s, huge LTA ships called **dirigibles** carried passengers and cargo around the world. They had rigid metal frames and were filled with hydrogen gas. The largest was the *Hindenburg*. More than 800 feet long, it could carry 100 people. Hydrogen burns easily, and many large dirigibles exploded, among them the *Hindenburg*. Thirty-six of the ninety-seven people aboard were killed.

Today, LTA ships called **blimps** use helium gas. Helium is heavier than hydrogen, but it doesn't burn. Blimps are not rigid LTAs. They are used for advertising, for some kinds of cargo-lifting, and as platforms for cameras.

LTA vehicles use **passive lift**. They float in the air because of their volume and weight. **Active-lift** vehicles create lift by their movement through the air. They are said to be in **powered flight** because they must have power to fly.

Active-Lift Aircraft

The first powered flight was made by Orville Wright on December 17, 1903. Orville and his brother Wilbur had been experimenting with gliders (unpowered planes). They added a 12-horsepower engine driving two propellers to a glider to build the first airplane.

Early planes were made of wood and cloth. They had two or three wings to increase lift. As engines became more powerful, heavier, stronger materials were used. The number of wings was reduced to one. Passenger service started in the United States in 1914. The U.S. Post Office started delivering air mail in 1919.

World War II brought many advances in airplane design and manufacture. Airplanes were mass-produced. Airframe design and electronics were improved. A very important advance was the jet engine. Jet engines were used on military planes right after World War II. They were not used on passenger planes until 1952.

The first person to fly faster than the speed of sound was Chuck Yeager. He flew an X-1 experimental plane to more than Mach 1 (about 700 miles per hour) in 1947. **Mach 1** is the speed of sound. **Mach 2** is twice the speed of sound. When an airplane gets near the speed of sound, the air it pushes ahead of it forms a shock wave called the sound barrier. The shock wave makes the plane hard to handle. Once the plane goes faster than Mach 1, the plane is easier to control.

The first passenger jets were the British DeHavilland Comets. Many of them crashed. Their problems were caused by

Planes like this one carry hundreds of people over thousands of miles economically. (Courtesy of Boeing Aerospace Company)

metal parts that failed because of the stress put on the planes when they climbed and accelerated. But the Comet was re-designed as a much safer plane. The planes that came after it were built better, as well.

Since the Comet, many other passenger jets have been built. The jets of today have an excellent safety record for many rea-

How Planes Fly

Four forces act on an airplane. They are weight, lift, drag, and thrust. **Weight** is caused by gravity. It is the force pulling the plane toward the earth. **Drag** is the wind resistance. This tends to hold the plane back when it moves forward. **Lift** is the upward force that must be created to make the airplane fly. **Thrust** is the forward force produced by the engines that move the airplane.

Airplane engines move airplanes forward at speeds of 100 to over 1,000 miles per hour. The wing is shaped so that air rushing over it produces a higher pressure on its bottom side than on its top.

Bernoulli's Principle describes this effect. It says that as air flows over a surface, its pressure decreases in places

where the air speed increases. The curve at the top of the wing makes the air rush over the top faster than under the bottom. Lower pressure is created on the top. The higher pressure on the bottom of the wing results in upward force on the wing, lifting it.

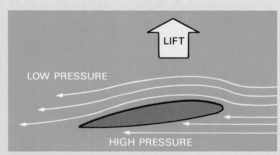

Airflow over an airplane wing.

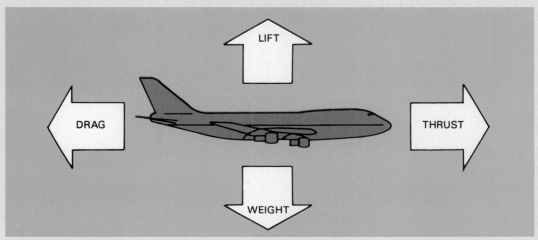

The four forces acting on an airplane.

The Wright brothers' first flight could have taken place in the cargo compartment of this C-5A. (Courtesy of Lockheed Corporation)

sons. Jet engines require less maintenance than internal combustion engines. Their flight crews are extensively trained. Modern air traffic control systems keep planes apart in the sky.

Many different kinds of planes are used today. Small private planes carry two to six people over short distances. Airliners that can land and take off on short runways handle commuter traffic. **Jumbo jets** carry hundreds of people at a time over long

Aircraft Engines

Many planes have used internal combustion engines that turn **propellers** to provide thrust. As airplane propellers turn, they cut into the air. They move the air from front to back. An airplane propeller can be used as a pusher, when mounted on the back of the wing. Or it can be used as a puller, mounted on the front of the wing. Most propeller-driven planes today use pullers.

The **jet** engine takes advantage of Newton's Third Law of Motion. Newton's third law states that **to every action there is an equal and opposite reaction**. For example, a balloon filled with air, suddenly let go, will fly around the room. The air rushing out of the balloon in one direction (the action) moves the balloon in the opposite direction (the reaction).

In a jet engine, a **compressor** forces air into a combustion chamber. There the air is mixed with a fine spray of fuel and burned. The burning mixture expands and rushes out. The force of the gases rushing out the back of the engine pushes the plane in the opposite direction, forward. The burning gases also turn a turbine as they leave the engine. The turbine turns the compressor and other devices such as electrical generators. In a **turbo-prop** engine, the turbine drives a propeller.

A jet engine uses oxygen in the air to burn the jet fuel. A **rocket** carries its own oxygen. It can fly where there is little or no oxygen, at high altitudes or in space. The rocket does not have a turbine, since there is no compressor.

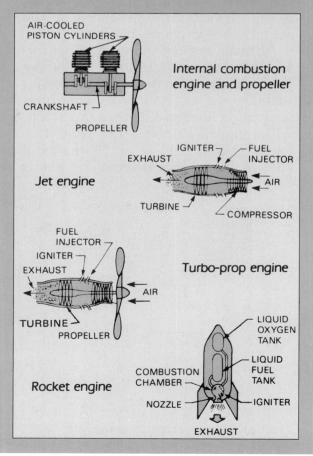

316

Transportation Safety

Most transportation systems that carry people put safety first. New ways for making transportation safer are constantly being designed and tested.

Sometimes new safety measures come about because of terrible accidents. In 1912, the largest steamship in the world, the R.M.S. *Titanic*, hit an iceberg on her first voyage. She sank in the North Atlantic. Over half of the 2,207 aboard were killed. As a result, new rules were set regarding lifeboats, lifejackets, and radios. These rules have made travel by ship much safer.

Many safety features are designed into transportation systems. Most vehicles must meet safety standards before being sold or used. Features such as air bags and improved bumpers on cars provide safer vehicles but also increase the cost to manufacture them. Designers, law makers, and public interest groups are constantly discussing the tradeoffs between costs and improved safety.

Some of the most important safety measures depend on the people who use transportation systems. For example, it is estimated that 17,000 fewer people would die in traffic accidents each year if people used seat belts. Only about 14 percent of adults, however, do use seat belts. Always buckle up!

Planes are dropped from this test stand to test for crash worthiness. (Courtesy of NASA)

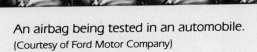

An airbag being tested in an automobile. (Courtesy of Ford Motor Company)

Always buckle up. (Courtesy of Ford Motor Company)

This plane has just been dropped from the test stand. (Courtesy of NASA)

Rockets are used to attain the high speeds needed to achieve orbit. (Courtesy of Lockheed Corporation)

distances. The British and French **Concorde** carries passengers at twice the speed of sound. Military planes carry large cargoes and refuel in-flight. They can travel anywhere without having to land to refuel. Small jets that travel at more than Mach 2 protect our borders.

SPACE TRANSPORTATION

The world entered the space age in 1957, when the U.S.S.R. launched *Sputnik*. *Sputnik* was a satellite, circling the earth once every 90 minutes. Yuri Gagarin of the U.S.S.R. made the first manned flight in 1961. Two American astronauts followed him later that year. During the 1960s, both the U.S. and the U.S.S.R. made more space flights. In July 1969, Americans Neil Armstrong and Edwin Aldrin became the first people to set foot on the moon.

Space vehicles must carry oxygen to burn their fuel because space is airless. So far, all space vehicles have been launched using rocket engines. The engines have used both liquid and solid fuels. Ideas for other kinds of engines are being studied.

To reach orbit, a speed of more than 17,000 miles per hour must be reached. An additional thrust is needed to accelerate the vehicle to 25,000 miles per hour, the speed needed to escape earth orbit.

Great advances had to be made in many areas of technology during the *Apollo* moon-landing program. Many of

CONNECTIONS: Technology and Science

- Heat causes molecules of a substance to move faster. The higher the temperature of a substance, the faster the molecules move. As a result, the molecules spread farther apart. Hot air is less dense than cold air since hot air has fewer molecules per unit of volume than cold air. Hot-air balloons can be lifted because, according to Archimedes Principle, the weight of the displaced "cooler" air is greater than that of the warmer air.

- Isaac Newton's First Law of Motion, also known as the Law of Inertia, states that resting objects tend to remain at rest and moving objects tend to remain in motion until a force acts upon them to change these situations. Seat belts are used because of inertia. When a moving automobile comes to a sudden stop, the passengers' inertia causes them to continue moving in the direction the car was moving. The passengers' motion will stop when a force such as a windshield or a seat belt opposes their inertia.

The space shuttle can travel in space, glide back to earth like a plane, and be used again for another trip. (Courtesy of NASA)

these advances had spin-offs that are now used in everyday life. They include the intensive care unit in hospitals, new fire-fighting equipment, and the integrated circuits used in many electric appliances.

Space travel poses many problems never faced before. In space, there is no air. Space travelers must carry their atmosphere with them. Once away from the earth's gravity, space travelers experience weightlessness. This is fun and exciting, but also presents problems such as how to take a shower or drink a liquid.

Space exploration brings up another very important problem. The great distances involved mean months and even years are needed for a journey. It took *Voyager*, an exploratory space vehicle, twelve years to travel from the earth to Neptune.

The space shuttle is the first reusable space vehicle. It is used as a space truck, carrying objects and people back and forth from the earth to low earth orbit. The shuttle uses three engines and two solid fuel rocket boosters to reach escape velocity. When it returns, it uses its wings to glide to a landing.

NONVEHICLE TRANSPORTATION SYSTEMS

Some transportation systems use vehicles to carry people or cargo. Vehicles can be trains, boats, planes, or space ships. But other transportation systems move materials and people without vehicles. These transportation systems include pipelines and conveyor belts.

Most transportation systems use vehicles to carry people or goods, but some systems do not use any vehicles.

Pipelines

Pipelines are used for moving crude oil or natural gas. They extend from the field where these resources are pumped to the

When completed, this pipeline will carry natural gas
from the wells to a major city.
(Courtesy of Perini Corporation)

Materials are automatically moved along an assembly line by a con
system. (Courtesy of Cincinnati Milacron, Inc.)

place where they are refined or loaded aboard trucks or ships.
The Trans-Alaska pipeline moves crude oil from northern
Alaska to tankers at Valdez, a port on Alaska's southern coast.
The pipeline is heated and insulated. This keeps the oil liquid
enough so pumps can move it even in cold weather. Great care
must be taken when designing and maintaining pipelines so
that they do not leak any oil or gas.

Conveyors

In most assembly lines, parts are moved from workstation
to workstation by **conveyors**. Most cars are assembled this way.
Large metal parts are cut, drilled, and machined at stations
along conveyor belts.

Small objects can be moved along a guide path with their
motion driven by vibrating the path. Coal and electronic parts
are often moved this way.

PEOPLE MOVERS

Some transportation systems move people over short distances.
These **people movers** include **escalators**, **elevators**, and **personal
rapid transit systems (PRTs)**.

Freight elevators had been used for some time before 1852,
when Elisha Otis invented a device to keep them from falling.
Otis's invention, and other advances, helped to make high-rise
buildings possible. Without elevators, cities would be more
spread out. They would not have central clusters of skyscrapers.

Elevators move people up and down. PRTs move people
horizontally from one place to another. PRTs use small cars

Escalators provide quick, easy movement from one level to another. *(Courtesy of Otis Elevator Company)*

that people stand in. They move along tracks from one part of an airport to another or from one part of a city to another. They are controlled automatically.

Elevators and PRTs use vehicles. Escalators and moving sidewalks do not. They can move people a little more quickly than walking or climbing stairs. They are often used where people are likely to be carrying packages or luggage, such as in department stores or airports.

SUMMARY

People have used transportation systems since ancient times. They have moved themselves and goods, and they have explored their world.

Transportation technology made it possible for people living far apart to exchange goods and ideas. This exchange has made countries economically dependent on each other. Transporta-

---

CONNECTIONS: Technology and Mathematics

■ Conveyor belts are used to move things from one place to another. A conveyor belt is a loop that has *two* sides, an inside and an outside. A mathematician, August Moebius, took a rectangular strip of paper, gave one end a half-turn, and then joined the two ends. He created a loop that had only *one* side. It has become known as the "Moebius strip." A conveyor belt with this "half-twist" lasts longer because it is a continuous loop and the entire surface "wears" equally.

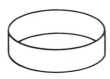 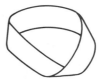

---

tion technology has changed our way of life by making travel an important aspect of modern life.

Most transportation systems use vehicles to carry people or cargo. An engine or motor changes energy to motion.

People are an important part of a transportation system. They do many jobs in managing the system. Many transportation systems include roadways or ports built with public money. Vehicles are often owned privately. Schedules are important in transportation systems.

Transportation systems include land, sea, air, space, and nonvehicle transportation. These systems are alike in some ways and different in others. The way in which subsystems are connected is determined by use.

Early vehicles that were mechanically powered were powered by steam. Railroads changed the way people lived. American railroads opened the West to settlement.

Automobiles have become necessary to many Americans. Cars are usually powered by the internal combustion engine.

The first mechanically powered ships were steamships. Passenger travel by ship has decreased because planes are faster. But ships are still used to carry freight and for vacation cruises.

Intermodal transportation combines different forms of transportation. Freight is packed into containers, and the containers are moved by different means. They arrive without having been opened.

The first powered flight in 1903 began air transportation. Air transportation vehicles were first powered by internal combustion engines. Now they are also powered by jet engines, turbo-prop engines, and rocket engines. Jet engines and rocket engines enable planes and rockets to fly faster than the speed of sound.

Safety is important in transportation systems. Safety measures come about as a result of accidents or through careful design.

Space transportation systems must solve problems of great distance, weightlessness, and airlessness. Vehicles can orbit the

(Courtesy General Motors Corporation)

earth if they can reach escape velocity—more than 17,000 miles per hour.

Some transportation systems do not use vehicles. They include pipelines and conveyors. They are used to carry oil from oilfields to ports and materials from station to station on assembly lines.

People movers carry people for short distances. Elevators and PRTs use vehicles. Escalators and moving sidewalks do not.

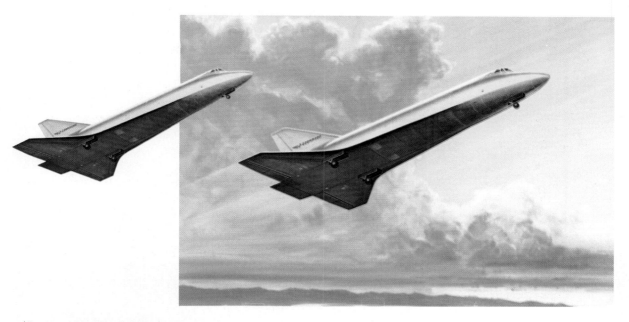

(Courtesy of Lockheed-California Company)

(Courtesy of United Nations/Photo by A. Holcomb)

1. Describe how transportation and communication systems have made countries interdependent.
2. In what way(s) have modern transportation systems affected the way your family lives?
3. Is an automobile a necessity or a luxury in your family? Why?
4. Why were steam engines replaced by internal combustion engines in cars?
5. Why were steam engines replaced by diesel and electric engines in trains?
6. What role did trains play in the settling of the American West?
7. Why are electric engines used on railroads that travel into major cities?
8. Describe why a boat made out of steel and cement can float.
9. Describe how intermodal transportation works and what its advantages are.
10. What shapes do submarines and submersibles use? Why?
11. Describe how a wing enables a plane to fly.
12. Give at least two reasons why jet engines have replaced internal combustion engines on commercial passenger planes.
13. How is a rocket different from a jet engine?
14. Why are rocket engines used for space vehicles?
15. Name two kinds of transportation systems that don't use vehicles. Describe how they work.
16. Name a human-powered vehicle that is widely used for sport or leisure traveling in this country, but is used as basic transportation in other countries.

(Courtesy of Mark Guran/Island Windsurfing)

(Photo by Jonathan Plant)

CHAPTER KEY TERMS

Across

3. An upward force exerted on an object by a fluid.
8. Similar to a gasoline engine, except it has no spark plugs.
9. A ship used to carry trailer-truck-size boxes.
11. This type of engine forces air into a combustion chamber, mixes the air with a fuel spray, and ignites the mixture.
12. This device can power craft in places where there is no oxygen since it carries its own supply.
13. To travel on a regular basis from home to work.
15. The gears or belts that connect a motor to the wheels.
17. Transportation systems that use different modes for different parts of a trip.
18. A type of external combustion engine used in early vehicles.

Down

1. The type of engine most often used to power automobiles.
2. A non-vehicle transportation system used to move crude oil or natural gas.
4. This propels an airplane or a ship.
5. The forward force (produced by an engine) that moves a vehicle.
6. A container that carries people or cargo.
7. An aircraft engine in which a turbine drives a propeller.
10. A non-vehicle transportation system used to move parts from one workstation to another.
14. The upward force that must be created to get an airplane to fly.
16. The wind resistance that tends to hold back an airplane when it moves forward.

Buoyancy
Commute
Container ship
Conveyor
Diesel
Drag
Engine
Intermodal
Internal combustion engine
Jet
Lift
Piggyback
Pipeline
Propeller
Rocket
Steam engine
Thrust
Transmission
Turbo-prop
Vehicle
Weight

SEE YOUR TEACHER FOR THE CROSSTECH SHEET

DESIGNING A SAILBOARD

Problem Situation

Modern sailboats depend on the wind and sails to help power the boat in a variety of directions. This is done through the use of triangular sails that can be rotated around a sailboat's mast. By allowing the sails to engage the wind at different angles, the sailboat can be propelled at almost any angle to the wind. Using a maneuver called tacking, sailboats can even travel into the wind by following a zigzag course and always keeping the sails at an angle to the wind.

Design Brief

Research, design, and construct a working sailboard that will perform a variety of maneuvers when placed in a tank of water and engaged by a wind from a 20" table fan.

Procedure

1. Research various hull, keel, and sail designs you might want to model or incorporate into your design.
2. Make some sketches and full-size drawings of a design you might want to model. Concentrate on how the sail will be maneuvered to control the ship.
3. Select appropriate materials to construct your model.
4. Using the safety procedures described by your instructor, construct your model.
5. Test its maneuverability in the water tank.
6. Make changes if necessary.

Suggested Resources

Appropriate modeling materials (balsa wood, pine wood, Styrofoam, etc.)
Sail materials such as fabric or plastic bags
Materials processing equipment and tools
20" table fan
Large fish tank or container for testing
Reference books on sailing

Classroom Connections

1. Objects float if their buoyancy is greater than their weight.
2. When wind fills the sail, it creates the shape of an airfoil. A difference in air pressure is created, causing a suction and pulling the ship through the water.
3. Geometry and vectoring can be used to graph the effects of the forces acting on the sail as it cuts into the wind at different angles.
4. The ability to travel longer distances because of better sailing technologies opened the earth to further exploration.

Technology in the Real World

1. New keel designs on racing yachts incorporate a ribbed design to help gain stability as the craft moves through the water.
2. Sails made of synthetic materials are lighter and much stronger than cloth or canvas sails.

Summing Up

1. Explain how a sailboat can sail into the wind.
2. Sketch your sailboard and label the keel, bow, stern, mast, and sail.
3. Explain Bernoulli's principle.
4. What function does the keel serve?

ACTIVITIES

DESIGNING A PASSENGER RESTRAINT SYSTEM

Problem Situation

The popularity of the automobile has brought with it some undesirable effects. Among these are air pollution, traffic jams, and traffic fatalities. Each year, thousands of lives are lost because of automobile accidents.

Experts have concluded that the use of passenger restraints is the one factor that could reduce automobile fatalities dramatically. Seat belts, shoulder harnesses, and airbags keep passengers from being thrown from the vehicle or bounced around the vehicle's interior or otherwise injured. The value of a well-designed restraint system has been demonstrated many times in the professional auto racing field. Race-car drivers have survived crashes at speeds of nearly two hundred miles per hour.

Design Brief

Design and construct a passenger restraint system that will keep an egg from being damaged while traveling in a model car down a ramp and into a concrete block. The ramp will be placed at a 30° angle. All students will begin with the same model vehicle.

Procedure

1. Using the safety procedures described by your teacher, construct the model transport vehicle as shown in the drawing.
2. Brainstorm possible restraints as well as shock-absorbing additions to the vehicle. Keep in mind that the goal is to protect the egg from damage upon collision.
3. Apply your solutions to the vehicle by constructing your restraining devices.
4. Test your vehicle with the wooden test egg. Observe the reaction of the egg to the collision. You may wish to videotape the collision and play it back in slow motion. Make changes if needed.
5. Replace the test egg with a raw egg and test your system.

Classroom Connections

1. Outputs of technological systems can be desired or undesired, expected or unexpected.

Suggested Resources

2" × 4" stock for vehicle, ¼" masonite for wheels, ¼" dowel for axle
Show cardboard
Styrofoam
Rubber bands
12' ramp
String
Assorted fasteners
Wire
Balloons
Raw eggs
Wooden practice egg
Springs, material processing tools and machines

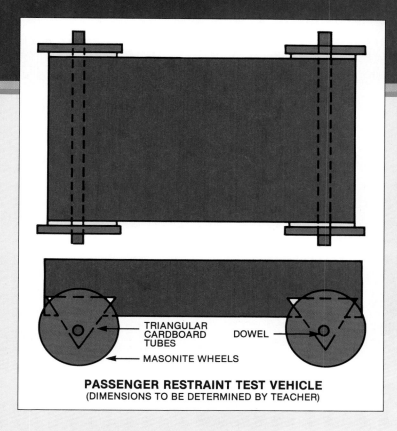

PASSENGER RESTRAINT TEST VEHICLE
(DIMENSIONS TO BE DETERMINED BY TEACHER)

Labels in diagram: TRIANGULAR CARDBOARD TUBES, DOWEL, MASONITE WHEELS

2. Technology produces many positive outputs and solves many problems. Sometimes, however, negative outputs create new problems.
3. Newton's laws of physics explain how a body reacts in a collision.
4. Alcoholism, especially among teenagers, has become a serious societal problem and is responsible for many automobile fatalities.

Technology in the Real World

1. Some automobile manufacturers now include an air bag as standard safety equipment in their vehicles. The air bag expands instantly in a collision, absorbing the impact of the driver being thrust forward upon collision. The air bag becomes a shield between the windshield and the driver.

Summing Up

1. What causes whiplash in an automobile accident?
2. List at least three safety devices found in the average vehicle.
3. What techniques and building concepts can be used to absorb the shock of a collision?
4. Are there any disadvantages to wearing seat belts and harnesses?

BIOTECHNICAL SYSTEMS

MAJOR CONCEPTS

After reading this chapter, you will know that:

- Biotechnology involves the use of living organisms to make commercial products.
- Fermentation is a biological process in which live organisms convert materials and produce by-products like alcohol and carbon dioxide.
- Modern biotechnology includes genetic engineering, bioprocessing, and antibody production.
- The most important resources in biotechnology are information (knowledge) and specialized tools and materials.
- Biotechnical systems such as agriculture, food production, and medical technology directly improve the quality of life.
- Agriculture refers to technological systems that produce plants and animals and process them into food, clothing, and other products.
- Agriculture has gone through three eras in the twentieth century: the era of mechanization, the chemical era, and the era of biotechnology.
- Modern medical technology cures disease, provides replacement parts for the human body, and improves medical diagnoses.

INTRODUCTION

Biotechnical systems are technologies that involve living things. People have used these systems since prehistoric times. For example, Stone Age people raised reindeer, goats, pigs, cows, and sheep. These animals were used both for food and for materials to make shelters, clothing, and tools.

Today we know more than ever before about living organisms. We are using that knowledge in farming, food production, and medicine.

Biotechnology involves the use of living organisms to make commercial products.

Problem Situation 1: While growing flowering plants, you find a flower with a color you like a lot. You are sure it will be a best seller on Mother's Day, which is only a few weeks away.

Design Brief 1: Using the techniques of asexual plant reproduction, design a system to increase the number of plants that look exactly like the original flower.

Problem Situation 2: In hydroponic gardening, the plants absorb the nutrients in a liquid state without having to search for them in the soil.

Design Brief 2: Design and construct a hydroponic greenhouse which can grow tomatoes without soil.

Problem Situation 3: A gardener's seed tray measures approximately 10″ × 14″ and is about 2″ deep. Seedlings are normally planted in the tray in 5 rows each containing 8 plants.

Design Brief 3: Design and construct a device which will enable the gardener to make at least one row of holes at a time, thus speeding up the transplanting process.

BIOTECHNOLOGY

The word **biotechnology** combines the words **biology** (the study of living things) and **technology**. Biotechnology is using technology to process living things into products or new forms of life.

Biotechnology From the Past

Fermentation is a biological process in which live organisms convert materials and produce by-products like alcohol and carbon dioxide.

Yogurt, cheese, beer, wine, and bread have been made for thousands of years using biotechnology. In the cheese-making process, living bacteria make milk sour and turn it into cheese. In making yogurt, milk is first heated, then bacteria are added. They make acids, which give the food its taste and thick, smooth texture.

Beer, wine, and bread are made through **fermentation**. In fermentation, living organisms feed on sugar and starch, breaking them down and producing alcohol and carbon dioxide gas.

In beer making, both alcohol and carbon dioxide are made by living cells called yeasts. Yeast cells feed on grain. The result is an alcoholic drink with a foamy head. Wine is made in much the same way. Yeast grows in fruit juice, turning it to wine.

CONNECTIONS: Technology and Science

■ Living things are composed of cells. The human body contains trillions of cells. Cells are the smallest units that can carry on activities of life. Cells vary in size, shape, and structure. Every living cell is a highly complex structural and chemical system.

■ Bacteria are small one-celled organisms found everywhere. They carry on activities that may be helpful and harmful to human beings. Some beneficial bacteria are

1. Effective at decomposing a variety of substances into simpler materials that can then be used again by plants.
2. Used in sewage treatment plants to destroy sewage.
3. Used to produce useful chemicals such as alcohols, antibiotics, and vitamins.
4. Used in food production (cheese, yogurt, and sauerkraut).

Some harmful bacteria:

1. Cause food spoilage.
2. Cause various diseases such as pneumonia, diphtheria, typhoid fever, and tuberculosis.

Grapes for fine wines are grown in vineyards such as this one in southern France. (Photo by Michael Hacker)

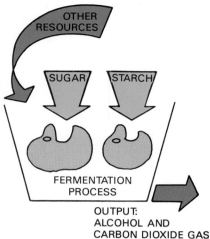
The fermentation process

Alcohol kills yeast. When the alcohol reaches a certain level in the wine, the yeast dies.

In bread making, yeast cells are added to the bread mixture. The yeast cells digest the sugar and starch in the mixture. They produce carbon dioxide and alcohol. The carbon dioxide makes the dough rise. The alcohol adds to the bread's flavor.

In each case, a biological process (fermentation) takes place in which living things (yeast cells) act on other materials. The yeast cells get energy from these materials. They reproduce and grow in number during fermentation.

Farmers have used biotechnology for centuries. Early farmers found they could get better crops if they saved seeds from the best plants to use at planting time. The valuable traits of the best plants were passed on in the seed. The passing of traits from one generation to the next is called **heredity**.

Crossbreeding is another form of biotechnology that has been used for a long time. Crossbreeding combines the traits of one plant with those of another. For example, one kind of wheat may have large seed heads but weak stems. Another kind may have strong stems but small seed heads. By crossing the plants, a new kind of wheat with strong stems and large seed heads might be produced. Crossbred plants or animals are called **hybrids**. They have the traits of both parents.

Luther Burbank used crossbreeding with plants in the late 1800s. He produced over 800 new kinds of flowers, plants, fruits, and vegetables. He was the developer of the Idaho potato.

George Washington Carver also made use of crossbreeding. He crossbred hundreds of plants to develop new and improved varieties. He also developed many new uses for the peanut, including coffee, ink, and soap.

Animal breeding was another early form of biotechnology. People chose animals with certain valuable traits to breed. Today, many animals are selectively bred. They include dogs, cattle, and racehorses.

George Washington Carver at work. (U.S.D.A. photograph)

Modern Biotechnology

Modern biotechnology includes genetic engineering, bioprocessing, and antibody production.

Modern biotechnology blends scientific knowledge with engineering techniques. There are three different areas of modern biotechnology. In **genetic engineering**, new forms of life are created. In **bioprocessing**, tiny living cells called **microorganisms** are used to process materials. In **antibody production**, substances are produced that may cure disease.

Genetic Engineering

Our genes determine how our bodies look and work. Our genes make us short or tall, light skinned or dark skinned, blue-eyed or brown-eyed. Having (or missing) certain genes results in genetic diseases that can make people sick or even kill them.

In genetic engineering, a gene is taken from one cell and **spliced** (moved) into another cell. The second cell has traits carried by the spliced gene. Someday it may be possible to cure genetic diseases by gene splicing.

Gene splicing can also be used with plants. For example, a gene for a substance that kills pests was put into tobacco and cotton plants. It is hoped that these new kinds of plants will repel insects. Farmers will not have to use so many dangerous and costly chemicals.

In one experiment, genes are being spliced into bacteria. Normal bacteria cause crystals of ice to form on the leaves of crops during frosts, killing them. Gene-spliced bacteria prevent this from happening. It is thought that this new kind of bacteria will save crops during cold spells.

Bioprocessing

Bioprocessing is a modern version of the ancient fermentation process. In bioprocessing, living cells do useful work. They break down materials, changing them into other materials. For example, garbage is broken down through bioprocessing. Microorganisms feed on plant and animal waste, giving off a gas called methane. This gas is much the same as the gas we use for heating and cooking. Bioprocessing can turn even garbage into a useful energy source.

Bioprocessing can be less costly than chemical processing. Living cells can produce useful materials through fermentation processes. This is carried out in huge vats, called **fermentors**.

One new bioprocess is using microorganisms to recover oil from old oil wells. When oil wells are first drilled, underground gases push oil to the surface. When the gases are gone, there is

These seeds are being coated with genetically engineered bacteria. The treatment is designed to improve the yield of the plant. (Courtesy of Cetus Corporation. Photo by Chuck O'Rear)

DNA: The Basis of Life

All living things are made up of cells. Inside each cell are threads called chromosomes. **Chromosomes** are made up of long molecules called **DNA**.

A DNA molecule has two connected spiral strands. The strands are twisted, forming a double helix. DNA strands are made up of sugar and a salt called phosphate. They are connected to each other by groups of three kinds of nitrogen compounds. These groups of three are called **codons**.

Groups of codons are genes. They combine in different ways to give organisms different traits. Genes determine whether a life form is a plant or animal. They determine what it will look like and how it will function. All life forms are based on the double-helix structure of DNA.

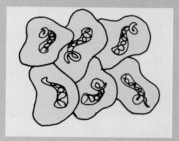

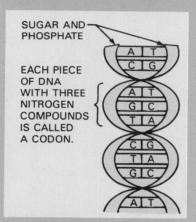

SUGAR AND PHOSPHATE

EACH PIECE OF DNA WITH THREE NITROGEN COMPOUNDS IS CALLED A CODON.

Designer Genes: High-Tech Style

Gene splicing may help in fighting cancer. Cancer occurs when cells in the body multiply rapidly and out of control. A substance called **interferon** kills cancer cells without hurting healthy cells. Interferon is made in the body. It is part of our natural defense against disease.

Right now, most interferon comes from human blood cells. But it takes a lot of blood to make even a little interferon. Therefore, the cost of treating one patient can be as much as $30,000.

The gene that makes interferon has been found. It has been spliced into a kind of bacteria, *E. coli*, that live in the human body. These bacteria divide every twenty minutes. After a short time, billions of cells have produced large amounts of interferon. This reduces the cost of treatment with this life-saving substance.

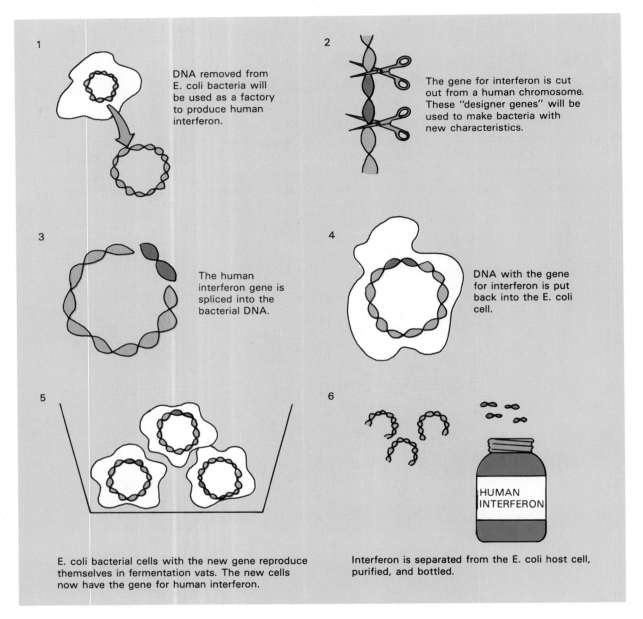

1 DNA removed from E. coli bacteria will be used as a factory to produce human interferon.

2 The gene for interferon is cut out from a human chromosome. These "designer genes" will be used to make bacteria with new characteristics.

3 The human interferon gene is spliced into the bacterial DNA.

4 DNA with the gene for interferon is put back into the E. coli cell.

5 E. coli bacterial cells with the new gene reproduce themselves in fermentation vats. The new cells now have the gene for human interferon.

6 HUMAN INTERFERON

Interferon is separated from the E. coli host cell, purified, and bottled.

no way to bring the oil up. Microorganisms can help do the job. Pumped into the oil-carrying rock along with sugar and water, they make gases that force the remaining oil to the surface. In the United States, about 300 billion barrels of oil cannot be recovered from old oilfields. Bioprocessing could help bring some of this oil to the surface so it can be used.

Bioprocessing is also part of sewage treatment. Sewage contains human and factory wastes. Treatment plants use microorganisms to break down the solid matter in sewage. They help clean the water so that it can be recycled.

Antibody Production

Antibodies are proteins in the blood that fight disease. They seek out invading disease-causing cells and kill them. Each antibody kills only certain kinds of invader cells (called **antigens**). With biotechnology, we can produce antibodies in huge amounts. Someday, we may be able to use antibodies to fight or even prevent diseases like cancer and AIDS.

To make antibodies this way, an antibody-producing cell is combined with a cell that multiplies quickly. The cell that results makes antibodies every time it divides. In a short time, billions of antibodies can be produced.

RESOURCES FOR BIOTECHNOLOGY

Each of the seven technological resources is used in biotechnical systems. People, information, tools, and materials, however, are most important.

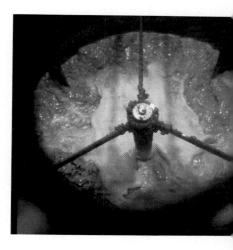

This mixer contains a fermentation "soup" of microorganisms. The microorganisms are being reproduced in great quantity.
(Courtesy of National Cancer Institute)

People

Scientists study living things to learn more about them and how they may be used. These people must be educated in the fields of biology and chemistry. **Engineers** design the tools needed to turn research ideas into large-scale production. **Technicians** run equipment and help the scientists and engineers in their work. **Nontechnical workers** such as managers, salespeople, lawyers, and financial experts are also needed.

Information

Information is the most important of all the resources in biotechnology. New discoveries are being made all the time.

People with scientific and engineering skills are hard at work developing new uses for biotechnology.
(Courtesy of Cetus Corporation)

Workers must keep up with them. They must also bring to their work much knowledge about biology and chemistry.

Materials

Many of the materials used in biotechnology are living cells or come from them. They can be human, plant, or animal cells. Cells can be used to make new life forms or process other materials into useful products.

Tools and Machines

The most important resources in biotechnology are information (knowledge) and specialized tools and materials.

Living things are often used as tools in processing materials. For example, a special type of substance called a **restriction enzyme** acts like a pair of scissors. This cell is used to cut a gene from a piece of DNA. The gene is then spliced into another piece of DNA. Another substance called a **ligase** is used to stick pieces of DNA together.

Other biotechnical tools include laboratory equipment, computers, and other electronic tools. These tools are used to do research and to monitor biotechnical processes.

Energy

Energy is part of bioprocessing. Living organisms get the energy they need to survive from the materials they feed on. Yeast cells, for example, use starch to get energy when they are used in bread making.

Capital

In biotechnology, capital is used to buy equipment and build laboratories. Some capital comes from investors. These people

At work in a biotechnology laboratory. (Courtesy of Cetus Corporation)

hope to make a profit when the company comes up with a new product or process. Capital also comes from government. Government wants to encourage new ideas that will improve people's health and living standards.

Time

In most factories, products can be made in minutes or hours. Computers process data in less than a second. In living things, the process is carried out while the organism is alive. This time period is called the **life cycle** of the cell. Some research can take a great deal of time. People have been searching for a cure for cancer for many years.

IMPACTS OF BIOTECHNOLOGY

Biotechnology is useful in many ways. We can use it to make large amounts of substances needed for human health. We can get rid of waste materials. We can change plants and animals to make them more useful. We can recover oil from deep underground that would otherwise be lost.

Biotechnology can be used to change life forms. Genes of farm animals are being altered so that the animals produce more meat. Plants are being altered to increase the nutritional value of their fruit and seeds. Bacteria that grow on plant roots are being changed so the plants can take nitrogen from the air. The nitrogen then acts as a **fertilizer** for the plant. Experimenters have even placed a gene from a rat into a mouse egg. The gene was one that directs rat growth. The result was a giant mouse.

Genetic engineering lets us create new forms of life. Is this a good idea, or a bad one? When scientists change bacteria to help plants grow, most people think it's good. But is it? What happens when we put a new kind of bacteria in the environment? Could it harm plants or trees? Could it cause a new disease, one that we couldn't cure?

What about genetic engineering of people? In time, we may be able to create new kinds of people. Some people even think that we should try to improve the human race by changing our genes. Others think that this should never happen. In the end, our values will decide how we use genetic engineering.

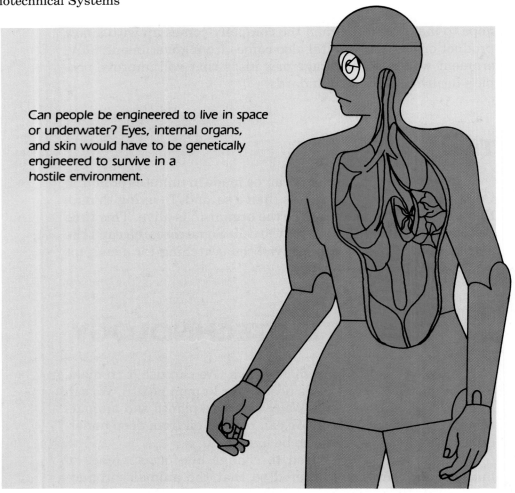

Can people be engineered to live in space or underwater? Eyes, internal organs, and skin would have to be genetically engineered to survive in a hostile environment.

CONNECTIONS: Technology and Social Studies

- Food production has been increased through the use of chemicals and pesticides. What impact do these items have on the health of the people who use them?
- Advances in medical technology have improved the health of many people in our society. But, at the same time, they have raised many difficult questions. For example, who should have access to these modern "miracles"? Since many are very costly, who should pay for them?
- People have become more health-conscious as scientists and doctors have been able to determine how certain foods and behaviors impact a person's physical well-being. This has led to the establishment of a multi-million dollar industry to meet these consumer demands.

OTHER BIOTECHNICAL SYSTEMS

Biotechnology involves living cells. Other technologies such as agriculture, food production, and medicine also involve living things.

Biotechnical systems such as agriculture, food production, and medical technology directly improve the quality of life.

AGRICULTURE

Agriculture refers to technological systems that produce plants and animals and process them into food, clothing, and other products.

People have used living things to satisfy their needs for as long as they have existed. At first, people traveled from place to place. They hunted animals and gathered plant materials for food. Around 8000 B.C., this began to change. People developed ways to plant crops and raise animals. Living in one place, they could produce their own food.

From these early beginnings until the late 1700s, little changed. Human and animal muscles provided the energy needed for farming. Tools were mostly hand tools. Some of these tools, like the digging stick and the sickle, required farmers to bend over while they worked.

The Industrial Revolution brought changes. New mechanical devices helped farmers do their work. Plows were made from cast iron and steel. A horse-drawn reaper (a machine that cuts wheat) was invented in 1831 by Cyrus H. McCormick.

Agriculture refers to technological systems that produce plants and animals and process them into food, clothing, and other products.

During the time of the Civil War, an average farm worker could produce enough food for a family of four. By the middle 1940s, an average farm worker could produce enough food for thirteen people. Today, the typical U.S. farmer can supply fifty people with food for one year. (Courtesy of State Historical Society of Wisconsin)

The green revolution improved food production all over the world. (Photo by Michael Hacker)

Chemicals are used to control pests in this soybean crop. (Courtesy of E.I. DuPont de Nemours & Company, Inc.)

In the twentieth century, farming has moved through three technological eras. In the **era of mechanization**, from 1900 to 1950, farmers changed from animal power to engine power. In the **chemical era**, from 1950 to today, farmers used chemicals to kill pests and make plants and animals grow better. Far more food could be produced. This era is sometimes called the **green revolution**. The **era of biotechnology** is the third era through which farming has moved in this century.

The green revolution was brought about both by the use of chemicals and by crossbreeding. Better kinds of wheat, rice, and corn have been created. These hybrids yield more grain per acre than older strains of plants. Today, farmers raise five times the amount of corn per acre that they did fifty years ago. Dwarf varieties of fruit trees make caring for orchards and harvesting fruit easier and cheaper.

Animal crossbreeding has also had an effect. In 1900, there were 20 million cows in the United States. Today there are only 11 million, but they produce twice as much milk.

Agriculture has gone through three eras in the twentieth century: the era of mechanization, the chemical era, and the era of biotechnology.

The Mechanization of Agriculture

It is said that the plow is responsible for the first great change in the way humans lived. People stopped wandering and could stay in one place and grow food to feed themselves.

Early plows were simple devices, pulled by people or animals. The earliest plows, first used about 5000 B.C., were wooden forked sticks that cut grooves into the ground. In Egypt in 3000 B.C., a person could farm only about half an acre using the simple tools of the time. Today, the average U.S. farmer farms over 400 acres.

Modern plows are called **cultivators**. They use steel blades

How Plants Grow

A plant uses sunlight to change carbon dioxide and water into sugar and oxygen. The sugar is used by the plant as food. The oxygen goes into the atmosphere and becomes part of the air we breathe. This process is called **photosynthesis.**

Plants get some of their nutrients from the soil. Water from the soil brings needed minerals to cells throughout the plant. The pressure of water on the cell walls keeps the plant standing upright. Without water pressure, a plant wilts.

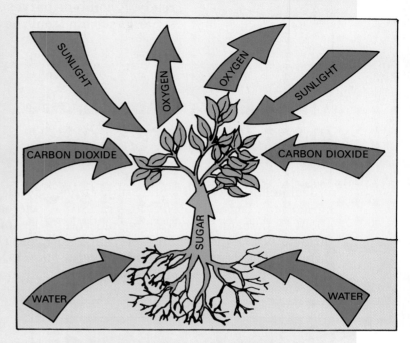

During photosynthesis, plants take in sunlight, carbon dioxide, and water. The plants produce oxygen and sugar.

Each of these 19 combines cuts a 30-foot path as it moves through the fields. In one hour, these machines harvested enough grain to fatten nearly 1,000 beef cattle to market size. (Courtesy of Deere & Company)

to break up the soil. They are pulled by tractors. Today, planting is carried out by machines. Machines loosen the earth, drop in the seeds, and spread soil over the seeds, all in one operation.

Harvesting is the process of gathering crops when they are ready. A combine harvestor is a machine that cuts the crop and separates grain from the straw.

Three Eras of Agriculture in the Twentieth Century

Technology has had a huge impact on agriculture. In the United States in 1820, about 70 percent of working people were farmers. Today, less than 3 percent are farmers. They not only provide food for the U.S. population but also produce food for export. This food-producing increase is due to changes during the **era of mechanization**, the **chemical era**, and the **era of biotechnology**.

When the mechanization era began around 1900, there were 6 million U.S. farms. Fifty years later, there were only 5.4 million. Farmers could farm more land with machinery. But not all farmers could pay for it. Smaller farms began to disappear.

During the chemical era, farmers began to produce more food per acre. They used chemicals to control pests. Some bought spraying equipment to do a better job. Others hired airplanes to dust their crops. Again, farmers who couldn't afford these improvements could not stay in business. Today there are less than 2 million farms in the United States.

The biotechnology era will change farming, too. By the year 2000, only about 1 million farms will be left. They will be large farms that use the most modern farming methods. Just 50,000 U.S. farms will produce 75 percent of the food supply.

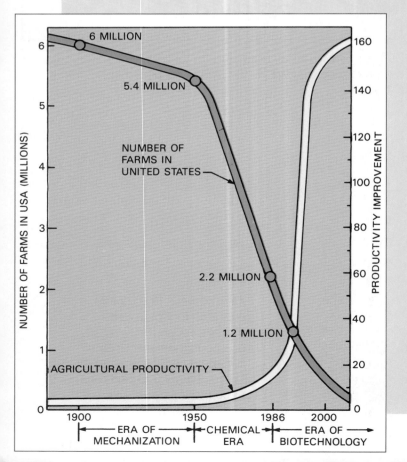

Center-pivot sprinklers.
(Photos courtesy of NASA)

Irrigation

Irrigation is supplying water to the land where crops are grown. In ancient times, people carried buckets of water to the land. Later they dug channels to bring it from rivers and lakes to their crops.

In some parts of the world, water is scarce. Irrigation methods must conserve water. In Israel, for example, **drip irrigation** is used. Each plant receives its own trickle of water through plastic tubing at root level. No water is wasted.

Center-pivot irrigation is used to supply water to large land areas. Water is sprayed through pipes that rest on wheels and travel around a center point. Large systems can be a quarter-mile long. They can water over 100 acres of land.

CONNECTIONS: Technology and Mathematics

■ Efficiency in packaging has not equalled efficiency in production because aesthetics, convenience, and safety have all become factors in how well a product sells.

For example: A rectangular box of raisins with dimensions 14cm by 10cm by 4cm has a volume of 560cm³. Its surface area is 472cm².

A cylindrical container holding 569cm³ has a surface area of only 398cm². Although the cylindrical shape uses less packaging, most raisin boxes are still rectangular.

Finding environmentally safe and recyclable packaging materials is a problem that must be solved with new technologies.

Over 250 tons of hydroponically produced cucumbers are harvested each year at the Environmental Research Laboratory of the University of Arizona. These plants grow in pure sand. (Courtesy of Environmental Research Laboratory, University of Arizona)

Jarred tissue culture. (Courtesy of United States Department of Agriculture)

These vegetables were grown in a controlled-environment greenhouse. (Courtesy of Environmental Research Laboratory, University of Arizona)

High-Tech Agriculture

Tissue culture is one way to produce new and better plants. A piece of leaf tissue is grown in a test tube. The millions of new cells are exposed to many different conditions, like heat or salty soil. The cells that survive are used to grow new plants.

The new plants have the same traits that let the parent cell survive the harsh conditions. A new life form with useful traits has been produced. This process of producing many identical organisms from one is called **cloning**.

The word **hydroponics** comes from Greek words that mean water and farming. Hydroponic farming is done without soil. Plants grow in water or in sand or gravel. Needed nutrients are supplied in water to the plants' roots. Hydroponic farming is done in greenhouses where weeds and pests can be eliminated. Air temperature and humidity can be kept at the best levels for growth.

FOOD PRODUCTION

People must eat to live. We eat a variety of foods prepared in many different ways because for most of us, eating is a pleasure.

What Is Food?

The human body needs six kinds of substances, called **nutrients**. These six are carbohydrates, fats, minerals, protein, vitamins, and water.

The way a food smells and tastes affects our enjoyment of it. It must also be preserved to keep it from spoiling. Food can be processed to make it smell and taste good, and preserved to keep it safe for eating.

Food Processing

Early people ate their food raw. Fruits, nuts, leaves, roots, animal flesh—all were eaten raw. About a million years ago, people learned how to build and tend fires. They began to cook their food. Cooking was one of the first ways food was processed.

Cooking makes food softer and easier to digest. The heat of cooking kills microorganisms that might cause food to spoil.

Most modern cooking appliances use heat from electric elements or gas flames. Microwave ovens, however, cook with radio waves. They send radio waves through the food, causing particles of water in it to vibrate and heat up.

Some foods are sold ready to eat. Others are partly prepared. Cake mixes, TV dinners, fish sticks, frozen pizza, and chicken pot pies are examples. They all must be cooked before they can be eaten.

Flour is processed industrially by a process called milling. **Milling** is grinding grain into fine particles. Then the particles are sifted. Only the finest are packaged for sale.

Shredded wheat is made by steaming grains of wheat. After the grain cools, it is passed through rollers. The rollers shred the grains. Blocks of shredded wheat are cut into small cakes

Chickens are processed in poultry plants. The live chickens are knocked out by an electric charge. They are then killed, cleaned, packaged, and sent to the store. (Courtesy of NASA)

Cooking causes the production or release of chemical substances in the food. These chemicals affect the food's flavor. (Courtesy of Occidental Petroleum Corporation)

Microwave cooking is very fast. The heat does not have to be conducted from the outside layer of the food to the innermost layers. In microwave cooking, food is cooked from the inside out. (Courtesy of Sears Roebuck and Co.)

for breakfast cereal. Cornflakes are made from grain that is flattened by rollers and then toasted.

Fruits and vegetables must be peeled before processing. Peel may give a bitter taste to cooked foods. Steam is often used to peel foods. When applied to the peel, steam breaks down the layer of tissue just below the peel. The peel comes off easily.

Fruit is sometimes crushed to make fruit juice. The water may be removed, leaving a concentrate. Juice concentrate is cheaper than fruit juice to transport from the factory to stores. The buyer need only add water to the concentrate to make the juice.

Some processed foods contain additives. **Food additives** are chemicals that improve the food or keep it from spoiling. These include preservatives, flavorings, and dyes. People have become cautious about buying foods with additives because some have proved to be harmful to people's health. Some people buy **natural foods**, which contain no additives.

Long ago, substances like sawdust were added to some foods. This increased the weight of the package and gave more profits to the food producer. Today the United States Food and Drug Administration monitors the content of additives in foods. A food package must list the additives in it and the amount of each.

Food Preservation

Most food spoils after a while. Bacteria, mold, and insects feed on it. Ways have been found to preserve food that prevent spoilage. Some of these methods have been in use for thousands of years.

Preserving Food with Heat

Some bacteria are useful. They can help preserve food, as in the case of buttermilk, cheese, and yogurt. Most, however, cause food to spoil. Bacteria must be killed to preserve food. Heat is one way to kill them. When food is heated to a high enough temperature, bacteria die.

In the 1850s, the French scientist Louis Pasteur discovered that heating milk for about half an hour could destroy harmful bacteria. The process was called **pasteurization**.

Today, all milk is pasteurized. First, the milk is heated to 160°F for 20 seconds. It is then quickly cooled to about 38°F. Pasteurized milk will keep for several days if refrigerated. After milk is pasteurized, it is pumped under pressure through a tube with a tiny hole in it. This mixes the milk and cream thoroughly. The process is called **homogenization**.

Canned foods are convenient, can be stored easily, and will last for a long time.
(Courtesy of Campbell Soup Company)

Another way to heat-treat milk is **UHT** (ultra heat treatment). In UHT, milk is heated to about 175°F. Then steam is used to heat it to 300°F. That temperature is held for a few seconds until all bacteria are killed. The milk is then cooled, homogenized, and poured into containers.

An unopened container of UHT milk can sit unrefrigerated for months without spoiling. People in Europe like and use it. But Americans have not yet accepted the idea that milk can be kept without refrigeration. This shows how people's attitudes can affect the growth of a new technology.

Canning is the most common method of preserving food. Food is steamed for a few minutes to kill bacteria. Then it is put into cans made of steel with a tin coating. Some items are packed in water or salt water. A few fruits, like pineapple, pears, and apricots, cause the inside of the can to corrode. The insides of cans used for these fruits are coated with lacquer to prevent corrosion.

When the cans are filled, any remaining air is removed. The lids are put in place. The cans are then put into an oven called an **autoclave**. The autoclave heats the can and destroys any remaining bacteria in it.

Preserving Food with Cold

Cold makes bacteria stop growing. People used to bury food in ice or snow to keep it fresh. In 1928, Clarence Birdseye discovered that meat frozen by Eskimos tasted better than refrigerated meat. He started the Birdseye food company, and the frozen food industry was born.

Below 14°F, bacteria stop growing but they are not killed. Before freezing, food is steamed to destroy all bacteria. After it is thawed, bacteria can grow again in food preserved by freez-

Frozen food retains much of its flavor and nutritional value. Because of the ease of processing and packaging, frozen foods are a favorite on supermarket shelves.
(Courtesy of Campbell Soup Company)

350 Chapter 12 Biotechnical Systems

ing. In time, the food will spoil.

Vegetables, fish, and poultry are frozen by dipping the packaged food into a tank of freezing salt water. Canned juices are frozen the same way. High-priced foods like shrimp are frozen by a spray of liquid nitrogen at $-320°F$. This is a costly way to freeze foods. Meat is frozen by **blast freezing**. The meat travels through a tunnel as fans blow air at $-40°F$ on it.

Preserving Food by Drying

When foods are dried, most water is removed. The concentration of salt and sugar increases. This salt and sugar concentration kills any microorganisms. Early people dried fish, meat, and fruit under the sun. Today, many foods are dried to preserve them. This process is called **dehydration**. Powdered milk, soups, potatoes, orange juice, and many other foods are dehydrated. To serve them, you just mix them with water.

Instant coffee is made by **spray drying**. Liquid coffee is sprayed into a container through which hot air blows. The hot air evaporates the water and turns the coffee to powder.

Nitrites are used to preserve luncheon meats like bacon, ham, and bologna. (Courtesy of Oscar Mayer Foods Corp.)

Preserving Food with Chemicals

For years, foods have been preserved using salt, sugar, and vinegar. Using salt or sugar as a preservative is called **curing**. The salt or sugar dissolves in the water in the food. The concentration of salt or sugar kills bacteria. When vinegar is used to preserve foods, the process is called **pickling**.

Today, new chemicals are being used as preservatives. Among them are nitrites and sulfur dioxide. Sulfur dioxide preserves fruits used to make jam, fruit juice, and dried fruits and vegetables. Such chemicals can be harmful if used in large amounts. The amount of chemical preservatives in foods is controlled by the Food and Drug Administration.

Preserving Foods by High-Tech Methods

X-ray radiation can be used to kill organisms in food. This process is called **irradiation**. It can be carried out after foods have been packaged. This is a great advantage. On the other hand, many people are not sure whether irradiation is safe. Large amounts of radiation must be used. It is known that radiation can have harmful effects on people.

Ultrasonic vibration is an experimental method of preserving food. This process makes the molecules of food vibrate. The vibration shakes the bacteria so hard that they die.

MEDICAL TECHNOLOGY

Thousands of years ago, medical care was primitive. People thought that illness and injury came from the gods, and that the gods would heal them. The Greeks and Romans based many of their medical cures on the healing properties of plants. Most people died of injuries or disease at an early age. If people got well, it was without much help from medical technology.

Early people wondered about the human body. The Egyptians thought that all blood vessels started at the heart. The Greeks thought the heart was the center of the soul and intelligence. In the seventeenth century, British doctor William Harvey discovered how blood flowed through the body.

In the centuries that followed, much was learned. Today, smallpox is gone. In most of the world, plagues and epidemics are rare. We know that microorganisms cause disease. We have learned about cell structure. Surgery can be performed on almost every part of the body. Diseased body parts can be repaired or replaced. Powerful drugs kill microorganisms and relieve pain.

Modern medical technology cures disease, provides replacement parts for the human body, and improves medical diagnoses.

New Medical Tools

Many new tools are helping in medical care. Instead of using scalpels (small, sharp knives) in operations, doctors are using **lasers**. A laser beam can be used to remove tissue and weld broken blood vessels. Lasers can also be used to unclog arteries that are blocked by fatty deposits. A few years ago, a person might have had to have a leg amputated because of poor blood circulation. Now, the circulation can often be restored through laser surgery.

Electronic tools have also greatly improved medical care. Heart and lung machines keep people alive during surgery. The heart and brain can be monitored by electronic devices. Doctors use fiber optic tools and cameras to see inside the body without cutting it open. Battery-operated pacemakers can be placed under the skin to control the heartbeat. Computers can be used in diagnosing disease.

Today, much research is being done on heart disease and cancer. The cure rate for these diseases is rising. The U.S. Department of Health and Human Services reports that more children with cancer are getting well. Less than 10 percent survived in the 1960s. Over 50 percent are surviving in the 1980s.

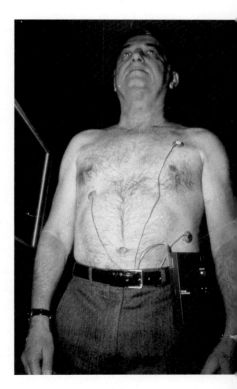

A portable heart monitor.
(Courtesy of NASA)

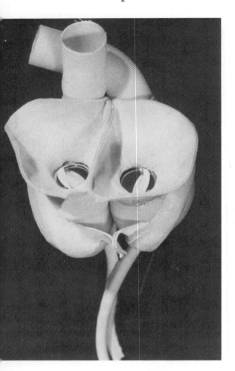

An artificial heart. (Courtesy of Texas Heart Institute, Houston, Texas)

Spare Parts for the Body

Around 500 B.C. a Persian soldier escaped from stocks by cutting off his foot. He replaced it with one made of wood. This is the first known use of a replacement for a human body part.

Today, many body parts can be replaced by natural parts. Organs like the heart and kidneys can be transplanted. Each day in the United States, an average of one heart, 20 kidneys, and 65 corneas (the clear covering of the eyes) are transplanted. More and more people are donating body parts when they die.

Much of the human body can be recycled. Hearts, kidneys, livers, bones, and lungs can be transplanted. Corneas can be transplanted to restore sight. Skin can be transplanted to the bodies of people who have been badly burned. Blood from one person can be used to replace blood lost by another. Even the lining of the brain can be reused.

For a long time, transplants were hard to do. People's bodies reject tissues taken from other people. Their immune systems react as they would to bacteria. New drugs suppress the immune reaction. More transplants are successful.

Some body parts can be made artificially. The technology of making body parts is called **bionics**. Hands, legs, bones, joints, teeth, and hearts can all be made artificially. Materials can be used to make body parts that do not cause an immune reaction. They include plastic, and alloys of titanium and cobalt.

Some artificial body parts function almost like those they were made to replace. The **myoelectric arm** is one such device. It is connected directly to nerve endings. It can be controlled by the user's brain.

Windows Into the Body

The development of the X ray let doctors look inside the human body without having to operate. However, X-ray radiation can be harmful to the body in large amounts. Also, X-ray images are not always clear enough to be useful. Now, new technology is providing a better window into the body than

This craftsperson has myoelectric arms. (Courtesy of Daily Telegraph Colour Library)

ever before. Three new methods use the computer to show the tissues inside the body.

A **CAT scanner** uses a rotating X-ray device that takes 288 pictures as it moves around the patient. The image is a cross section of the body, shown on a computer screen. It is much clearer and more complete than an ordinary X-ray image. Also, it uses lower radiation levels. Still, CAT scanners may be harmful under certain conditions.

Nuclear magnetic resonance, or **NMR**, is another body imaging system. NMR uses powerful magnets to make atoms in the body line up. High-frequency radio waves make these atoms vibrate. The length of time it takes for the atoms to stop moving determines the kind of image that shows up on the computer screen. The image tells a great deal about body tissues.

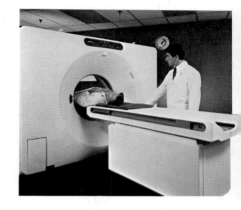

A CAT scanner in use. (Courtesy of General Electric Company)

Technology to Help People Walk

Biotelemetry is a space-age technology. It is used to monitor from the earth the heartbeat, pulse rate, and other body functions of astronauts in space. Now it has a new use. Doctors are using biotelemetry to study the problems of children who have difficulty walking. These are often children who have cerebral palsy, muscular dystrophy, or injuries.

Tiny sensors, each about the size of a half-dollar, are attached to the skin over the muscle being studied. Signals from muscle activity are sent to a computer. Doctors study this information to decide whether surgery or physical therapy will help the child.

The image of the walking pattern helps doctors decide what kind of physical therapy or surgery is needed. (Courtesy of NASA)

Ultrasound uses sound waves that are too high for humans to hear. These sound waves are passed through body tissues. The echo that is received as the sound wave bounces off body tissue is changed into electrical signals.

These signals create an image on the computer screen. Ultrasound is often used by doctors to learn about a fetus inside its mother's body.

SUMMARY

Biotechnical systems are technologies that relate to living things. Biotechnical systems include biotechnology, agriculture, food production, and medical technology.

Biotechnology includes genetic engineering, bioprocessing, and antibody production. In genetic engineering, a gene is taken from one cell and spliced into another cell. Genetic engineering is used to create improved kinds of plants and animals. Bioprocessing uses living organisms to process materials. Bread, wine, beer, and cheese are ancient examples of bioprocessed products. Modern examples include waste disposal and oil recovery. Antibody production is done by making hybrid cells that produce antibodies every time they divide.

The seven technological resources are needed for biotechnology. The most important are information and specialized tools and materials. Often the tools are living organisms that do the processing.

Biotechnology's biggest impact is in the ability to create new kinds of life. It has yet to be decided whether these techniques will be used on people.

Other biotechnical systems include agriculture, food production, and medical technology. Agriculture is a very old technology, and one that was slow to change. Until the late 1700s, human and muscle power provided the energy for agricultural tasks. In recent times, the mechanical, chemical, and biotechnical eras have sped agricultural change. Today, farmers can produce far more food than they could years ago. The number of farms is dropping, but the farms have gotten larger.

Food production is another biotechnical system. Most foods must be processed industrially or at home. We process food by cooking it. Cooking softens food and kills microorganisms that would make it spoil. Processed foods sometimes contain food additives. These are chemicals used to improve food and keep it from spoiling. Natural foods contain no food additives.

Food is sometimes preserved by using heat to kill bacteria. In pasteurization, milk is heated for a short time then cooled. Cold temperatures also preserve food by slowing down bacterial growth. Freezing food is a common way to preserve it. Food can also be preserved by drying, curing, and pickling. High-technology processes like irradiation and ultrasonic vibration also preserve food.

Early medical practices were primitive. Many cures involved the use of plant and animal parts. Modern medicine has wiped out diseases like smallpox. Progress is being made in finding cures for cancer and heart disease.

Body parts may be transplanted from one person to another. Artificial parts can be manufactured to take the place of missing body parts. Computers have made it possible to see inside the human body with devices like the CAT scanner, nuclear magnetic resonance (NMR), and ultrasound. Electronic tools like the laser have made surgery safer and better.

1. What three technologies make up the field of biotechnology?
2. What are some ancient examples of biotechnology?
3. Define fermentation.
4. What is a hybrid?
5. Name three applications of bioprocessing.
6. State two possible positive applications and one possible negative application of genetic engineering.
7. Draw a diagram of DNA.
8. Describe the process of antibody production.
9. What is a possible use of antibodies produced in the laboratory?
10. Explain how the tools used for processing materials in biotechnology differ from tools used in other technologies.
11. Describe the process of genetic engineering.
12. Why are agriculture, food production, and medical technology referred to as biologically related technologies?
13. Explain photosynthesis.
14. What are the three eras of agriculture in the twentieth century?
15. How can agricultural productivity be rising in the United States even though the number of farms is decreasing?
16. What are the six basic nutrients humans need to survive?
17. Identify four methods used to preserve food.
18. What are some of the impacts of modern medical technology on society?
19. Explain how technology can provide substitutes for body parts. Draw a picture of a human being with the body parts marked that can be replaced by artificial parts.

(Courtesy of Campbell Soup Company)

(Courtesy of United States Department of Agriculture)

CHAPTER KEY TERMS

Across

2. The process of creating new forms of life by splicing genes.
3. Using living cells to process materials.
5. Technology applied to medicine.
7. Uses cells like yeast to process materials and produce byproducts like alcohol and carbon dioxide.
9. Includes genetic engineering, bioprocessing, and antibody production.
11. A high-tech method of using X rays to sterilize foods.
12. Producing one or more identical copies of an organism from a common ancestor.
13. Has a shape like a twisted spring or a double helix.
14. A substance normally found in human blood, now produced by genetic engineering.
16. A very tiny living being.

Down

1. Producing proteins in the blood that defend against disease.
4. The technology people use to produce plants and animals and process them into food, clothing, and other products.
6. An improved species produced by crossbreeding.
8. The technology of producing artificial body parts.
10. A type of camera that moves around the body and takes X-ray pictures.
15. Nuclear magnetic resonance (abbreviation).

Agriculture
Antibody production
Bionics
Bioprocessing
Biotechnology
CAT scanner
Chromosome
Cloning
Crossbreeding
DNA
Fermentation
Food processing
Genetic engineering
Hybrid
Hydroponics
Interferon
Irradiation
Irrigation
Medical technology
Microorganism
NMR (Nuclear magnetic resonance)
Ultrasound

SEE YOUR TEACHER FOR THE CROSSTECH SHEET

ACTIVITIES

ERGONOMICS: DESIGNING WITH PEOPLE IN MIND

Problem Situation

Ergonomic design can improve products used by a particular group, such as the elderly. As we approach the twenty-first century, the number of people over age 65 will continue to increase. Improved medical care helps people remain healthier and live longer. However, as people age they undergo changes that affect their vision, strength, and mobility. Products designed to help people compensate for these changes can improve the lives of the elderly.

Design Brief

An electronics firm has decided to introduce a new line of products designed specifically for the elderly. You have been assigned to work on a team of two or three people to design a new AM/FM clock radio that will be attractive and easy for the elderly to use.

Procedure

Use the safety procedures described by your teacher.
1. Make a list of the features that your radio will include, such as controls that are easy to locate and easy to push or turn, extra-large time display, and adjustable volume for the alarm.
2. Do at least six rough sketches to develop your ideas. Select your best design and present it to your team.
3. Combine the ideas developed by your team members into a final sketch.
4. After your sketch is approved by your teacher, use the available materials to make a model of your product.
5. Use markers to add color and transfer type to identify the controls.
6. Modeling clay can be used to make control knobs. Examine the dashboard of a new automobile to get ideas for ergonomically designed controls.
7. Write a user's guide that is brief and clearly explains how to operate the radio. If possible, use word-processing software and extra-large type.
8. Show your model to an elderly person. Ask for suggestions to improve both the product and the user's guide.
9. Make changes if necessary and then present your product to the entire class.

Suggested Resources

Drawing paper
Illustration board
Foam board
Color markers
Transfer type modeling clay
White glue
Hot glue and glue gun
Computer
Word-processing software

Classroom Connections

1. The field of ergonomics applies technology to the needs of human beings.
2. Clearly written directions help communicate desired information.
3. Color and graphics can help communicate a message.
4. All products should be designed so that they are easy and safe to use.

Technology in the Real World

1. Glasses and hearing aids are products of technology that help people compensate for changes that are part of the aging process.
2. Wheelchairs, canes, and walkers help elderly people remain mobile.

Summing Up

1. Use your own words to define the term ergonomics.
2. Identify a product that could be improved to make it easier to use. Suggest several changes.
3. Explain why the AM/FM clock radio you designed will be easy for the elderly to use.

ACTIVITIES

CREATING THE BIONIC PERSON

Problem Situation

Modern technological processes and advances in medical techniques have made possible the technology of creating human body parts, known as bionics. New plastics and alloys of steel, titanium, and cobalt are being processed into artificial arms, hands, knees, hips, teeth, and many other body parts.

Many of these artificial parts are applications of the six basic simple machines that are controlled by electric circuits. The fields of microcomputers, electronics, pneumatics, and hydraulics have also had a large impact on the bionics industry.

Design Brief

Design and model an artificial body part that can be used to replace a skeletal joint, organ, or other body part. The model should be made of available materials and be an actual working part.

Procedure

1. Identify the major body joints from pictures or models of the human skeleton. Examine how they work and how they can be modeled artificially. Do the same as above for body organs.
2. Make sketches showing how the body part can be duplicated artificially and still perform the same work.
3. Select the appropriate materials to use in your model.
4. Using the safety procedures described by your teacher, process the materials into a working artificial body part.
5. Test the bionic part and make modifications if necessary.

Classroom Connections

1. Modern medical technologies cure disease, provide replacement parts for the human body, and improve medical diagnoses.
2. Bionics is the study that involves the design and manufacture of artificial body parts. The study of the human body—how it develops, works, and reacts to different conditions—is vital to the study of bionics.
3. The knowledge of simple machines and their ability to increase force and power helps in the design of artificial parts.
4. The development of new alloys and materials that can function within the body without rejection is vital to bionics.

Suggested Resources

Plastic sheets, bar stock, round stock

Sheet metal, bar stock, round stock

Balsa wood, particle board, solid wood, plywood

Plastic tubing and syringe plungers for pneumatics

Show cardboard, foam board

Assorted fasteners, adhesives

Material processing equipment

5. The manufacture of artificial body joints requires accurate measurements—sometimes to thousandths of an inch.
6. The engineer and bionic designer must express technical ideas graphically so technicians and medical people can understand how the device works and how it must be installed.

Technology in the Real World

1. Diabetes patients who depend on daily injections of insulin to maintain a proper blood sugar level may now have an insulin pump inserted into the body that will automatically release insulin into the bloodstream.
2. Dental implants screwed into the jawbone have become permanent teeth replacements.

Summing Up

1. Name five parts within the human body that can be replaced with bionic parts.
2. What properties must a material have for it to be considered in the manufacture of bionic parts?
3. Which simple machine or machines operate like a hip joint?
4. List three energy forms that can be found in the human body.

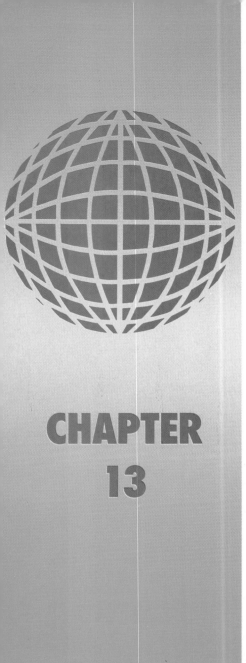

CHAPTER 13

CONTROLLING THE SYSTEM

MAJOR CONCEPTS

- Open-loop systems cannot correct for changing conditions. Their outputs usually change when surrounding conditions change.
- Closed-loop systems can correct for changing conditions to maintain the desired output.
- All feedback loops have a monitor, a comparator, and a controller.
- Sensors monitor a system's output.
- Decision makers (comparators) compare the system's output (actual result) to the system's input (desired result).
- Controllers turn a system's process on or off, or adjust it in some way.
- Within certain limits, systems give different outputs if the command input is changed.
- Computers can be used to control both open-loop and closed-loop systems.

In this chapter, you will learn how technological processes are controlled.

Control is important for three reasons. First, we want the goods, services, information, or energy we produce to be exactly right. Second, we need to be sure a technological system does not harm people or the environment. Last, we want to make sure that resources are used in the best way possible.

Open-loop systems cannot correct for changing conditions. Their outputs usually change when surrounding conditions change.

Problem Situation 1: In a trout hatchery, the young fish have to be kept in constantly flowing fresh water.

Design Brief 1: Design and construct a switch mechanism to operate a pump and partially drain the tank, which is constantly filling. In this way overflowing is prevented.

Problem Situation 2: The task of getting a vehicle to go up a hill and then descend smoothly on the other side presents a wide range of problems to the designer.

Design Brief 2: Design and construct a model vehicle which will climb to the top of a small incline about one foot high, stop, and then reverse back down the slope to its original starting place.

Problem Situation 3: Closed-loop systems are controlled by sensors that monitor such things as temperature, light, smoke, noise, heat, moisture, and pressure.

Design Brief 3: Design and construct a system using sensors that is controlled by a computer. You may use plastic modeling equipment like Lego® or Fishertechnik® if you wish.

OPEN-LOOP AND CLOSED-LOOP SYSTEMS

Many technological systems are controlled automatically by feedback. These systems adjust themselves if the conditions around them change. This is an important feature of a system with feedback.

Feedback comes in many forms. It can come from a human. Your parents may yell at you to turn down the stereo. Feedback can also come from a mechanical device. A thermostat, for example, tells the furnace that the room is too cold. These systems are called **closed-loop systems** because the feedback "closes the loop" from output to input.

Some systems are operated without feedback. These are **open-loop systems**. A model railroad is an open-loop system when no one adjusts the speed control. The train runs at a fixed speed on a level track. Uphill, it slows down. Downhill, it speeds up. The speed is not maintained at the desired level. With no feedback, the system output (the speed) changes when conditions change.

A person operating the train can provide feedback. The operator sees the train climbing the hill and changes the speed setting to increase the train's speed. When the train goes down the hill, the operator slows its speed. The operator has made the system a closed-loop system.

Closed-loop systems can correct for changing conditions to maintain the desired output.

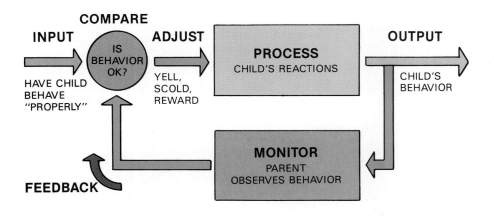

Example of a closed-loop system.

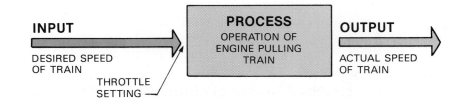

An open-loop system cannot correct for changing conditions.

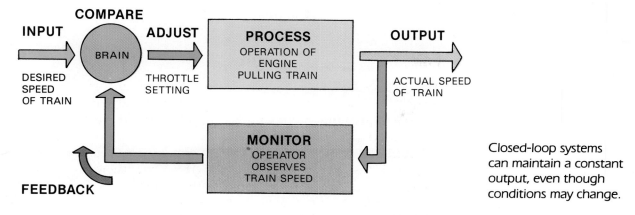

Closed-loop systems can maintain a constant output, even though conditions may change.

SYSTEM	OPEN LOOP	CLOSED LOOP
BICYCLE STEERING	Rider blindfolded	Rider watches road
CAR SPEED	Driver doesn't look at speedometer	Driver looks at speedometer on a regular basis
MODEL TRAIN SPEED	Speed control is set once and not readjusted	Operator watches speed of train and adjusts control to maintain constant speed

A comparison of open-loop and closed-loop feedback systems.

In this case, feedback comes from the operator's eye, or **monitor**. The operator's brain **compares** the actual speed to the desired speed. The operator's hand then **adjusts** the speed setting. Feedback lets a system maintain the desired output, even though conditions change.

USING FEEDBACK

A jet can fly thousands of miles almost on its own. The light, temperature, and humidity of a greenhouse can be automatically controlled. An AM radio can automatically search for a station, stop at it, and play it. Robots can produce goods in a factory with only a few people. All of these are possible because they use **feedback loops**.

The Three Elements of a Feedback Loop

Feedback loops have three necessary parts. The three parts are sometimes combined into one mechanism, but all three must

All feedback loops have a monitor, a comparator, and a controller.

Each of these uses several feedback loops. (top) A jet airliner in flight. (Courtesy of Boeing Corp.) (bottom) A tuner with scan/seek feature. (Courtesy of Panasonic) (right) An automatic guided vehicle moving through a factory. (Courtesy of Cincinnati Milacron, Inc.)

be present for the loop to work.

A **sensor** monitors the output. The sensor sends information to a **comparator** or decision maker. The comparator compares the actual output with the desired output. A **controller** uses the comparison to change the process to make the actual output closer to the desired output.

Sensors are made to monitor almost any kind of output. Sensors that you know include your eyes, a thermometer, an electric meter, and a compass. Many systems use several kinds of sensors at the same time.

Comparators compare the actual result to the desired result. They compare the output, as monitored by the sensor, to the system input or desired result. If the two are not the same, the comparator indicates that there is a difference. For example, a device called a thermostat in a home heating system decides whether the room temperature matches the setting, or command input.

Some comparators tell how different the actual and desired results are. For example, a computer in a car engine monitors gas. It tells whether enough gas is going into the engine. It can decide how much to add to reach the amount needed. Comparators can be simple. Or they can be complex devices that contain computers.

Controllers are the devices that turn the process off or on or that change it. The controller chosen depends on the job to be done. In an electrical circuit, the controller may be an on/off switch. The flow of water is often controlled by a faucet or a valve. In a camera, the light hitting the film is controlled by the shutter.

Sensors monitor a system's output.

Decision makers (comparators) compare the system's output (actual result) to the system's input (desired result).

Controllers turn a system's process on or off, or adjust it in some way.

Types of Control Loops

Some controllers can only turn a process on or off. They are called **bang-bang controllers**. A feedback loop senses whether the output is different from the desired result. The bang-bang controller then turns the process on or off. Home heating systems use bang-bang controllers. The heater is either on or off.

Another kind of controller changes the process by small amounts. This is called a **proportional controller**. If the output is far from what it should be, the controller makes a big change in the process. If the output is close to what it should be, the

How Fast Does Feedback Have to Be?

In Chapter 6, a car accelerating from a stop to 30 miles per hour was shown as a system diagram. In this example, the driver's eyes and the speedometer are the sensors. The driver's brain is the comparator. The driver's foot on the gas pedal is the controller.

Suppose a driver travels on a high-speed highway. What would happen if the driver looked at the speedometer only once an hour? The car might travel faster than the speed limit. Or it might slow way down and the driver would not know it. For good system control, the feedback must be timely. Some systems act in millionths of a second. Other systems act over the course of years.

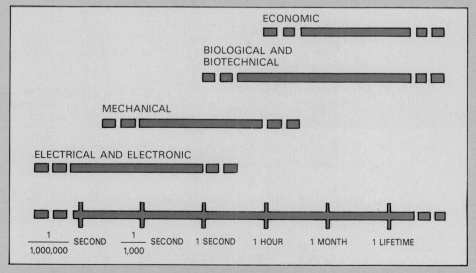

Some systems act in millionths of a second, while others act over the course of years.

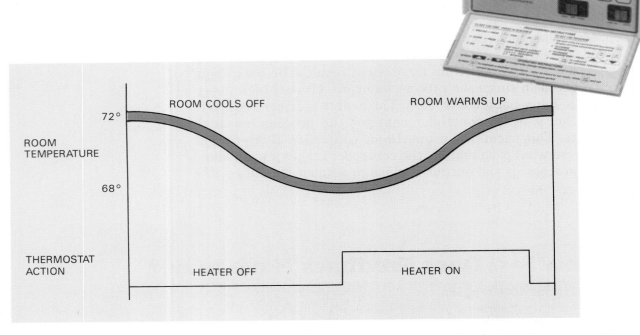

A home heating system thermostat is an example of a bang-bang controller.
(Photo courtesy of White-Rodgers Division, Emerson Electric Company)

The rider and steering on this vehicle are an example of a proportional
controller. (Photo courtesy of Envision Communications Consultants/Michael Hochanadel)

controller makes a small change. Proportional controllers give
better control, but they are more complex and often more costly.

Controllability

A system is **controllable** if we can make the results (outputs)
what we want them to be. We can use a thermometer to moni-

tor the temperature in a house. We can also use the information the thermometer provides to control the room temperature. We can use a thermometer outside, but we can't control the outside temperature. Monitoring a system doesn't always mean we can control it.

Most systems are controllable over a small range of outputs. We can keep temperature inside a home between 65 and 85 degrees with an air conditioner. We need only set the thermostat. We cannot, however, set the thermostat to lower our home temperature to 25 degrees. The lowest and highest temperatures that the air conditioner can maintain are called its **limits** of controllability. A system operator must be aware of these limits and not go beyond them.

Within certain limits, systems give different outputs if the command input is changed.

SENSORS

A sensor monitors output. It sends information to the comparator. Some sensors send this information directly. Others change it into another form first. Information is most often

An Accurate Control System

A powerful space telescope is now in orbit around the earth. Astronomers use it to observe the universe without having to look through the atmosphere. It will expand the size of our known universe by 350 times.

The space telescope must be pointed at the object being studied. It must be held very steady in position, or the image will blur. Commands sent from the earth will point the telescope using a feedback control system. The system will keep the telescope pointed at its target. The system is so accurate that if the telescope were located over New York, and it were commanded to point at a butter knife over San Francisco, it would have to know whether to point at the handle or the blade.

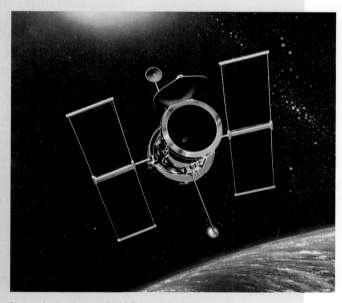

An artist's view of NASA's space telescope.
(Courtesy of NASA)

changed into electrical signals. These signals can be sent easily and quickly.

Devices that change information in one form of energy to information in another form are called **transducers**. Some change information into electrical signals. Pressure sensors can change mechanical pressure into electrical voltage. In a stereo, tape heads are transducers. So are phonograph cartridges and speakers.

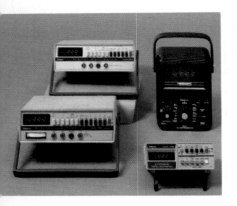

Meters give a visual indication of electric current and voltage. (Courtesy of Simpson Electric Company)

Electrical and Electronic Sensors

Sensors can be used to monitor electrical quantities such as voltage and current. Voltage and current are measured by **meters**. A meter displays the amount of current flowing through it. A needle may point to a number on a scale or the meter may have a readout like a computer display. This kind of meter must be read by people. It is used as a monitor when people are part of the feedback loop. Other kinds of meters can send electronic signals to computers. These are used in fully automatic systems.

Mechanical Sensors

Mechanical sensors measure position, force, and motion. One device that measures **position** is a float that monitors the level of a liquid.

A control system using a float was invented in Russia by I. Polzunov in 1765. The system controlled the water level in a steam engine's boiler. A float operated a valve, letting water in or shutting it off.

Today, floats are used in toilet tanks to sense when the water

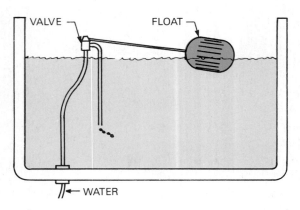

A float in this toilet tank measures the position of the surface of the water.

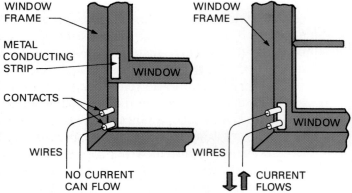

One of the simplest electrical sensors is a conducting strip that touches two contacts, interrupting or completing a circuit.

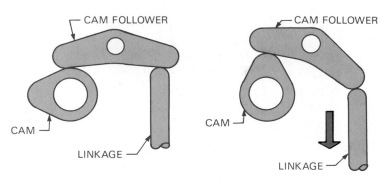

Cams are used in many different kinds of mechanical control systems.

Cams are used in many different kinds of mechanical control systems.

level is high enough. The position of the float causes a valve to close, stopping the flow of water. Floats are also used in automobile gas tanks to sense the amount of gas in the tank. As the level falls, the float causes a dashboard gauge to indicate the new level.

Another kind of position sensor is used on windows in home security systems. An electrical conducting strip is attached to the window. Electrical contacts are placed on the window frame. When the window is closed, the conducting strip completes the circuit between the two contacts. If the window is opened, the circuit is broken and an alarm sounds.

A **cam** and **cam follower** are mechanical sensors in wide use today. A cam is an irregularly shaped part mounted on a shaft. As the shaft turns, the cam turns. The cam pushes on the cam follower, which moves another device. Cams are used in car engines to control the intake and exhaust valves.

Pressure is a measure of the amount of force applied against a surface. It is the force per unit area. It can be measured in pounds per square inch. Devices that monitor pressure are called pressure sensors.

Pressure sensors are often transducers that change mechanical energy into electrical energy. The oil-pressure sensor in a car measures the oil pressure. Then it sends an electrical signal to a meter or light on the dashboard. If the pressure reading is too low, the driver adds oil or has someone check the engine.

Pressure sensors can be used to measure water level. You remember from Chapter 11 that pressure under water depends on water height. A pressure sensor on the bottom of a water tower monitors the water pressure, and therefore the water height. Water is pumped in to keep the water at the right height.

Pressure sensors are used to measure the pressure of liquids being canned or bottled. They are used to measure the pressure of fluid going through a pipeline. They are also used to control tank levels in sewage treatment plants.

A speedometer on a bicycle is a sensor that measures **motion**. A disk is attached to the front wheel of the bicycle. The disk is

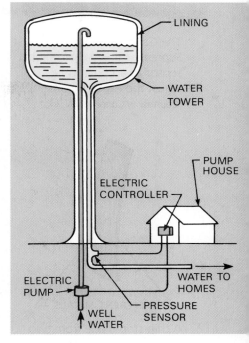

A pressure gauge can be used to sense the level in a water tank. This information can be used in a control loop to open a valve when the level is low, controlling the level of water in the tank.

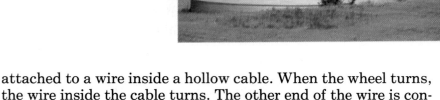

Pressure sensors are used in many different kinds of systems. (top left) A polishing and grinding line. (Courtesy of Allegheny Ludlum Steel Corp.) (top right) Trans-Canada pipeline. (Courtesy of Perini Corp.) (bottom) A wastewater treatment plant. (Photo by Jonathan Plant)

The Canon Sure Shot Supreme camera uses a light sensor to automatically adjust the exposure time. (Courtesy of Canon U.S.A., Inc.)

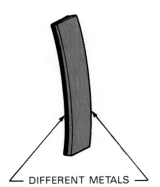

DIFFERENT METALS

A thin strip made from two different kinds of metal will bend in one direction when it is heated, and in the opposite direction when it is cooled.

attached to a wire inside a hollow cable. When the wheel turns, the wire inside the cable turns. The other end of the wire is connected to an indicator that shows how fast the wheel is turning.

Optical Sensors

The optical feedback system you know best is your eye. In bright light, your pupils are small. Only a little light enters your eye. In dim light, your pupils enlarge to let in more light. You can see this by looking in the mirror in dim light and then turning on a light.

A photocell is an optical sensor that monitors light. A photocell produces an electrical voltage when struck by light. The brighter the light, the greater the voltage. Automatic cameras use photocells to measure light and a microprocessor to adjust the camera. Some TVs use photocells to brighten or darken the picture based on the amount of room light.

Thermal Sensors

A thermal sensor is often no more than two thin strips of different metals joined together. This is a **bimetal strip**. Metal expands when heated and contracts when cooled. No two metals expand or contract at exactly the same rate. Because the two metals are joined, a bimetal strip bends as one metal expands more than the other. The amount of the bend indicates the temperature.

Another kind of thermal sensor is a **thermistor**. A thermistor is a resistor whose resistance changes with temperature. It is used to change temperature measurements to electrical signals.

372

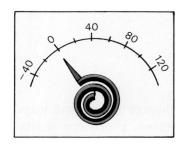

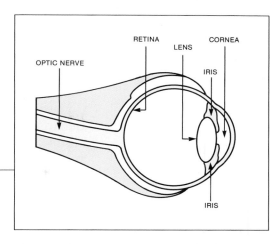

A thermometer is made if one end of a bimetallic strip is held firm and the other end is attached to a pointer.

CONNECTIONS:

Technology and Science

- The human eye has structures controlling how well we see. Light enters the eye through a clear membrane called the cornea. It then passes through the pupil, a hole in the iris. The iris controls the size of the pupil. The pupil is smaller in bright light than it is in dim light. The light then passes through the lens, which focuses the light on the retina. The lens shape changes with the distance of objects one is viewing. The retina contains nerves like the optic nerve which carry messages to the brain.

- Electromagnets are temporary magnets that can be made by wrapping a coil of wire around a soft iron core and then running a current of electricity through the wire. The strength of the electromagnet can be increased by increasing the amount of electrical energy in the wire or by increasing the amount of wire wrapped around the iron core. Electromagnets are useful because they can exert large magnetic effects. Also, they can be turned off simply by turning the electric current off.

Magnetic Sensors

If a magnet is moved past a piece of wire, a small voltage is produced in the wire. If wire is wound into a large coil, a larger voltage is produced when the magnet is moved through it. Generators work on this principle. The same idea can be used to make a transducer that changes magnetic information into electrical information.

This principle is used in recording tape and computer disk heads. It can also be used to make sensors that measure rotation. A magnet is attached to the spinning object. A wire coil is mounted in a fixed position near it. As the magnet spins, the voltage changes indicate how many times the magnet has spun past the coil.

A magnetic sensor can also be used to detect the opening of a door or window in a security system. A thin magnet is used to open or close an electrical circuit when a magnet is brought near it.

CONTROLLERS

Controllers turn a process on or off or change it in some way. Because there are many kinds of processes, there are many kinds of controllers. Many modern systems are controlled by an electrical signal. The controllers in these systems change the electrical signals into mechanical, hydraulic, pneumatic, and other forms of action.

RADAR: Monitoring Objects at a Distance

Some sensors are used to gather information about objects or people at a great distance. These sensors use radio, light, infrared (invisible red light), and other waves.

Radar (**RA**dio **D**etection **A**nd **R**anging) sends out a radio wave. An object in the wave's path reflects some of it back. A radar receiver is used as a sensor. It either displays radio echoes on a screen, or the echoes may go to a computer that analyzes them. The echo signal indicates whether the object is moving, and in what direction.

Radar is useful for detecting planes and missiles from hundreds of miles away. Some radar systems have been used to map the surfaces of other planets. Radar is used to detect rain, thunderstorms, and tornadoes. Police use small hand-held radar units to find out a car's speed.

Shipboard radar antenna.
(Courtesy of U.S. Navy)

Air traffic control center.
(Courtesy of Sperry Corporation)

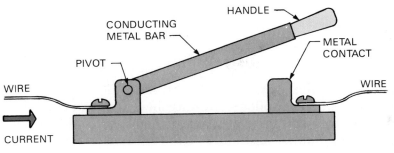

HANDLE

CONDUCTING METAL BAR

METAL CONTACT

PIVOT

WIRE

WIRE

CURRENT

A switch is a mechanical device for interrupting electric current flow.

Switches come in a variety of sizes and styles. (Courtesy of C & K Components, Inc.)

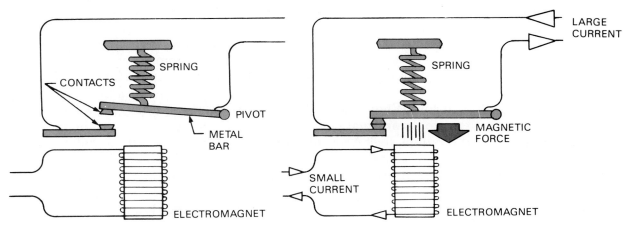

SPRING

CONTACTS

PIVOT

METAL BAR

ELECTROMAGNET

LARGE CURRENT

SPRING

MAGNETIC FORCE

SMALL CURRENT

ELECTROMAGNET

A relay is a remotely controlled switch.

Electrical and Electronic Controllers

The simplest kind of electrical controller is the **switch**. A switch lets a person apply electricity to a circuit or turn it off. Switches are often used in systems in which people are the comparators.

A **relay** is a switch that can operate remotely. A relay contains a coil that builds up a magnetic field when a current is passed through it (an electromagnet). The magnetic field pulls a metal bar toward the coil. A contact on the metal bar touches another contact, closing the circuit. The current used in the relay coil is small, but it may control a large current flowing through the contacts. Using a relay, a small coil current can control a large current through the contacts.

Through the use of transistors (see Chapter 7), relays, and switches, many kinds of machines can be controlled. For example, an on-board computer can control most systems on an airliner. A plane's flight can be largely automatic.

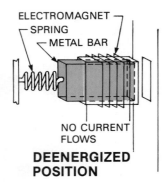

ELECTROMAGNET
SPRING
METAL BAR

NO CURRENT
FLOWS

**DEENERGIZED
POSITION**

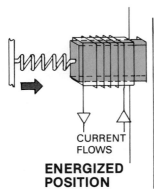

CURRENT
FLOWS

**ENERGIZED
POSITION**

A solenoid is used to remotely control the position of a metal bar. This solenoid is a remotely controlled lock for a door.

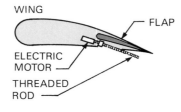

WING

FLAP

ELECTRIC
MOTOR

THREADED
ROD

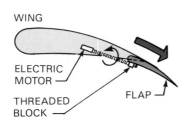

WING

ELECTRIC
MOTOR

THREADED
BLOCK

FLAP

The position of an airliner's wing flap can be controlled by a combination of mechanical and electrical parts.

Mechanical and Electromechanical Controllers

Through the ages, people have made many different kinds of mechanical controllers. Some are levers, rods, and wheels. Others have been more complicated. Controllers range in size from a watch stem to large parts used on locomotives.

A car with a manual (standard) transmission has a controller called a **clutch**. The clutch is used to connect and disconnect the rotating shaft of the engine to and from the wheels. When the driver pushes on the clutch pedal, the clutch is disengaged. The engine is disconnected from the wheels. The car then either sits still or, if it was moving before the clutch pedal was pushed, it coasts.

One electromechanical controller used in mechanical systems is a **solenoid**. A solenoid is an electromagnetic coil wrapped around a metal bar. The bar is held in position by a spring. When a voltage is applied to the coil, the bar is pulled into the coil. When the voltage is removed, the spring pulls the bar back to its original position. The bar's movement can open or close a valve, lock and unlock a door, or perform some other function.

Electric motors are often connected to threaded shafts or other mechanical parts that change rotary motion to other kinds of motion. On an airplane, flaps on the wings that control the airplane's flight are often moved up and down by a motor with a threaded rod.

Pneumatic and Hydraulic Controllers

Pneumatic controllers use air pressure. Air under pressure comes from a compressor and is carried by hoses. The air always moves from a place of high pressure to a place of low pressure until the pressure is the same everywhere in the system. The movement of air can provide a very large force. The flow of air is controlled by valves in the hoses.

Hydraulic controllers use water, hydraulic oil, or other liquids. Liquids do not compress under pressure. Hydraulic systems use the principle that the pressure is the same at all points throughout a liquid. If you press on a liquid in one place, the force will be transferred to another place.

Brakes on a car are hydraulic controllers. While the car is moving, springs hold the brake shoes away from the brake drum, which connects to the wheels. When the driver presses the brake pedal, brake fluid is forced into the system. The fluid presses the shoes against the drum wall, stopping the wheels and the car.

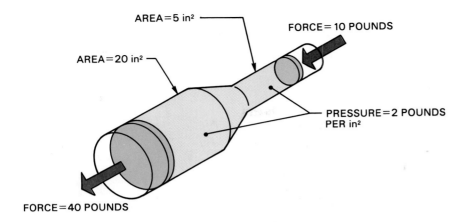

AREA=5 in²

AREA=20 in²

FORCE=10 POUNDS

PRESSURE=2 POUNDS PER in²

FORCE=40 POUNDS

As the pipe diameter becomes larger, the force applied by the fluid becomes greater.

How Clutches Work

A clutch is made up of two circular plates that are pressed together by a strong spring. When they are pressed together, they turn together. If they are separated by a force pushing against the spring, usually the driver's foot on the pedal, the rotation of one does not cause the other to turn.

Clutches can also be made with pulleys and a belt. Here, the engine shaft has a pulley that drives a belt. The idler pulley pushes down on the belt, making it tight around the pulley wheels and turning the load (or wheels, as in the case of a go-cart). When the idler pulley moves back, the belt slips around the pulley, and it cannot turn the load.

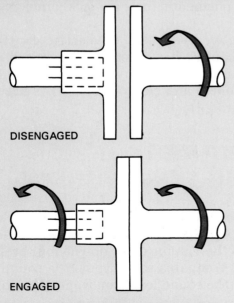

DISENGAGED

ENGAGED

A clutch is used to interrupt the transfer of power from one rotating shaft to another.

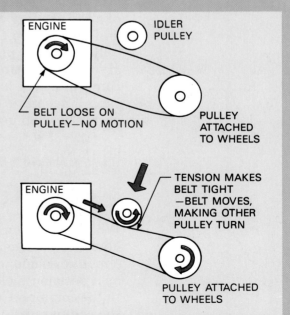

ENGINE

IDLER PULLEY

BELT LOOSE ON PULLEY—NO MOTION

PULLEY ATTACHED TO WHEELS

ENGINE

TENSION MAKES BELT TIGHT —BELT MOVES, MAKING OTHER PULLEY TURN

PULLEY ATTACHED TO WHEELS

An idler pulley and belt can perform the same function as a clutch.

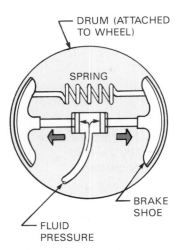

Hydraulic pressure is used to activate the brakes on a car.

Robots with hydraulic controllers can hold heavy loads with greater accuracy than robots with pneumatic or electric controllers. (Courtesy of Prab Robots, Inc.)

Pneumatic and hydraulic controllers are used in factories. They control motion and tools on assembly lines. Valves automatically open and close to control the liquid or air pressure.

Some tools are pneumatic tools. A trigger controls the flow of air to the tool. Some pneumatic tools are jackhammers, drills, and nut drivers.

Robots use pneumatic, hydraulic, or electric controllers. Hydraulic controllers can lift heavy loads. The fluid does not compress, so a load can be held steady in one spot for a long time. Pneumatic controllers and electric controllers are not as accurate when used in robots.

DECISION MAKERS: MAKING THE COMPARISON

When you listen to a radio, you are part of a feedback loop. Your ears act as sensors, monitoring how loud the radio is. You may decide to turn the volume up or down. Your brain acts as a decision maker. It compares the output (the sound level) to the command input (the sound level you want). Your brain is the comparator.

There are many ways of comparing the actual output to the command input. Many comparators are mechanical or electrical or both.

Guidance Systems

Many vehicles have **guidance** systems that keep them on course. One of the simplest is used to steer a sailboat. A windvane senses the direction of the wind. A linkage connects it to the rudder under water. The system keeps the boat headed in the chosen direction.

Computers are often used in guidance systems. Planes and missiles use moving wing sections called **control surfaces** to change direction. The control surfaces may be adjusted by a computer on a set course.

A land vehicle that has no driver has been developed by the Department of Defense. It uses TV cameras and a laser to guide it.

Guided missile. (Courtesy of Lockheed Horizons)

Autonomous Land Vehicle (ALV).
(Courtesy of Martin Marietta)

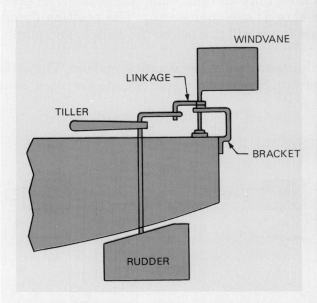

This guidance system steers a sailboat.

■ In technology one studies about magnetic sensors that can measure rotation. In mathematics classes one studies angles as rotations and learns how to use a protractor to find the degree measure of an angle. In measuring angles students often use the wrong scale of the protractor and give the measure of an acute angle of 60° as 120°. To measure angles correctly, students have to continually monitor their work. They have to ask themselves, "Am I measuring an acute angle or an obtuse angle," compare their answers with the diagram, and make corrections as needed.

Electrical and Mechanical Comparators

A comparator can be made with a mechanical **linkage** and an electric switch. A linkage is a rod that transfers position information or movement from one place to another. A sump pump uses such a comparator.

A sump pump is a pump that keeps water from rising in the basement. As water rises in a well in the basement floor, a float rises. The float is attached to a collar with a linkage that pushes a switch. When the float gets high enough, it turns on the switch and starts the pump. When the water has been pumped out, the float falls and the pump turns off. The position of the collar acts as the comparator. The position determines how high the water will rise before the switch, and pump, go on.

Sometimes computers are used as decision makers. The software is the comparator. Special electronic circuits are often used as comparators.

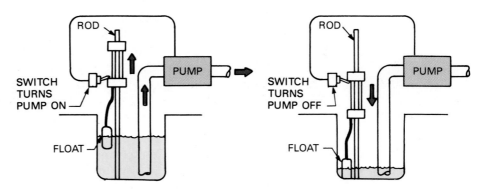

The sensor, comparator, references, and controller for a sump pump.

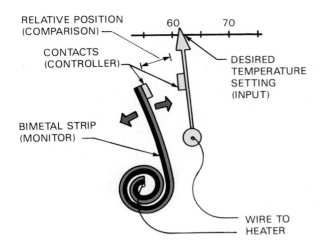

RELATIVE POSITION (COMPARISON)

CONTACTS (CONTROLLER)

DESIRED TEMPERATURE SETTING (INPUT)

BIMETAL STRIP (MONITOR)

WIRE TO HEATER

60   70

The monitor, comparator, input (temperature setting), and controller in a thermostat.

COMBINING THE THREE ELEMENTS

Sometimes one device contains all three feedback elements. A **thermostat** is an example. The sensor, comparator, and controller are all in the thermostat. A bimetal strip is the sensor that monitors room temperature. A wire connected to one end of the strip and electrical contact at the other make it a switch, or controller. The position of the contact can be changed to set the desired temperature, providing a comparator. In a thermostat, one mechanism serves as monitor, comparator, and controller.

TIMERS AND OTHER PROGRAM CONTROL

Timers control some systems, making them perform actions in sequence. A timer can be used to turn lights on for security in a house when people are not at home. The lights might be set to go on at 7 p.m. and off at 11 p.m. A traffic light is controlled by a timer. It can be made to turn green for two minutes for the main road, and then green for one-half minute for the cross street. These timers are useful because a person does not have to make regular changes in the system. They are an example of open-loop control called **program control**.

But these timers cannot adjust to changing conditions. The timer set for house lights in January will not do the job right in June. The traffic light cannot adjust itself to give a longer green light to heavy traffic on the cross street.

This student is operating a computer numerical controlled (CNC) machine.
(Courtesy of EMCO)

Timers can be made with motors, cams, and switches, but computers provide the best control. They can be reprogrammed and provide much more precise control than timers that use cams. Today they are doing more and more timing control jobs.

Programming the computer can be done in several ways. Often, simple programming languages are used. They can be understood more easily by people who work with the machinery that is to be controlled. This is true, for example, with the numerically controlled tools described in Chapter 9.

Robots are sometimes programmed by having a person show the robot how to move. A person actually moves the robot arm through the required motions. The robot's memory records the motions. This is referred to as "teaching" the robot.

CONNECTIONS: Technology and Social Studies

- The importance of monitoring sensors is apparent when you consider the oil flow through the Alaskan pipeline. The protection of the environment depends on the safeguards and monitoring devices that have been built into this system to prevent oil spills.
- Sensing devices have now been created that enable blind people to "read" printed material. By passing a device over a printed page, the blind person hears the words reproduced orally.
- There are many factors that indicate the relative health of an economic system. The unemployment rate, movement in the stock market, level of foreign trade, and gross national product are all examples of measures used by economists to determine the strength of an economy.

COMBINING FEEDBACK WITH PROGRAM CONTROL

Timer control and other kinds of programmed control can change an operation based on expected conditions. But they won't change when conditions change. To solve this problem, some systems combine feedback with programmed control.

At some traffic intersections, electronic sensors have been buried in the road. The sensor provides feedback to the traffic light control system, which has a timer in it. A microprocessor uses information from both the timer and the sensor to control the changing of the lights.

Computers can be used to control both open-loop and closed-loop systems.

SYSTEM	OPEN LOOP	CLOSED LOOP	PROGRAM CONTROL	PROGRAM + FEEDBACK CONTROL
TRAFFIC CONTROL	Stop sign	Traffic light with sensor in street	Timed traffic light	Timed traffic light with sensor in street
COOKING	Put food in oven, forget to set timer, don't test	Cook food until it is done according to taste, touch, or temperature	Cook food for specified time	Cook food for specified time, then test to see if done
LEARNING	Random reading	Reading followed by questions and discussion	Reading certain subjects at certain times	Reading certain subjects at certain times, followed by questions and discussion

A comparison of different types of systems.

CONTROL OF NONTECHNICAL SYSTEMS

You have been learning how technological systems are controlled. But nontechnical systems must be controlled, too. Very often, information used to make decisions is processed by computer. Some systems are so complex that information is incomplete. People must decide based on experience or guesses.

A nation's economy is controlled in part by the amount of money in circulation. A government may decide to print and distribute more money. This may make the economy grow. Here, the controller is the printing of more money. The sensor is a set of economic indicators that tell how the country's economy is

SYSTEM	SENSOR	COMPARATOR	CONTROLLER
HEATING SYSTEM (THERMOSTAT)	Bimetal strip (thermometer)	Relative position of contacts	Contacts
BICYCLE (STEERING)	Rider's eye	Rider's brain	Handlebars
TOILET TANK (WATER LEVEL)	Float	Relative position of float arm and valve lever	Valve
WATER TOWER	Pressure sensor	Microcomputer	Valve
TRAFFIC LIGHT	Sensing loop in road	Microcomputer	Relay (changes light)
AIRPLANE AUTOPILOT	Gyrocompass, radio navigation equipment	Autopilot computer	Transistors, relays, hydraulic valves

Sensors, comparators, and controllers for different systems.

doing. The decision maker is the group of people who study the indicators and decide whether to print more money.

In a company, managers are the decision makers. They study company data. The data include profits, debt, market forecasts, and many other factors. They make long- and short-term decisions. Should the company hire new people? Should it spend more money on research? The controlling factors in a company are the number of people hired, the amount spent on research, and new products developed, among others.

Laws and regulations also provide controls. Laws are made because of feedback. For example, tests showed that our air was becoming polluted by trucks, cars, and buses. New laws were made to require changes in fuel and car design that reduced pollutants. The laws were used as a controller. Other laws govern toxic waste disposal, nuclear power plants, and genetic research.

SUMMARY

Open-loop systems provide a desired output under a certain set of conditions. The output may vary as conditions change. Closed-loop systems adjust to changing conditions. They maintain the desired output.

The three elements in a feedback loop are the sensor, which monitors the output of the system; the comparator, which compares the actual output to the desired output (command input); and the controller, which changes the process in some way.

Controllers that turn a process on and off are called bang-bang controllers. Proportional controllers adjust the process by just the right amount, based on how far the actual output is from the desired output.

A system is controllable if we can make the outputs what we want them to be. All systems are controllable only within limits. People must know a system's limits and operate the system only within them.

The feedback loop must work quickly enough to control the process. If the feedback is too slow, the system cannot adjust itself, and will not respond quickly enough to changes.

Sensors must be chosen to monitor the system output and to signal the comparator. Sensors can monitor electrical, mechanical, optical, thermal, magnetic, and other system outputs. Sensors that convert information in one form of energy to information in another form are called transducers.

Most controllers used in technological systems are operated mechanically or electrically. Electrical controllers include transistors, switches, and relays. Mechanical controllers include linkages, clutches, and solenoids. Most comparators used today are electrical, mechanical, a combination of the two, or computer software.

Mechanisms that combine two or all three of the elements can be made. Many systems use separate sensors, comparators, and controllers.

Timers and other programmed controllers are examples of open-loop systems that are changed in response to expected conditions. Without feedback, these systems cannot respond to changes.

While many machines have been made with mechanical programmed controllers, computers are now in widespread use for this task. Computers have more precise control over the process. They can be reprogrammed easily. Computers also make it easier to combine feedback with programmed control, using both ways to control a system.

Nontechnical systems also have feedback loops with sensors, comparators, and controllers. These are most often dependent on information. People usually make the decisions, often with the help of computers.

(Courtesy of Allen-Bradley Company)

Feedback

1. A printing system is set to print dark black letters on posters. The system has no feedback loop. After thirty minutes of printing, the ink becomes thinner, and the letters are not as dark. Can the system automatically compensate for this and make the letters dark again? Why or why not?
2. A boy riding a bicycle changes gears as he pedals up a hill. Are his actions part of a closed-loop system? Why or why not?
3. When a person drives a car, there is a closed-loop system made up of the person, the gas pedal, the engine and wheels, and the speedometer. Identify the different parts of the system. Is this a bang-bang control system or a proportional control system?
4. Is a home heating system a controllable system? Why or why not?
5. Many sensors used today change mechanical, optical, thermal, or magnetic information into electronic pulses. Why is this true?
6. A timer used to turn a light on at dusk is set in June. In December, it gets dark much earlier. Since the timer is an open-loop system, it will not adjust for the earlier dusk. How can you change the system so that it automatically compensates for changes in the time of dusk?
7. Why are computers used in so many feedback control systems?

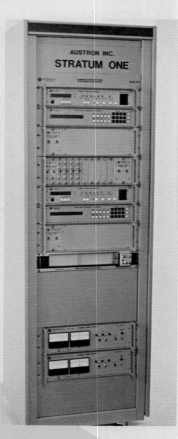

(Courtesy of Austron, Inc., Austin, Texas. Stratum One Communications Network Synchronization System)

(Courtesy of Ford Motor Company)

(Courtesy of Martin Marietta)

CHAPTER KEY TERMS

Across
1. This kind of control adjusts a process in small amounts.
3. A device that monitors the output of a system.
8. This type of control is provided by a feedback loop.
10. The decision maker that compares the output (actual result) to the command input (desired result).
14. This high-tech tool can use pneumatic, hydraulic, or electric controllers.
15. Control by means of a timer, as in a washing machine that goes through a wash and rinse cycle.
17. This consists of three parts: a sensor, a decision maker, and a controller.
18. These systems use air pressure to control the process.
20. The range of outputs over which a system is controllable.

Down
2. Control without feedback.
4. Radio detection and ranging; used to detect planes, ships, and missiles at a distance.
5. This type of control only turns a process on or off.
6. This part of the feedback loop is used to turn the process on or off or to adjust it in some way.
7. A sensing device that changes one form of energy to another form.
9. This controls a home heating system. It combines a sensor, comparator, and controller in one unit.
11. Water or other liquids are used to control these systems.
12 An open-loop controller that causes systems to perform actions in a sequence.
13. A device that can be used to connect and disconnect the engine from the wheels of a vehicle.
16. A type of electrical switch that can control large amounts of current from a distant location.
19. An irregularly shaped part mounted on a rotating shaft.

Bang-bang control
Cam
Closed-loop control
Clutch
Comparator
Controllable
Controller
Feedback loop
Hydraulic
Limits
Open-loop control
Pneumatic
Program control
Proportional control
Radar
Relay
Robot
Sensor
Thermostat
Timer
Transducer

SEE YOUR TEACHER FOR THE CROSSTECH SHEET

387

ACTIVITIES

BUILDING A SMART HOUSE

Problem Situation

One of the newest applications for electronic sensors can be found in new homes and buildings. Smart buildings, as they are called, sense the environment and adjust to conditions within the building. For example, sensors monitor and adjust a building's heat, air conditioning, and humidity level. Intruder alarm systems can sense body heat, motion, sound, and light, and notify people in charge about intruders. Fire alarm systems can detect smoke and heat and then notify the fire department. Light sensors turn on outdoor lights as the sun goes down. Door locks that used to be opened with keys have been replaced by sensors that read coded information on a personalized plastic entrance card. Touch sensors on walls and appliances turn the power on and off at the touch of a hand. In smart buildings, doors can open and close automatically after they detect a person's presence in the area.

Design Brief

In a team of two or three students, design and build a model of a smart building that will monitor, react to, and/or adjust at least three different conditions within the building.

Procedure

Use the safety procedures described by your teacher.
1. Use the cardboard and markers to simulate the front of your house or building. Make the doors so they open and close, and the windows so they move up and down. Keep the design simple and in a large enough scale to work with easily. Place all your components behind the building's front; there will be no sides to the building.
2. As a team, decide on a minimum of three systems to use within your smart house. Be sure to consider your power source, sensor, and output device.
3. Each member of the team will select at least one of the systems and assemble it.
4. Install the systems in the model house and demonstrate them to the class.

Suggested Resources

Show or stiff cardboard and markers
Electronic components:
  switches
  photo resistors
  thermistors
  DC motors
  1.5-volt bulbs and sockets
  buzzers
  relays
  batteries
  moisture-detector circuit
  wire
Soldering equipment and supplies

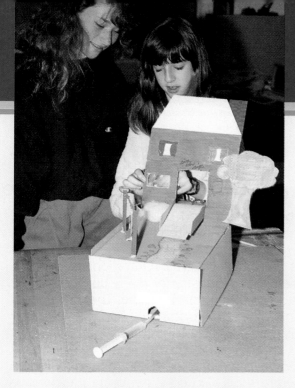

Classroom Connections

1. Sensors are used to detect existing conditions or the output of a system.
2. Feedback loops contain monitors (sensors) that sense changes in a condition, comparators that compare the system's output with its input, and adjustors that adjust the system.
3. Sensors can be electronic, mechanical, or electromechanical.
4. Ohm's Law can be used to determine voltage, current, and resistance.
5. A knowledge of schematic drawing is necessary to communicate electronic ideas.

Technology in the Real World

1. Many public restrooms now utilize water faucets that turn on when you break an invisible beam of light by passing your hands underneath. The sensor prevents people from leaving the water running when they leave the restroom.
2. Money machines with touch-sensitive screens are being installed in buildings and shopping malls all over the world.

Summing Up

1. List the names and functions of three different types of switches.
2. What electronic sensor might be used to turn a fan on when the lights in a room go out?
3. What type of switch would be found on a refrigerator door that will turn the refrigerator light on and off?
4. What types of electronic sensors would be used to detect human motion?

ACTIVITIES

PNEUMATIC CONTROL SYSTEMS

Problem Situation

Systems controlled by air are called pneumatic systems. Air-powered jackhammers are used in construction work to break up concrete. Pneumatic wrenches are used by mechanics to remove and attach the lug nuts that hold wheels on automobiles.

Pneumatic devices are used in industry for many purposes. They are used to move parts on or off an assembly line. The movements of some robots are pneumatically controlled. For example, the Jarvik-7 artificial heart is pneumatically controlled.

There are two kinds of pneumatic control systems: open and closed. Open systems require a compressor. They are used for industrial control and to operate tools. Closed systems are inexpensive and easy to make. A closed system can be assembled using a plastic bottle, PVC tubing, a rubber band, and a balloon.

Design Brief

In this activity you will assemble and test a closed pneumatic system. Afterward, you will use the system to control a device that you design and construct.

Procedure

Use the safety procedures described by your teacher.
1. Insert the tubing into a ³⁄₁₆" hole in the bottle top. Make sure that there is a tight fit between the bottle top and the tubing.
2. Use a rubber band to attach the balloon to the other end of the tubing. When the bottle is squeezed, the balloon should inflate. Seal any leaks in the system.
3. Test the system by placing one or two books on top of the deflated balloon and squeeze the bottle. What happens? Do the results surprise you?
4. Use the pneumatic system to control a device that you construct, such as a model dump truck, a vehicle with pneumatically operated doors, a model elevator, railroad crossing gates, or a jack. Make a series of sketches showing possible devices.
5. Select one idea and prepare a detailed sketch. Obtain approval from your teacher.
6. Construct your device using available modeling materials.
7. Test your device and make improvements until it operates properly.
8. Demonstrate how the pneumatic system controls your device.

Suggested Resources

Squeezy plastic bottles
2' to 3' of aquarium tubing
Balloons
Rubber bands
Illustration board
Balsa wood
½" × ½" pine strips
White glue
Hot glue
Glue gun

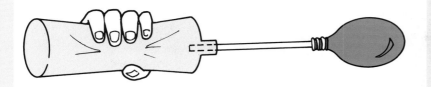

Classroom Connections

1. Controllers can be used to adjust a process, or turn it on and off.
2. Systems can be controlled by many different controllers.
3. Air always moves from a place of high pressure to a place of low pressure.
4. Air can be compressed. Liquids do not compress under pressure.
5. Pressure is measured as force per unit area. Pounds per square inch is a common measure of pressure.

Technology in the Real World

1. Pneumatic systems and hydraulic systems are similar. Pneumatic systems use air while hydraulic systems use liquids. Because liquids do not compress, they provide more accurate control.
2. Pneumatic systems are used to operate tools and for industrial control.
3. In the future a pneumatically controlled artificial heart may be used to keep patients alive while they wait for human hearts.

Summing Up

1. List three applications of pneumatic systems.
2. Compare open and closed pneumatic systems.
3. Give one advantage of hydraulic control over pneumatic control.
4. Describe two problems that you had to solve while constructing your device.

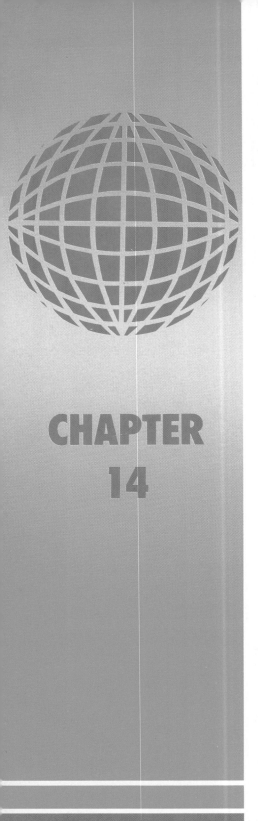

CHAPTER 14

IMPACTS AND OUTLOOKS

MAJOR CONCEPTS

After reading this chapter, you will know that:

- Outputs of a technological system can be desired, undesired, expected, or unexpected.
- People determine whether technology is good or bad by the way they use it.
- Technology produces many positive outputs and solves many problems. Sometimes, however, negative outputs create new problems.
- Technology must be fitted to human needs.
- Technology must be adapted to the environment.
- Existing technological systems will act together to produce new, more powerful technologies.
- Using futuring techniques, people can anticipate the consequences of a new technology.

INTRODUCTION

Long ago, travel across the United States took weeks or even months. The pony express was the quickest way of delivering mail. People had to make many products by hand. Today, we can buy the goods we need, goods produced by manufacturing technology. Jets take us across the country in hours. Technology makes it possible to move information at nearly the speed of light.

There is another side to technology. In December 1984, poison gas leaked from a chemical plant in Bhopal, India. More than 2,000 people were killed. In January 1986, the space shuttle *Challenger* exploded seconds after launch. All seven astronauts were killed. In April 1986, an accident occurred at a nuclear power plant in Chernobyl, in the Soviet Union. People died from the radiation given off in the accident.

Some people say that we have lost control over our technology. They say that technology results in tragedies like the explosion of the *Challenger*. They say we should stop building nuclear power plants because of the dangers of radiation. They blame technology for pollution, noise, and crowded cities and highways.

Is technology good or evil? Would we want to go back to the days when we had simpler technology—and could we, if we wanted to? What effects will technology have on our world in the future? These questions are the subject of this chapter.

The Challenger crew included (clockwise from top left) Ellison S. Onizuka, Sharon Christa McAuliffe, Gregory Jarvis, Judith Resnick, Ronald McNair, Francis R. Scobee, and Michael J. Smith. (Courtesy of NASA)

IMPACTS OF TECHNOLOGY

In Chapter 6 we learned that there are four possible outputs from any technological system. It is the unexpected and undesirable output that is of greatest concern to us. If we cannot plan for an event, we may not be able to deal with it.

The accident at Chernobyl was one such event. It took the Soviets several days to evacuate the people who lived near the plant. Moreover, all the harmful effects of the accident will not be known for a long time, since radiation can cause cancer deaths years later.

Still, technology is neither good nor evil. It can be used for good, or it can hurt us. How we control it determines whether it is useful or harmful. For example, the U.S. Nuclear Regulatory Commission (NRC) sets standards for nuclear power plants. A plant must have a 9-inch-thick steel shell and a 3-foot-thick concrete building around the nuclear core. The

Outputs of a technological system can be desired, undesired, expected, or unexpected.

People determine whether technology is good or bad by the way they use it.

393

Problem Situation 1:   People with arthritis in their fingers have great difficulty in gripping small items.

Design Brief 1:   Design and construct a device which will help a person with arthritis to use a house key more easily when opening a door. The device must be small enough to fit into a purse or jacket pocket.

Problem Situation 2:   The normal (qwerty) typewriter keyboard was designed with the most frequently used letters in awkward positions to slow early typists down and prevent keys in mechanical typewriters from jamming. Computer keyboards do not jam, and can accommodate faster typists.

Design Brief 2:   Redesign the computer keyboard with the most frequently used letters under the strongest fingers. Use a page or two of printed text to determine which letters are most frequently used.

Problem Situation 3:   Some environments, like the moon, the ocean, and underground are hostile to human life.

Design Brief 3:   Design and construct a model of an environment that could be built in one of these locations and could support human life.

Chernobyl reactor was built without such protection. Of course, an accident is possible even in NRC-approved plants in the U.S. But control of technology lowers the risks.

Today, **pollution** is a worldwide issue. Millions of automobiles and trucks give off dangerous gases. These gases pollute our air. Some factories dump their wastes into rivers and oceans. These wastes have polluted our water supply.

Technology itself is not harmful. Rather, it is the way we use technology that causes problems. And, just as technology can cause problems, it can solve them. For example, pollution control devices on cars and scrubbers on factory smokestacks are reducing pollution.

Technology produces many positive outputs and solves many problems. Sometimes, however, negative outputs create new problems.

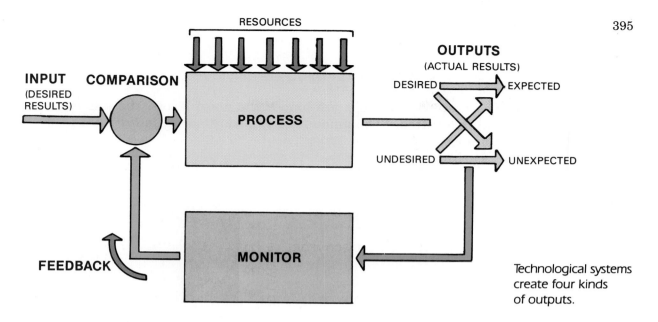

RESOURCES

INPUT
(DESIRED
RESULTS)

COMPARISON

PROCESS

OUTPUTS
(ACTUAL RESULTS)

DESIRED → EXPECTED

UNDESIRED → UNEXPECTED

MONITOR

FEEDBACK

Technological systems
create four kinds
of outputs.

This automated control room
in a nuclear power plant
is watched over by human
workers. (Courtesy of Rochester
Gas & Electric Co.)

Sanitation man sweeping
manure in New York City.
(Courtesy of the Museum of the
City of New York)

Should We Go Back to Nature?

If we went back to a simpler life, would problems like pollution disappear? No. Pollution has been a problem since people first began to live in cities. The rivers around Rome were so full of wastes that people were forbidden to bathe there. Before the automobile, one hundred years ago, 150,000 horses in New York City produced over 500,000 tons of manure a year.

We face other problems. Our environment is noisy. People who live near airports hear jets taking off and landing day and night. Subways roar. Road traffic makes constant noise. Factory workers listen to the pounding of machinery all day long. Even loud music may damage our hearing.

CONNECTIONS: Technology and Mathematics

■ The graph shows that the louder the noise levels, the less exposure permitted. (The amount of exposure varies *indirectly* to the noise level.)

If limits are exceeded, efforts must be made to reduce noise. If this is not possible, workers must be given earmuffs or plugs.

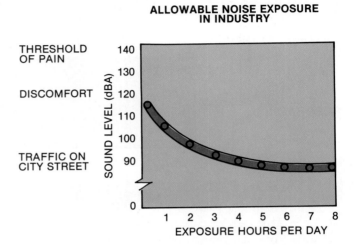

In earlier days, there were other noise problems. Factories were even more noisy than they are today, when we have laws that protect workers. Trains rumbled through cities, shaking the buildings as they passed. That's only the beginning. Medical care was crude. People lived in unsanitary conditions and in unsafe buildings.

The fact is, we depend on technology. Our lives and our routines are built around it. We could not go back to a life without automobiles, flush toilets, telephones, and modern medical care. Technology is here to stay. But it is up to us to use it well. We can pass laws that protect us and our environment. We can control technology.

Matching Technology to the Individual

Technology must be fitted to human needs.

Products can be designed to make our lives more comfortable. The field of **ergonomics,** or **human factors engineering,** uses technology to meet the physical needs of people. Chairs fit the shape of the back. Drinking glasses are easy to hold. Tables and desks are the right height. Auto seats make driving more comfortable.

Computer keyboards are a good example of ergonomic design.

One company tried out different designs on hundreds of secretaries. The secretaries typed on the keyboards and provided feedback on how comfortable they were. The shape of the keys, the pressure needed, and the way they sprang back were things that determined how "user friendly" the keyboard was.

Making products safe is another way ergonomics helps. The blades of kitchen machines such as food processors must be guarded for safety. Automobiles are designed for people's safety.

Special needs can be met. People with disabilities can choose from many new products. Bicycles, cars, and skis have been designed for people who do not have the use of their legs.

Handbikes and Sunbursts

Bicycling can help people learn good balance and it is also an excellent form of exercise. Some people, however, are unable to pedal a bike. To meet their needs, an arm-powered bike has been designed at the Veterans Administration Rehabilitation Research and Development Center in Palo Alto, California.

The Handbike, as it is called, comes in adult, child, and racing versions. The bike combines the best ideas from bicycles, people-powered vehicles, and sport wheelchairs.

The Sunburst Tandem is a spin-off of the Handbike. The front rider pedals with arms or legs or both. The back rider steers. The Handbike and Sunburst show how technology can help the physically challenged to enjoy activities that would otherwise not be possible.

Macy Tackitt on her Handbike. Macy was partly paralyzed as a result of an auto accident. (Courtesy of VA Rehabilitation R & D Center, Palo Alto, California)

Designer Douglas Schwandt, a biomedical engineer, and friend Dianna Gubber riding the Sunburst. (Courtesy of VA Rehabilitation R & D Center, Palo Alto, California)

Matching Technology to the Environment

Technology must be adapted to the environment.

Technology has been damaging the environment for thousands of years. Forests were cut down for fuel in the 1600s. Garbage was burned in open fires. Human wastes were poured into rivers. However, all this occurred on a small scale. Today, technology could harm the environment on a worldwide scale.

We burn over 700 million tons of coal each year to provide energy, producing pollutants like **sulfur dioxide**. In the United States, nearly 170 million vehicles burn about 125 billion gallons of fuel each year, adding huge amounts of pollutants to the air.

Chemicals given off by automobiles and industrial plants can rise into the atmosphere and return to the earth as acid rain. **Acid rain** has killed forests in some parts of the United States, Canada, and Scandinavia. It has killed lake fish. In New York State, fish have been completely killed off in about 200 lakes. The same thing may happen in hundreds more.

Over the last thirty years, new chemicals have come into use in farming. They control pests and disease. But they may remain in the soil and air. They can get into the food we eat and the water we drink.

Some factories make chemicals that can harm people. Many companies try to limit this harm by doing **technology assessment studies**. Laboratory animals are exposed to the chemicals. The results help determine the effect chemicals will have on people. Such studies can save lives. In Niagara Falls, New York, chemicals dumped by local industries caused illness and death. About 800 families living in an area called Love Canal had to leave their homes.

Our **environment** must be treated with care. We must use resources wisely and look for alternatives for scarce resources. We must develop technologies that do not cause environmental harm. For example, solar energy can provide for our energy needs without harming the environment.

Today, federal laws like the Clean Air Act ban industries from polluting the environment. *(Courtesy of Carnegie Library of Pittsburgh—Pittsburgh Photographic Library)*

SEEING INTO THE FUTURE OF TECHNOLOGY

How will technology affect the way we live in the future?

In recent years, it has brought about enormous changes. Twenty-five years ago there were no video games, home computers, or industrial robots. Artificial hearts had not been used. We didn't have genetic engineering. Only in the last decade

CONNECTIONS: Technology and Social Studies

■ Sometimes correcting the problems caused by technology is not easily done. Acid rain that comes from smokestacks in the Midwest damages forests and streams in the Northeast. But there are people whose economy depends upon those smokestacks. It is difficult to set regional interests aside for the benefit of the common good.

■ As new technological devices are developed, people seek some assurance that the devices will really benefit society. Did the artificial heart that was implanted in a number of human beings really improve their quality of life? Some people argue that much was learned from these experiences, but there are others who question the use of people to test a device of questionable value.

■ With the rapidity with which technological advances are taking place, a concern is whether people can keep up with the new developments. Even people who wish to stay on top of the field may find it impossible because they cannot find the time to learn about all the current happenings.

have we had space shuttles, test-tube babies, and fully automated factories.

Technology is growing at an ever-increasing rate. Think about the changes that have happened in the last twenty-five years. What might happen in the next twenty-five? Let's look ahead and see what technological systems of the future might be like.

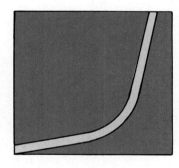

Technology is growing at an exponential (ever-increasing) rate.

COMMUNICATION IN THE FUTURE

In the future, we will be able to "talk" to machines. Already, computers can speak like humans. The technology that makes this possible is called **speech synthesis**.

You may already know about synthesized music. This is music made by electrical devices rather than by acoustic instruments. Telephone companies use speech synthesis. When you call directory assistance to ask for someone's number, a human operator switches you to a computerized device. The device gives you the number, using a computer-generated voice.

Soon **voice recognition** technology will be in widespread use. Computers will be able to respond to human voice commands. The telephone operator may no longer be needed. These systems may be used in typing. You will be able to speak to a

This typewriter can recognize 1,000 spoken words.
(Courtesy of Kurzweil Applied Intelligence, Inc.)

typewriter, and it will print whatever you say. The disabled may be helped by this technology. Computers, robots, and wheelchairs will be controlled by voice. A machine for the blind that can read the words of a printed page aloud has already been developed.

Communication at Home

Homes will have robots with whom it will be possible to communicate. People can order their household pet robots to fetch a glass of water or shine their shoes. Robots will help with housework, watch children, and aid the elderly and disabled. Robots will be made so they seem more like people.

We will be able to communicate with appliances. We will tell lights to turn on and television sets to turn off. Our telephones will talk to us. When we come home, a telephone will tell us who called, at what time, and give us the message. Telephones will become portable. We will be able to carry one in a pocket and talk anywhere or anytime we wish. They will also become more powerful. We will be able to see the person we are talking to in 3-D. As space travel becomes more common, we will talk to friends in space.

WABOT II, The Keyboard-Playing Robot

Robots of the future will not only be factory workers but will provide personal services for people. Service robots will look and act more like people.

WABOT II is an intelligent robot musician. It can talk. It can read sheet music and play an electric organ, using both hands and legs. The robot was constructed by a research group at a Japanese university.

The fingers of both hands play keyboards. The feet work the bass keyboards and the pedals. A video camera picks up the musical score. WABOT II can accompany a singer, playing in tune with the singer's pitch.

(Courtesy of Ichiro Kato)

Cameras will be video machines that use video disks, or cassettes no larger than an audio cassette. We will view our pictures on a home TV screen or hook up a printer to print them on paper.

Television sets will be bigger and better. With a new technology called HDTV (high-definition television), pictures will be clearer. On today's TV screens, the picture is made up of 525 fine horizontal lines called **scanning lines**. HDTV will have 1,125 scanning lines, and provide more detailed pictures.

Holographic telephones will create 3-D images. *(Courtesy of General Electric Company)*

Communication at Work

Satellites are used today to send telephone conversations, sporting events, and military information around the world. A technology called **video conferencing** is helping people to communicate by TV.

Meetings between people in different places are possible by way of a TV screen. An electronic blackboard allows people to draw diagrams and send them instantly to a screen in another country, or even in space.

We may someday be able to put a microchip in the brain, and connect people up to computer data bases. In the distant future, we may be able to record our thought patterns for future generations. Some people fear that this technology will be used to monitor a worker's job performance. Employers might be able to control the thoughts and feelings of workers by sending signals to the microchip.

Would this ever happen? Some unions, such as the Newspaper Guild, the United Auto Workers, and the Communication Workers of America, have asked for antimonitoring clauses in their contracts with employers.

Only uses of technology that enhance the quality of life should be encouraged. Computers, for example, have made telephone operators more productive. However, the operators report that they are now expected to answer more calls per minute than ever before. This has made their jobs more stressful.

Communication and the Consumer

Computers are being connected to telephone lines, changing the way people shop. You may use your computer to search through lists of merchandise, order, and pay without leaving home. This method of communicating will become more common in the future. You will be able to pay bills and move money from a checking account to savings. You will be able to reserve a flight and buy airline tickets. You will be able to use video

Personal computers can communicate with mainframe computers over telephone lines. *(Courtesy of AT&T, Bell Laboratories)*

This woman is using a computerized mirror to "try on" a new outfit. (Courtesy of L.S. Ayres)

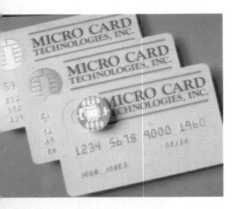

One of the new "smart cards." (Courtesy of Motorola, Inc.)

disk images to choose a vacation place, a hotel, and the sights you want to see.

You'll use the same technology to choose a new hairstyle or a new outfit. Your hair stylist will combine your image with that of a new hairstyle on a computer screen. In the same way, you'll be able to see yourself wearing new clothes in a computerized mirror.

Computers and credit cards could turn our country into a cashless society. Credit cards will have miniature computers in them. You'll be able to use them to buy anything you need.

These **smart cards** have an integrated circuit memory chip within them. The chip keeps track of the consumer's bank balance. After a card is used, the chip's memory stores the new balance. Smart cards will be used to pay for telephone calls and pay TV. They will store information about you, such as your blood type, drug allergies, and medical insurance information.

MANUFACTURING IN THE FUTURE

Robots will do more and more manufacturing jobs. More of our factories will operate almost without people. Today, robots do simple jobs such as welding and painting parts, and moving them from one place to another. Putting products together most often requires people, because this requires human skills. Future robots will be smarter. They will be able to see and feel and learn. They will be able to put products together as well as make parts. Many people will lose factory jobs. Those jobs that remain will go to people in countries where labor is cheap. Those whose jobs have been taken by robots will have to be retrained to learn new jobs.

A hundred years ago, factory workers in textile plants worked 75 hours a week. Today, the average worker puts in 42 hours a week. Fifty years from now, some people forecast, the work week will be only 20 hours. There will be "parents' hours" jobs, from 9:00 A.M. to 3:00 P.M. There will be jobs with "students' hours," from 3:00 to 6:00 P.M. What will people do with all this leisure time? Will the extra time parents can spend with their children result in happier, stronger families?

Manufacturing in Space

Some products can be manufactured in space more easily than they can be on earth. Low gravity is one reason.

Gravity pulls the molecules of materials down, making it difficult to separate and purify materials. Suppose you wanted to take certain cells out of a group of living cells so you could find a cure for a disease. On earth, because of gravity, only a tiny number of the cells could be removed at a time.

There is almost no gravity in space. Materials "float." They can be given an electric charge and pulled apart by an electric field. The pure materials can be collected. In space, it is possible to purify 700 times as much of certain materials as on earth.

Making some kinds of electronic chips requires a very pure form of a material called gallium arsenide. Right now, large numbers of these chips cannot be used because the material is not pure. In space, very pure crystals of gallium arsenide can be grown.

Optical glass, used in microscope lenses, must be pure. To make this kind of glass on earth, sand and limestone must be heated to about 3100°F. At these high temperatures, the glass is corrosive. It will corrode its container, so little bits of the container get mixed with the glass. The glass is no longer pure.

In space, glass can be melted without containers. The molten glass is contained by sound waves. The glass produced in this way is pure.

This symbolic photograph shows a group of pure, needle-shaped urea crystals. The crystals were grown aboard the orbiter "Discovery" in a gravity-free environment. (Courtesy of NASA)

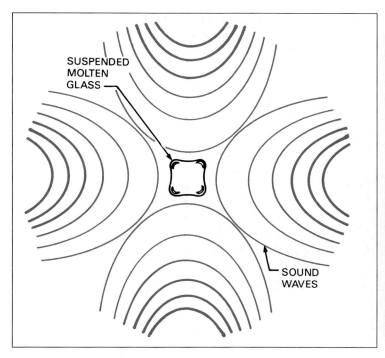

SUSPENDED MOLTEN GLASS

SOUND WAVES

Acoustic waves (sound) hold molten glass during gravity-free processing in space.

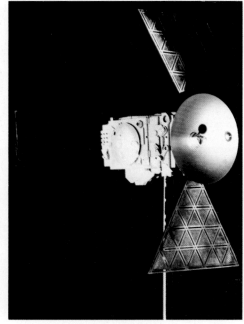

Model of a space factory.
(Courtesy of NASA)

It costs a great deal today to make products in space. We make cost-benefit trade-offs. We choose whether to make very pure optical glass in space at high cost or less pure glass on the earth at lower cost.

The first Industrial Space Facility (ISF) will be launched by a space shuttle in the 1990s. It will serve as a platform to make pure optical glass and develop new kinds of alloys. It will also be used to grow pure crystals for use in electronic devices.

As more low-gravity processing is done, space factories will be started. They will use minerals available on the moon and nearby planets. They will use solar energy for power. Although people will be able to visit these factories, they will operate automatically. The moon's soil is rich in minerals like aluminum, silicon, and titanium. A factory on the moon could use these materials to build other factories.

CONSTRUCTION IN THE FUTURE

Space construction will become a big business. Space stations will orbit the earth, serving as bases for people who come into space to work on satellites and space factories. Space stations will grow larger as more workers are needed in space. Eventually, they will become cities. People will live in space for months or years at a time.

A futuristic space station. (Courtesy of Boeing Aerospace Company)

The community of Arcosanti, located in the Arizona desert. (Courtesy of Ivan Pintar)

One design for a space city is a cylinder that spins slowly on its axis. Spinning would create an artificial gravity. People would live on the inside surface of the cylinder. A person could look up and see people and structures upside down on the opposite side of the cylinder.

On earth, more homes will be factory built and moved to a site to be put together. Today it takes five people two days to frame a wood-frame house. A factory-built house can be put up by two people in two days.

As energy costs go on rising, communities of energy efficient homes will be built. One such community is Arcosanti, in the Arizona desert. Arcosanti is a town of 5,000 people that contains high-rise apartment buildings and covers a small area of land.

Arcosanti has homes, businesses, and offices. There are spaces for parks, recreation, performances, and small meetings. The apartment buildings surround a shopping mall. From their apartments, people can look inside at the shops or outside

---

CONNECTIONS: Technology and Science

■ Gravity is a force of attraction between objects. Weight results from the pull of gravity. The greater the force of gravity, the greater the weight. The farther two objects are apart, the weaker the force of attraction between them. People are lighter on the moon because the moon is lighter than earth. As one moves farther from the earth in a spacecraft, one's weight rapidly decreases until one experiences weightlessness.

Artificial gravity could be achieved in very large space-craft by spinning the spacecraft. As a result of spinning in a weightless environment, objects come to rest on the inside walls. People and objects will push outward on the inside wall while the wall will push inward on them. The inside wall will become their floor.

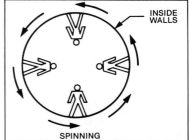

■ Ionizing radiation affects living organisms. It has enough energy to destroy or damage DNA molecules. If damaged, the faulty DNA may pass inaccurate information on to daughter cells, resulting in genetic effects. The amount of radiation affecting humans is measured in units of rems or sieverts. One sievert is equal to 100 rems. In the United States people receive, on the average, 100 millirems. A millirem is one thousandth of a rem.

A city of the future, complete with parking for your personal hovercraft. (Courtesy of General Electric Company)

toward the wilderness. In the community there are no roads or cars. People walk instead. Greenhouses allow residents to grow their own food year round. Garbage is used for energy. Water is recycled and purified. The architect, Paolo Soleri, said, "The need to plan for tomorrow is a part of all life."

Built-in computers will control many systems in buildings in the future. **Smart houses** will have bathrooms in which you can choose a shower, spring rain, or warm breeze. Bedrooms will have voice-activated radio and TV systems. Houses of the future will be built so people can live and work at home.

Office buildings will be "**intelligent buildings,**" wired with data communications lines. Each will have a central computer tying all the offices together. People will be able to communicate between offices using desktop terminals. Energy use will be computerized, as will security. Each building will have a word-processing center and a long-distance data communication system.

The lighter-than-air Magnus. (Courtesy of Magnus Aerospace Corp.)

TRANSPORTATION IN THE FUTURE

In 1889 Nellie Bly traveled around the world by coach, ship, train, and camel. It took 72 days to do so. In 1977 a Boeing 747 made the trip in 54 hours. Today it takes 3½ hours to travel from New York City to Paris on the Concorde, an aircraft that can fly at speeds of over 1,000 miles an hour. How fast will we travel in the future?

Air Transportation

New ways of moving people and goods will meet the needs of the future. One new vehicle is the LTA (lighter-than-air) airship. This airship will take off and land vertically and will be able to carry very heavy loads.

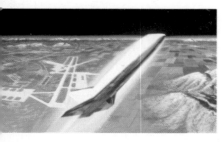

A transatmospheric aircraft. (Courtesy of Lockheed—California Company)

The Magnus LTA will be able to move over 50 tons, more than the largest helicopters, and do it more cheaply. It will be made of a 200-foot nylon sphere filled with helium. The cabin will carry two engines, which will provide power for takeoff. The ship will travel at a maximum speed of 70 miles per hour and will be able to travel 500 miles before landing.

Engines will also make the sphere spin. When a sphere spins in air, forces cause it to rise. This is known as the *Magnus effect*. It was discovered in 1853 by a German physicist named H. G. Magnus. The Magnus effect causes a spinning baseball to rise as it nears the plate.

Soon a transatmospheric aircraft will be built for military use. It will travel through the atmosphere, then move into low earth orbit. The aircraft will blast off like a rocket but travel like an airplane. In less than two hours, it will come down on the other side of the world.

Space Transportation

Future travel will take people far from earth. Already travel agencies are selling tickets for rides on the space shuttle. The tickets cost $50,000. Before the *Challenger* disaster, the first flight had been planned for 1992. It most likely will not take place that soon.

Someday tourists will vacation in space. They will visit space colonies, the moon, and nearby planets. Small vehicles will carry people in space to and from space stations. It is not known when it will take place.

A maglev train. (Courtesy of General Motors Corporation)

Ground Transportation

On the ground, vehicles will be faster and more energy efficient. Railroads will make use of **magnetic levitation**. Maglev trains, as they are called, will use strong magnets to repel the train from its track. This will make the train float just above the track, providing a fast, smooth ride.

Automobiles of the future will be sleeker, safer, and more energy efficient. Their streamlined shapes will save on gas. More automobile systems will be computer controlled. Cars will talk to drivers. A computer will tell us that the lights are on or ask who we are before the door can be unlocked. The driver's voice print will be on record in the car's computer and will have to match the speaker's voice before the door will open.

New electronic systems will be added. A sonar system will warn of objects behind the car as the car backs up. A four-wheel steering system will let the driver turn all four wheels

This satellite navigational system is combined with a laser disk map that indicates the car's position. (Courtesy of Chrysler Motors)

The Moller 400 is a flying car that carries four people, can be parked in a normal garage, takes off vertically, and flies at speeds of up to 400 miles per hour. (Photo courtesy Moller International)

toward the curb to make parking easier. Doze alerts on the steering wheel will sound an alarm if the driver gets sleepy and relaxes his or her grip on the wheel. Windshield wipers will start automatically as soon as rain falls on the windshield. A navigating system will show the car's location on a video screen map. It will also give the best way to get to the driver's destination.

Cars will use new energy sources. Electric cars will become popular. They will run on large banks of batteries that will last for several hundred miles. The driver will pull into a service station when the batteries run down to exchange them for a new set.

Cars will travel on "autopilot." Future superhighways will work with computers in cars, sending signals to keep the car on the road until the right exit is reached. The driver will then take over for travel on local roads.

Personal rapid transport vehicles (**PRTs**) will become popular. PRTs are automated vehicles that travel on tracks or guideways. They can be built either above or below the ground. Using PRTs, people will be able to avoid heavy highway traffic.

BIOTECHNICAL SYSTEMS IN THE FUTURE

New drugs and treatments will cure or prevent many diseases, such as cancer and heart disease. We may even learn what causes the aging process. We could then slow the progress of aging or even stop it. Already, scientists have had success in keeping older rats from aging. Their hearts and lungs are as strong as those of young rats.

A PRT system. (Courtesy of West Virginia University)

As we have been able to clone plants and some lower animals, we might begin to clone people. We could put the nucleus of a person's cell into an egg cell and let it grow. In this way, it would be possible to have a hundred or even a thousand people who were exactly alike. It may also be possible to engineer humans to live in hostile environments. For example, a fish-like gill could be put in a person to allow breathing under water.

Ethics

Ethics are the standards of right and wrong that people use to govern their behavior. The ethics of bioengineering will probably cause as much debate in the future as it does now. Should scientists grow human embryos to experiment on? Should we produce hundreds or even thousands of clones of a certain kind of person? Should we change people so they can live under the sea or in polluted or unsafe environments?

Who should make decisions about bioengineering? Should we leave it to scientists? To religious leaders? Should decisions be made by governments? Or by the United Nations? Could we reach worldwide agreement?

Food Production

How about a nice bowl of newspaper for breakfast? Or a plateful of maple leaves for lunch? It may not sound very good, but foods of the future may be taken from such sources. Chemists are learning how to break down materials into their basic

From the control panel on a desert farm, people will harvest crops with robotic harvestors. (Courtesy of General Electric Company)

Salt-tolerant plants, called halophytes, will be grown with seawater. (Courtesy of Environmental Research Laboratory, University of Arizona)

elements, such as carbon, hydrogen, and oxygen. They could combine these elements to make nutrients. The nutrients would then be made into artificial foods, just as healthful and maybe even as tasty as those we eat today.

Many of us would probably want an old-fashioned hamburger, but artificial foods could prevent starvation in some parts of the world. Feeding people in developing nations is becoming a problem. We could use technology to grow crops in the desert, irrigating them with salt water from the sea. These plants, called **halophytes**, would first be used to feed animals and as a source of plant oils. Later, they could be used to make foods.

Within 35 years, the earth will have a population of 8 billion people. We will need to find new ways to feed all of these people. Farmers will use **growth hormones** to produce larger, less fatty animals. Cows will produce more milk. Chickens might be developed that have no wings or feathers, making more meat with less chicken feed. Vegetables and fruit will be improved, too. For example, beet genes could be transferred to tomatoes to give tomatoes tougher skins and make them easier to ship. We may even be able to implant genes from cows into fish and raise sea steaks!

ENERGY IN THE FUTURE

As nations become more technological, more energy will be needed. Fossil fuels will run out. Renewable energy sources will fill most of our energy needs. Space-based power stations will send microwave energy to the earth.

Green algae plants may provide energy. Algae contain large amounts of oils and fats. These oils and fats are easily separated from the algae and can be changed into fuels. One acre of water

An artist's idea of an algae farm. (Courtesy of Solar Energy Research Institute)

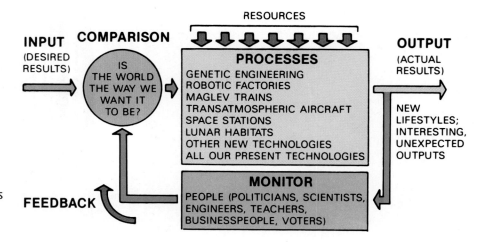

This system diagram shows how future technologies will affect our lives.

or ground covered with algae can produce as much as 100 barrels of oil a year, which could be used to make gasoline.

One new energy source may be **nuclear fusion**. In fusion, atoms join together, giving off energy. But fusion reactions need huge amounts of energy to start. Researchers are trying to find ways to get at least as much energy from the reaction as is needed to start it. This balance is called **breakeven**. Once the reaction makes more energy than is needed to start it, the excess energy can be used to generate electricity. Some researchers predict that fusion reactors will be built by the year 2020. Fusion is better than fission because it produces clean energy. There is little or no nuclear waste or radiation.

FUTURING—FORECASTING NEW TECHNOLOGIES

The newest technologies are the most complex. Many different technological systems work together. Space shuttles, satellites, robotic factories, and maglev trains, for example, require the newest in transportation, communication, manufacturing, and

Existing technological systems will act together to produce new, more powerful technologies.

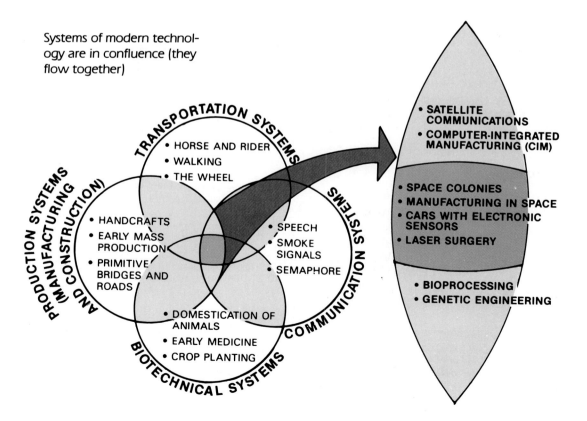

Systems of modern technology are in confluence (they flow together)

construction systems. This **confluence** (flowing together) of systems will mean that new technologies will not be separate and distinct. Many systems will be combined to produce new technologies.

Modern biotechnology is an example of the confluence of systems. Genetic engineering is really a manufacturing process. New kinds of organisms are produced. When doing genetic engineering, technologists and scientists make use of construction, manufacturing, and communication technologies. They construct structures (like fermentation vats) to house process-

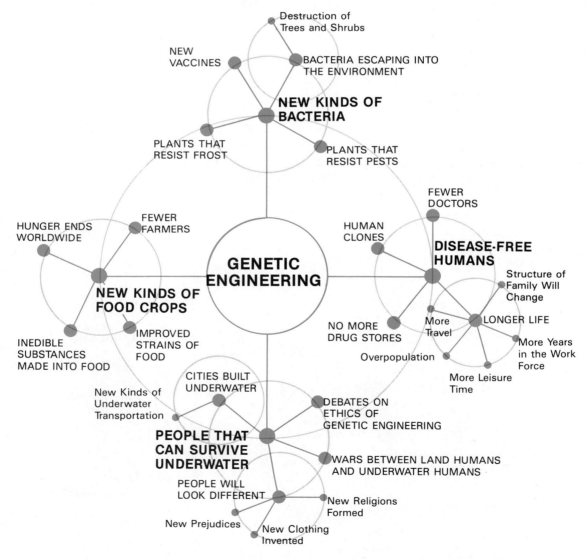

A futures wheel helps us to determine possible outcomes of a new idea or technology. In this case, outcomes of genetic engineering are being forecast.

ing operations. They use manufactured tools and instruments. They use communication and information technology when they do computer modeling.

Futuring Techniques

Futurists think about how technology will change our future. They can do this in at least four different ways.

One way to forecast the future is to use a futures wheel. A **futures wheel** is a diagram with one "big idea" at the center of a circle. The central idea leads to other ideas. Each of these ideas, in turn, has its own likely future outcomes.

Another way to forecast the future is a **cross-impact analysis.** Here, we look at many different future possibilities. We see what might happen when they are combined.

A third way to forecast the future is through **Delphi surveys.** In a Delphi survey, experts list ten kinds of changes they expect in the future. All their ideas are put on a list that

Using futuring techniques, people can anticipate the consequences of a new technology.

	GENETICALLY ENGINEERED HUMANS	CITIES ON THE MOON	LIFE ON OTHER PLANETS	CITIES UNDER THE SEA
AIR POLLUTION	• Humans will be able to breathe sulphur dioxide	• Moon societies will safeguard against the kind of pollution that occurred on earth	• People will vacation on nonpolluted planets	• Polluted air will be used as fuel by sea cities
RISING ENERGY COSTS	• Humans will be able to survive on fewer calories	• Energy will be beamed to earth from a moon base	• Energy sources will be exported from other planets to earth	• There will be new uses of sea water to provide energy
OVERPOPULATION	• Child-bearing will be forbidden • Only genetically engineered humans needed for specific purposes will be authorized • People will debate whether we should alter human life	• Societies will move to the moon	• Marriages will occur between earthlings and space beings • Attempts to transfer populations to other planets will meet with resistance; interspace wars will result	• People will work in new industries like mining sea beds
LONGER LIFE SPANS	• People will be altered several times during their lifetimes as conditions change on earth	• People will commute to "summer homes" on the moon • New travel agencies will specialize in moon-earth travel	• People will spend more time traveling to distant planets	• People will live part of their lives on land and part in sea cities to learn more about all forms of life

A cross-impact analysis.

(Courtesy of General Electric Company)

is sent to all of them. Each arranges the ideas in an order from most likely to happen to least likely to happen. A new list is made up of their lists. The new list is sent around and the experts rank the ideas again. The final list is their forecast of the future.

A **trend analysis** forecasts the future by looking at the past. For example, the cost of computers has dropped at an increasing rate over the last ten years. Forecasters might suggest that the price will continue to drop in this way. Computer prices in the year 2000 might be so low that they could be used everywhere and anywhere. On the other hand, just because things happened one way in the past does not mean that this trend will continue. Trend analysis does not always work.

SUMMARY

Some impacts of technology are helpful and some are harmful. It depends on how technology is used. Planning can limit the undesired outputs of technology.

We depend on technology. It would not be easy to go back to a simpler life-style. Technology must be fitted to the needs of people, and it must protect and work with the environment.

Technological change will continue to increase exponentially. Computers, telephones, satellites, and television will work together to provide new ways to communicate. Voice recognition technology will allow us to speak to our typewriters. We will hold videoconferences with people in distant lands. Shopping, working, and studying will depend on the computer.

In developed nations such as the United States, robots will be used in more and more manufacturing jobs. Only in less developed nations will a large part of the work force work on factory production lines. Robots will do more for us, on the job and at home. Factories will be built in space. New materials will come from the sea, the moon, and other planets.

Space stations will orbit the earth. They will send energy back, and serve as stopping places for people on the way to the moon or the planets.

Construction technology on the earth will be more energy efficient. New housing developments will take up less space, preserve the wilderness, and use less energy. We will live in "smart houses" and work in "intelligent buildings" in which computers control many of the systems.

Future transportation will be faster and use less energy. Maglev trains will "float" on their rails, giving a fast, smooth ride. Space vehicles will carry us around space stations and space colonies. On the earth, personal vehicle systems such as PRTs will move us around cities.

Biotechnical systems will do important jobs for us. They will help us cure and prevent disease, extend our life spans, and feed billions of people.

By early in the twenty-first century, we will use nuclear fusion to produce electricity. It will be a safe, cheap, and clean energy source.

Futuring techniques like futures wheels, cross-impact analysis, Delphi surveys, and trend analysis can help us forecast the changes technology will bring.

CONCLUSION

One thing is certain. Technology will continue to grow at an ever-increasing rate. Because of this, people must be ready to plan and direct it.

Life in the future will depend on how well we control technology. The quality of life on our planet should be our main concern. We must think in terms of the human race and less in terms of individual countries or groups. We must work together to ensure our own survival and that of the living things with which we share our planet.

Technology is not magic. As you learn more about science and mathematics, you will see that we can and must control technology, for our own good and for the good of future generations.

Feedback

1. What four possible combinations of outputs can result from technological systems?
2. Do you believe that technology is good, evil, or neutral? Explain why.
3. How can people control the development of a technology they feel may be harmful?
4. Give two examples showing that pollution is not just a recent problem.
5. Give an example of technology that is poorly matched to the human user.
6. Give an example of a problem technology has caused and a solution technology has provided.
7. Explain how technology can make life easier for a disabled person.
8. Draw a design for a futuristic communication system that would allow you to communicate with a class of students in Europe.
9. What would be two good products to manufacture in space? Why?
10. How will future travel differ from present-day travel?
11. Would a personal rapid transit (PRT) system be useful in your town or city? Explain why or why not.
12. What kinds of moral questions will arise if people are able to engineer human life?
13. Give an example of how a modern or future technology will require inputs from several existing technological systems. (Hint: Think of how biotechnical, communication, construction, manufacturing, and transportation systems act together to produce new technologies, such as manufacturing in space.)
14. Draw a futures wheel. At the center, place a skin cream that brings back youth. Predict the possible outcomes.

(Courtesy of International Business Machines Corp.)

(Photo by Jeremy Plant)

CHAPTER KEY TERMS

Across

3. Before implementing a technology, we must consider its effects on this.
5. Buildings with data communications lines in addition to electric wiring.
7. This futuristic communication technology will permit us to talk to a typewriter.
10. This technology allows us to hold face-to-face meetings with people in distant places.
11. The flowing together of technological systems.
13. Not just a modern-day problem.
14. Personal rapid transport (abbreviation).
15. A plant that can be irrigated with salt water.
16. Techniques for forecasting the future.

Down

1. A _____ forecasts the possible outcomes of the event at the center.
2. A clean, safe, atomic energy source for the future.
4. Maglev.
6. Artificial speech produced by a computer.
8. Human factors engineering that fits technology to human needs.
9. Pollution from automobile exhaust and burning coal causes this.
12. Effect of a technological system on people or the environment.

Acid rain
Confluence
Environment
Ergonomics
Fusion
Futures wheel
Futuring
Growth hormone
Halophyte
Impact
Intelligent buildings
Magnetic levitation
Manufacturing in space
PRT
Pollution
Smart houses
Speech synthesis
Sulfur dioxide
Videoconference
Voice recognition
Wheel

SEE YOUR TEACHER FOR THE CROSSTECH SHEET

MAGNETIC LEVITATION TRANSPORTATION SYSTEMS

Problem Situation

Congested highways and airports are a problem in many parts of our country. New transportation systems that reduce the number of people traveling by airplane and automobile are needed. Magnetic levitation (maglev) vehicles that travel twice as fast as conventional trains are a possible solution. Instead of engines, maglev vehicles use electromagnetism to levitate (raise) and propel the vehicle. Alternating current creates a magnetic field that pushes and pulls the vehicle and keeps it above the support structure, called a guideway. West Germany and Japan have successfully tested maglev vehicles. The prototypes that they have built resemble monorails and travel at speeds of up to 300 miles per hour. The West German maglev system uses conventional electromagnets. The Japanese prototype uses strong magnetic fields developed by superconductors.

Design Brief

Design and construct a model maglev vehicle that travels the length of an 8-foot magnetic guideway.

Suggested Resources

An 8-foot magnetic guideway
Four ½" × 1" magnets
2½" × 6" × ⅛" hardboard for vehicle base
Modeling materials:
  Illustration board
  Foam board
  Color markers
  Masking tape
  Hot glue
  Glue gun
Small DC motors for use as possible propulsion systems

Procedure

Use the safety procedures described by your teacher.

1. Use masking tape to temporarily attach one magnet to each corner of the vehicle base.
2. Test the vehicle base to determine if it will levitate above the magnetic track. Turn the magnets over and around, as needed until the vehicle base levitates about ⅜" above the track.
3. Remove the tape and fasten each magnet in place with a small amount of hot glue.
4. Test the vehicle base to see that it levitates and moves freely along the track. Make adjustments if necessary.
5. Design and construct a body for the vehicle using illustration or foam board. A streamlined design will reduce aerodynamic drag.
6. Attach the body to the base and test to make sure that the vehicle moves freely along the track.
7. Design a propulsion system for your vehicle.

Classroom Connections

1. Future railroads will make use of magnetic levitation.
2. Vehicles will become faster and more energy efficient.
3. Like poles of magnets repel each other. Unlike poles attract.
4. Superconducting electromagnets are much more powerful than ordinary electromagnets.
5. Maglev vehicles are quiet and don't emit pollutants. However, pollution may be emitted during production of the electricity that a maglev system requires. Some people are opposed to developing maglev systems. They feel that conventional trains and buses can be used to solve our traffic problems.

Summing Up

1. Identify one advantage and one disadvantage of maglev transportation systems.
2. Describe two problems that you had to solve while designing your vehicle.
3. Compare the propulsion system you used to the system that will probably be used in commercial maglev systems.

ACTIVITIES

WHERE DOES IT COME FROM?
WHERE DOES IT GO?

Problem Situation

Each day the average American creates about 3½ pounds of garbage. More than a third of this comes from bottles, cans, boxes, and wrappers that are no longer needed.

Well-designed packaging helps protect and sell products, but many items can be sold without packaging. Manufacturers can reduce the packaging for many products and use recyclable materials for others. Convenience items, including disposable cups, plates, razors, and diapers, also increase the amount of garbage we create.

Many communities have started recycling programs. Materials such as glass, metal, newspapers, and cardboard are usually separated for collection. In addition to reducing the volume of trash, recycling can help preserve natural resources and save energy. Trees are saved when paper products are recycled. Aluminum made from recycled materials uses one-half the energy that aluminum made from bauxite (aluminum ore) requires.

Design Brief

For this activity you will monitor the trash disposed of by your family. Each day for a week you will determine the weight of the garbage created by your family. Recyclable materials will be separated so that the percentage of these materials being disposed of can be calculated.

Procedure

1. Discuss this activity with other members of your family. Ask for their help. Find out if your community has a recycling program. Learn how the program works.
2. Locate several waste baskets or other suitable containers so you will be able to separate recyclable materials. Weigh each empty container. Record this information.
3. Weigh each collection container every day when it is full. Subtract the weight of the empty container to determine the weight of the garbage in it.
4. Keep a log and record the weight of recyclable and non-recyclable trash.

Suggested Resources

Weight scale
Paper bags
Waste baskets
Graph paper
Markers

5. At the end of the week, use the information you collected to make a bar graph showing the weight of all garbage and the weight of recyclable materials produced by your family each day.
6. Calculate the percentage of your garbage that is recyclable, and the average amount (in pounds) of garbage produced by your family each day.
7. Estimate how much of the garbage was packaging materials.

Classroom Connections

1. Waste disposal is a major technological problem.
2. When choosing materials for technological systems, we should think about how best to dispose of them when they are no longer needed. For example, the acid in flashlight batteries is a hazardous material. Batteries disposed of in landfills may contaminate groundwater.
3. Resources that break down into natural materials are called biodegradable.
4. Mathematical data can be expressed in graph form to communicate information.
5. Solid waste disposal is a local, national, and global problem.
6. Recycling programs can significantly extend the life of many landfills.

Technology in the Real World

1. New landfills are being designed so that contaminants do not leak into the groundwater.
2. The cost of waste disposal is increasing in many communities because government regulations require the closing of landfills.
3. To date there is no satisfactory short-term method to dispose of radioactive wastes.

Summing Up

1. Does your community have a recycling program? Briefly explain the program. If there is no program, how do you think one could be started?
2. Make a list of the kinds of materials that many communities recycle.
3. What can you do to reduce the amount of garbage produced by your family?

TECHNOLOGICAL TIMELINE

4000 B.C.	Earliest known cuneiform writing
3500 B.C.	Hieroglyphic writing
1500 B.C.	Syrian alphabet
550 A.D.	Babylonian Talmud written
650 A.D.	Writing of the Koran
750 A.D.	Arabs learn papermaking from Chinese prisoners
1024 A.D.	Chinese use first paper currency
1450 A.D.	Gutenberg invents movable type
1752 A.D.	Benjamin Franklin proves lightning is electricity
1760-1840 A.D.	Industrial Revolution in Britain
1776 A.D.	U.S. Declaration of Independence
1789 A.D.	French Revolution
1816 A.D.	Regular transatlantic sailing service between New York and Liverpool, England makes regular overseas written communication possible
1826 A.D.	First photograph (Niepce)
1837 A.D.	First telegraph in U.S. by Samuel F.B. Morse
1844 A.D.	Morse completes telegraph line between Baltimore and Washington, D.C.
1850 A.D.	First telegraph cable placed across English Channel
1856 A.D.	Western Union formed
1858 A.D.	First transatlantic cable (Cyrus Field)
1861-1865 A.D.	U.S. Civil War
1876 A.D.	Alexander Graham Bell invents the telephone
1887 A.D.	Energy transmitted through air (Hertz)
1888 A.D.	First Kodak handheld camera
1892 A.D.	First Strowger telephone switch in service
1896 A.D.	Marconi sends radio signals over two miles
1900 A.D.	EverReady flashlight invented
1901 A.D.	Marconi sends first transatlantic radio signals
1903 A.D.	Wright brothers' first flight
1905 A.D.	Albert Einstein proposes theory of relativity
1906 A.D.	First voice transmission by radio
1907 A.D.	Rubel invents offset printing
1909 A.D.	Beginning of Age of Plastics—Bakelite patented
1912 A.D.	Sinking of the Titanic—use of radio gains worldwide attention and mandatory use on all later ships
1915 A.D.	Sonar invented
1920 A.D.	First U.S. commercial broadcast at KDKA, Pittsburgh
1926 A.D.	First television demonstrated
1927 A.D.	Theory of negative feedback control systems developed
1931 A.D.	First regular TV broadcasts from experimental station W6XAO in California
1935 A.D.	Radar developed
1939 A.D.	World War II starts in Europe; Pan Am begins first regular commercial transatlantic flights
1942 A.D.	LORAN navigation developed; first color snapshots
1958 A.D.	First integrated circuit made by Texas Instruments; bifocal contact lenses; first U.S. satellite
1959 A.D.	Pilkington float glass process perfected; Sony produces first transistorized TV; Russia sends unmanned spacecraft to the moon (Luna 2); Xerox copier
1960 A.D.	Lasers; light-emitting diodes (LEDs); weather satellites; felt-tip pens

1962 A.D.	TV signals sent across the Atlantic via satellite; world's first industrial robot produced by Unimation Corporation; Tang orange juice
1963 A.D.	John F. Kennedy assassinated; cassette tapes; Instamatic cameras; the tranquilizer Valium produced
1964 A.D.	China tests nuclear bomb; IBM word processors; photochromic eyeglasses that change in response to the amount of sunlight
1965 A.D.	First space walk; Super 8 cameras; reports link cigarette smoking to cancer
1966 A.D.	Flashcubes; electronic fuel injection; Russians make successful soft landing on the moon (Luna 9)
1967 A.D.	Christiaan Barnard performs first heart transplant; energy-absorbing bumpers; Frisbees
1969 A.D.	American manned moon landing; first Concorde SST flight; artificial heart used during surgery
1970 A.D.	Russians land moon robot (Lunokhod); floppy disks; jumbo jets
1971 A.D.	Microprocessor developed by Intel; Mariner 9 orbits Mars
1972 A.D.	CAT scan; photography from satellites (Landsat); videodiscs
1973 A.D.	Genetic engineering; Skylab orbiting space station launched; supermarket optical price scanning; Selectric self-correcting typewriter; push-through tabs on cans
1974 A.D.	Toronto Communications Tower (tallest building in the world) opens
1975 A.D.	Liquid crystal displays; video games; disposable razors; cloning of a rabbit; Americans and Russians dock in space
1976 A.D.	Birth of the Apple computer; electronic cameras
1977 A.D.	Space shuttle; trans-Alaska pipeline system completed; neutron bomb developed
1978 A.D.	First test-tube baby born; programmable washing machines; computerized chess; auto-focus cameras
1979 A.D.	Skylab, the orbiting space station, falls back to earth; Rubik's cube
1980 A.D.	Solar-powered aircraft
1981 A.D.	Silicon 32-bit chip, nuclear magnetic resonance (NMR) scanner; first launch of a reusable space vehicle (the space shuttle *Columbia*)
1983 A.D.	Biopol (biodegradable plastic); carbon-fiber aircraft wing; 512K dynamic access memory chip
1984 A.D.	Compact disk player; genetically engineered blood-clotting factor; megabit computer chip
1985 A.D.	CD-ROM (compact-disk read-only memory); image digitizer; soft bifocal contact lens
1986 A.D.	Uranus moons photographs; DNA fingerprinting; diminished ozone shield
1987 A.D.	High-temperature superconductivity confirmed
1988 A.D.	Patented animal life
1990 A.D.	NASA's Hubble Space Telescope deployed by space shuttle *Discovery*

TECHNOLOGY EDUCATION STUDENT ASSOCIATIONS

Arvid Van Dyke

INTRODUCTION

Technology education student associations help you work with others in your technology class, your school, and your community. The technology student organization is the group in your school that uses technology in a variety of interesting activities, projects, and contests.

Think of the other groups in your school. Some, called "clubs," serve a special interest or offer after-school activities. Associations such as the Student Council serve all students in the school. The technology education student association, now called the Technology Student Association (TSA), serves all students taking Technology Education or Industrial Technology in their school.

Technology education student associations should meet the standards for technology education that have been established by an international association of teachers. These standards are helpful in making student associations serve the needs of students who must be technologically literate to live and work in our technological world.

Technology education student associations involve communication, construction, manufacturing, and transportation activities. Opportunities for creative thinking, problem solving, and decision making are also provided. Learning how technology works can be greatly enhanced by the student association's activities.

LEARNING TO BE LEADERS

Every group or organization has a set of leaders, called officers. Most officers are elected by group members. In a corporation, for example, a board of directors may appoint (or employ) a president to run the company. Vice presidents, treasurers, and other managers or leaders are the top people who make things happen in the organization.

Every student should serve at some time as an officer in the student association. You can volunteer to be a candidate or be nominated by someone else. Experience as an elected officer will improve your ability to lead and to work with others. Leadership skill is especially valuable when you are attending college or working. Employers look for people who are willing to learn and are able to get along well with others.

ASSOCIATIONS START IN CLASS

Many teachers recognize the benefit of student associations and involve all of their students in the associations. The students learn to lead and to use technology to make learning activities more significant. The officers and committees of an in-class student organization can help lead and manage activities during the class period. Each class may then want to take part in the school's technology student association activities.

Student organizations help build leadership and public speaking abilities. (Courtesy of Kramer Photos, Jerry Kramer, Box 87, 118 S. Main, Melvern, KS 66510)

THE SCHOOL ORGANIZATION OR CHAPTER

Most student groups in your school are organized to allow students to work together in activities related to a subject they take. A student association for technology education plans activities that relate to technology. By working together the group accomplishes more than each member could accomplish individually.

The association forms committees to allow more of its members to lead activities. Each committee plans and leads at least one activity relating to the school's technology education classes. The activity adds to students' understanding of technology and its impact on our world.

The student association in your school **affiliates** (joins together) with the state association so that all school chapters throughout your state are stronger and can share information. In the same way, your state office affiliates with all other state offices to form a strong national association. The national association offers many services to its member states, schools, and students.

ADVANTAGES TO STUDENTS AND THEIR SCHOOL

The technology education student association can help you continue your exploration in the field of technology. You can learn about career options and opportunities. By working in groups, you learn leadership skills and work cooperatively with other students and adults. Your school technology education program becomes better known because the student association attains recognition through the success of its members. You prepare for contests, follow the Achievement Programs, and travel to conferences in your state. Attending a national conference is an honor and a very educational experience.

LEARNING ABOUT TECHNOLOGY WITH THE TECHNOLOGY EDUCATION STUDENT ASSOCIATION

Your class and chapter officers can select several activities that will help students learn more about technology. Each activity allows students an opportunity to make technology work. Begin a brainstorming session, and find activities that are closely related to the technology area you are currently studying. Try to use technology in your student association activities.

CONTESTS TO MOTIVATE AND TEACH

Contests start in the classroom or laboratory, like many other activities of the technology education student association. The contest or project can be one of your class assignments. For example, all students might be required to make a safety poster during the laboratory safety unit of instruction. All posters are then graded by the teacher before a student committee judges the posters for in-class awards. The top posters are entered (registered) in the TSA Safety Poster Contest for recognition at the state or national conferences.

Competitive events motivate students to learn and to use technology for problem solving. Contests should offer a challenge to students' minds, hands, and attitudes. Some types of contest test a student's understanding of technology and its impact. Other TSA contests recognize leadership skills. Still others allow students to use communication skills in speaking or writing about technology and its significance.

Working on technology activities is an important part of your student organization membership.
(Courtesy of Kramer Photos, Jerry Kramer, Box 87, 118 S. Main, Melvern, KS 66510)

CAREER PAGE

EXAMPLES OF WORKERS WITHIN OCCUPATIONAL CLASSIFICATIONS

Production Workers

Laborers
Machine operators
Assemblers
Welders
Machinists

Management, Administrative, and Clerical Workers

Managers
Proprietors
Clerical workers
Computer operators
Retail trade workers
Finance workers

Service Workers

Fast food workers
Hospital workers
Security guards
Personal service workers (hair stylists, tour guides, etc.)

Technical and Professional Workers

Programmers
Engineers
Technicians
Teachers
Lawyers
Health care workers

The graph shows the continuing decrease in workers engaged in production, an increase in management and administrative workers, average growth for service workers, and a steadily increasing growth in technical and professional jobs. (Data from Dr. Dennis Swyt, National Bureau of Standards.)

TOMORROW'S JOBS

In Chapter 1, the shift from an agriculturally based to an industrially based to an information-based society was described.

During the industrial age, many jobs required people to do physical labor, such as working on factory assembly lines operating large, noisy machines. In the information age, many good jobs require people with a great deal of knowledge. The jobs that pay the most are those that require people to use their heads, not their muscles. For example, engineers and computer programmers are paid more than clerks and fast food workers.

WHAT THE FUTURE WORK FORCE WILL LOOK LIKE

Workers can be divided into four categories. These categories are production workers, management and administrative workers, service workers, and professional and technical workers. During the industrial age, more people worked in production than in any other category. Today, there are fewer production workers, but more management and administrative workers, and more professional and technical workers. These workers have jobs that require knowledge and a good education.

WHAT EMPLOYERS LOOK FOR

Besides looking for workers with knowledge and skills, employers want to hire people who have good work attitudes. A good work attitude includes the ability to get along with other people, the desire to do a good job, and the ability to get things done on time. Sometimes, employers feel that a good work attitude is even more important than good technical skills. Employers can often teach their workers how to do technical things, but an employee with a good attitude toward work will always be an asset to a company.

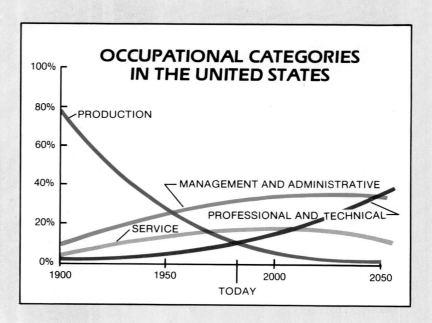

OCCUPATIONAL CATEGORIES IN THE UNITED STATES

PRODUCTION
MANAGEMENT AND ADMINISTRATIVE
PROFESSIONAL AND TECHNICAL
SERVICE
TODAY

CAREERS IN COMMUNICATION

The communication field includes a broad range of occupations having to do with research, writing, editing and production. They may be in the areas of education, journalism, publishing, television, business, advertising, public relations, photography, and speech.

OCCUPATION	FORMAL EDUCATION OR TRAINING	SKILLS NEEDED	EMPLOYMENT OPPORTUNITIES
ELECTRICAL ENGINEER AND TECHNICIAN—Electrical engineer designs, develops, tests, and supervises the manufacture of electrical and electronic equipment, such as radios, televisions, radar, industrial measuring and control devices, computers, and navigational equipment. Technicians build and service this equipment.	**Engineer:** Four-year college degree. **Technician:** Specialized training at technical institutes, junior and community colleges, and vocational and technical high schools.	**Engineer:** Excellent math skills, an analytical mind, and a capacity for detail. **Technician:** An aptitude for mathematics and science, and an enjoyment of technical work.	Much greater than average. There will be a strong demand for computers, robots, communications equipment, and electrical products for military, industrial and consumer use. Employment opportunities for engineers and technicians are expected to increase by 23% between the years 1988 and 2000.
PHOTOGRAPHER—Takes pictures of a wide variety of subjects. Still photographers specialize in portrait, fashion, or advertising. Industrial photographers provide illustrations for scientific publications.	Entry-level jobs for photographers have no formal educational requirements. Special courses are available in universities, junior colleges, and high schools. Over 100 colleges offer a bachelor's degree.	Good eyesight and color vision, artistic ability and manual dexterity. Photographers should be patient, enjoy working with detail, and have some knowledge of chemistry, physics and mathematics.	Above average. The demand for photographers will be stimulated as business and industry place greater importance upon visual aids in meetings, stockholders reports, and public relations work.
PRINTING PRESS OPERATOR—Prepares and operates the printing presses in a pressroom or print shop.	Apprenticeship or training on the job. Courses in printing, communications technology, chemistry, electronics, and physics are helpful.	Mechanical aptitude is important in making press adjustments and repairs. An ability to visualize color is essential for work on color presses.	Average. Increased use of color printing will contribute to the growth of new jobs, computer-operated equipment will reduce labor requirements.

Data from *Occupational Outlook Handbook, 1990-91*, U.S. Department of Labor

CAREERS IN PRODUCTION

Production technology includes construction and manufacturing. Workers in the construction trades build, repair, and modernize homes and other kinds of buildings. They also work on projects, such as airports, mass transportation systems, roads, recreation facilities, and power plants. Between the years 1988 and 2000, employment in the construction industry is expected to increase by 15%. Most manufacturing workers work in manufacturing plants, although some jobs involve sales and considerable travel.

OCCUPATION	FORMAL EDUCATION OR TRAINING	SKILLS NEEDED	EMPLOYMENT OPPORTUNITIES
ARCHITECT—Plans and designs attractive, functional, safe, and economical buildings.	College degree (bachelor's degree in architecture) as well as three years of experience in an architect's office.	Engineering design and managerial skills. Knowledge of building materials and modeling techniques.	Above average. Rapid growth in construction of non-residential structures will increase the demand.
JOURNEYMAN—A member of the building trades, such as: Carpenter—Builds framework, frames the roof and interior partitions. Concrete mason—Places and finishes concrete. Electrician—Installs, assembles, and maintains electrical systems. Plumber—Installs and maintains water and heating systems.	On-the-job training; two- to four-year apprenticeship programs. A high school education including courses in basic mathematics, applied science, electricity and electronics, mechanical drawing, and construction technology.	Sketching; reading drawings; must lay out, measure, cut, shape, and fasten materials; use hand and power tools safely. Must know about building codes and regulations.	Average. As the population grows, more journeymen will be needed to help build and maintain structures.
MANAGER AND ADMINISTRATOR—Plans, organizes, directs, and controls an organization's major functions.	College degree and management training. Top managers often have a master's degree in business administration.	Determination; self confidence; high motivation; strong decision-making, organizational, and interpersonal skills.	Opportunities will increase faster than the average occupations as business operations become more complex.
MANUFACTURING SALESPERSON—Most manufacturers employ sales people who market products to consumers in other businesses.	For technical products, college degrees in scientific or technical fields. For nontechnical products, college degrees in liberal arts or business administration.	A pleasant personality and appearance, and the ability to get along well with people are important.	Lower than average. Many large firms and chain stores buy direct from manufacturers, but this is a large occupational field with many yearly openings.

Data from *Occupational Outlook Handbook, 1990-91,* U.S. Department of Labor

CAREERS IN ENERGY, POWER, AND TRANSPORTATION

OCCUPATION	FORMAL EDUCATION OR TRAINING	SKILLS NEEDED	EMPLOYMENT OPPORTUNITIES
AIRCRAFT PILOT—Transports passengers, cargo, and mail. Some pilots dust crops, spread seed, test aircraft, and take photographs.	A commercial pilot's license issued by the Federal Aviation Administration (FAA) requires at least 250 hours of flight experience. Flying can be learned in military or civilian flying schools.	An understanding of flight theory and the ability to interpret data provided by instruments.	Above average. The growth in cargo and passenger traffic will create a need for more pilots, and more flight instructors.
PETROLEUM ENGINEER— Explores and drills for oil and gas. Determines the most efficient methods to recover oil and gas from petroleum deposits.	Four-year college engineering degree. Courses in energy technology, mathematics, physics, chemistry, mechanical drawing, and computers are useful.	Excellent mathematical and computer skills. Good background in chemistry.	Average. Oil and gas are becoming harder to find. More people will be needed to explore new sources, like the oceans and the polar regions.
TRUCKDRIVER— Transports goods from producer to consumer. Long-distance truck drivers spend most of their time behind the wheel. Local truck drivers spend much time loading and unloading.	In most states, a chauffeur's license is required. Employers prefer applicants with a good driving record. New drivers often start on small panel trucks and advance to larger trucks.	Ability to drive large vehicles in crowded areas and in highway traffic; knowledge of federal, state, and local regulations. Ability to inspect trucks and freight.	Average. However, this occupation is among the largest. The number of job openings each year will be very high.
VEHICLE MECHANIC— Repairs and maintains motor vehicles and construction equipment. Types of mechanics include aircraft, automotive and motorcycle, diesel, farm equipment, and heavy equipment.	Mechanics training is provided by the military, by private trade schools, or public vocational schools. Aircraft mechanics must be licensed by the FAA. Auto mechanics still can learn the trade by working with experienced mechanics, but formal training is becoming more important due to the complexity of new cars.	Knowledge of electronics and engine technology; knowledge of mechanical, hydraulic, pneumatic, and electrical systems. Good manual dexterity.	Above average. Rising incomes and a growing population will stimulate the demand for airline transportation. Expansion of the driving population, and more complex automotive systems will require more skilled mechanics.

Data from *Occupational Outlook Handbook, 1990-91,* U.S. Department of Labor

GLOSSARY

Acid Rain Rain containing chemicals given off by automobiles and factories, often hundreds of miles away. Acid rain can damage trees and other life forms.

Actual Results The outputs of the system.

Aesthetics The way something looks from an artistic point of view.

Agriculture The production of plants and animals, and the processing of them into food, clothing, and other products.

Alternative The choice of one thing from out of two or more possibilities.

Ampere The unit in which electric current is measured. An ampere is the amount of current that flows when one volt pushes electrons through one ohm of resistance.

Analog In electronics, a smoothly varying voltage or current that can be set to any desired value. Analog is opposed to digital, which can be set only to a limited number of values (e.g., 1s and 0s in binary form).

Antibody A cell (made of protein) that acts to defend a body against disease.

Architect A person who designs structures.

ASCII American Standard Code for Information Interchange; a digital code in which seven binary bits represent letters, numbers, and punctuation.

Assembly Line A manufacturing method in which each worker or machine does only a small part of the whole job. In an assembly line, the products being built move from worker to worker or machine to machine.

Automation The process of controlling machines automatically.

Bandwidth The measure of a communication channel's ability to carry information. The larger (wider) the bandwidth, the more information the channel can carry.

Bang-Bang Control The control of a system by turning the process on and off.

Bionics The technology of producing and using artificial body parts.

Bioprocessing The use of living organisms to process materials.

Biotechnology   The use of living organisms to make commercial products. Biotechnology includes antibody production, bioprocessing, and genetic engineering.

Bit   The smallest unit of digital information; a 1 or 0 in digital coding.

Brainstorming   A method of coming up with alternative solutions in which group members suggest ideas and no one criticizes any of them.

Brittle   Easily broken. A material that breaks more easily than it bends is said to be brittle.

Broadcast   The sending of a message to many receivers at the same time.

Bronze Age   The period (3000 B.C. to 1200 B.C.) in which people discovered how to combine metals to make alloys. The most common of these alloys was bronze, made from copper and tin.

Buoyancy   The upward force on an object in a liquid.

Byte   A group of bits, usually eight, that is used to represent information in digital form.

CAD   (1) Computer-Aided Design—the use of a computer to assist in the design of a part, circuit, building, etc. (2) Computer-Aided Drafting—the use of a computer to assist in the creation, storage, retrieval, modification, and plotting of a technical drawing.

CAM   Computer-Aided Manufacturing—the use of computers to control a manufacturing process.

Capital   One of the seven resources used in technological systems. Capital is the money or other form of wealth used to provide the machines, materials, and other resources needed in a system.

Casting   Forming a product by pouring a liquid material into a mold, letting the material harden into a solid, and removing the solid object from the mold.

CAT Scanner   A medical tool that takes X-ray pictures of a portion of a body from many angles. A CAT scanner produces a very detailed image of the area being studied.

Cement   A contruction material made of limestone and clay.

Channel   The path that information takes from the transmitter to the receiver.

Chromosomes Thin threads of DNA contained within cells. Chromosomes carry the genetic code for the characteristics of an organism.

CIM Computer-Integrated Manufacturing—the use of computers to control both the business and production aspects of a manufacturing facility.

Circuit A group of components connected together to perform a function.

Clone (1) An identical reproduction of a living organism, produced by placing the nucleus of a cell from the parent into an egg of the offspring (2) Cells produced from a common ancestor.

Closed-Loop System A system that uses feedback to adjust the process. Adjustment is made based on a comparison of the system's output(s) to its command input(s).

Code A set of signals or symbols that have some specific meaning to both the sender of the message and the receiver of the message.

Combining The joining of two or more materials in one of several ways, including fastening, coating, making composites, and so on.

Communication Successfully sending a message or idea from one person, animal, or machine (origin) to a second person, animal, or machine (destination).

Compact Disk A thin round disk on which information is stored digitally in the form of pits that reflect or absorb light from a laser.

Comparator The part of a closed-loop system that compares the output (actual results) to the command input (desired results).

Component A part that performs a specific function.

Composite A synthetic material made of other materials.

Compression A force that squeezes a material.

Concrete A construction material made of stone, sand, water, and cement.

Conditioning Changing the internal properties of a material.

Conductor A material whose atoms easily give up their outer electrons, letting an electrical current flow easily through the material.

Confluence The flowing together of systems and technologies.

Constraints The limitations that form boundaries around the possible solutions to a problem.

Construction The building of a structure on a site.

Container Ship A ship whose freight is prepacked in containers that are carried to and from the ship on railroad flat cars and trucks.

Controller The part of a system that modifies the process by turning it on and off or by adjusting it.

Crossbreed To combine the traits of one plant or animal with the traits of another.

Current The flow of electrons through a material.

Data Raw facts and figures. Data may be processed into information. (*See* Information.)

Design Brief A simple statement that spells out the proposed solution to a problem. The design brief should include the design criteria and the constraints.

Design Criteria The specifications that the solution to a problem must meet.

Design Folder A place where a record of all the ideas and drawings that lead toward a solution can be kept.

Desired Results A system's command input; a statement of the expected output(s) of a system.

Diesel A type of internal combustion engine that does not use spark plugs.

Digital In electronics, the use of a limited number of values of voltage or current to represent information (e.g., 1s and 0s in a binary system).

DNA Long, spiral-shaped molecules containing sugar and phosphates connected together by four different kinds of nitrogen compounds whose order of occurence determines the characteristics of a living organism.

Downlink The portion of a satellite communication link from the satellite to the ground station(s).

Drag The wind resistance that holds a plane back when the plane is moving forward.

Drilling A separating process that cuts holes in materials.

Ductile Easily formed. A material that bends more easily than it breaks is ductile.

Elastic Capable of returning to original size and shape after being deformed. A material that bends and then returns to its original shape is elastic.

Electron The negatively charged part of an atom that orbits the nucleus. The movement of electrons from one atom to another creates an electric current.

Energy One of the seven resources used by technological systems. The capacity for doing work. Energy takes many forms (thermal, electrical, mechanical, etc.) and comes from many sources (solar, muscle, chemical, nuclear, etc.)

Energy Converter A device that changes one form of energy into another.

Engineer A person with an engineering degree and/or a state license who designs buildings, roads, electronic circuits, airplanes, computers, and other technological systems.

Entrepreneur A person who forms and runs a business. An entrepreneur's business is often based on new ideas and/or inventions.

Ergonomics Fitting technology to human needs.

Exponential Increasing or decreasing at a changing rate (as opposed to linear, increasing or decreasing at a constant rate).

Extruding A method of forming parts in which a softened material is squeezed through an opening, giving the part the shape of the opening.

Fastening The process by which one part is attached to another.

Feedback The use of information about the output(s) of a system to modify the system's process.

Fermentation A process in which living organisms digest starch and sugar, producing by-products like alcohol and carbon dioxide gas.

Fiber Optics The passing of light through a flexible fiberglass or plastic light guide.

Finite Having limits.

Flexible Manufacturing A manufacturing process in which the tools and machines can be easily reprogrammed to produce different parts. Flexible manufacturing makes it economically feasible to produce small quantities of any given part.

Forming Changing a material's shape without cutting the material.

Forms In construction, molds that give concrete its shape while it is hardening.

Fossil Fuel   Fuel made from remains of animal or plant matter. Coal, oil, and natural gas are fossil fuels.

Foundation   In construction, the portion of a structure that supports the structure's weight; the substructure.

Frequency   The number of times an electromagnetic or other wave changes polarities in a specified period of time. Frequency is usually measured in cycles per second (Hertz).

Futures Wheel   A graphic method of forecasting the future. It is used to forecast second- and third-order impacts that result from a technology.

Fusion   A way of producing nuclear energy by causing atoms to combine to form a new atom plus energy in the form of heat.

Genetic Engineering   The use of gene splicing (moving genes from one cell to another) to change the characteristics of living organisms.

Geothermal Energy   Energy derived from the heat of the earth.

Grinding   Removing small amounts of a material by rubbing it with an abrasive.

Halophyte   A plant that can be grown using salt-water irrigation.

Hardness   The ability of a material to resist being dented or scratched.

Heat Treating   A conditioning process in which a material's internal properties are changed through the application of heat.

Hybrid   The plant or animal resulting from crossbreeding.

Hydraulic   Activated by fluid pressure.

Hydroelectricity   Electricity produced by turbines driven by falling water.

Hydroponics   Growing plants without soil by using water containing the nutrients required by the plants.

Impact   The effect of a technology on people, society, the economy or the environment.

Implementation   In problem solving, trying out a proposed solution.

Industrial Revolution   A period of inventive activity (beginning around 1750 in Britain) during which machines mechanized what had previously been manual work. The Industrial Revolution was responsible for many social changes, as well as changes in the way things were manufactured.

Input   The command input to a system; the desired results of the system.

Insight   A process of conceiving possible solutions to a problem. Solutions arrived at by insight leap into the mind of the problem solver.

Insulator   A material whose atoms hold their outer electrons tightly, resisting the flow of electrical current through the material.

Integrated Circuit   A complete electronic circuit built on a single piece of semiconductor material. Integrated circuits contain the equivalent of from dozens to over 500,000 transistors and other circuit components and can perform extremely complex functions.

Intelligent Building   A building with advanced communication capabilities built in to be shared by different tenants.

Interferon   A drug that appears to kill cancer cells, while leaving healthy cells unaffected.

Intermodal   A transportation system that uses more than one type (mode) of transportation.

Internal Combustion Engine   An engine that burns its fuel within a totally enclosed chamber.

Iron Age   The period during which people learned to extract iron from iron ore and to make tools with the iron. In the Middle East, this advancement occurred around 1200 B.C.

Irradiation   Sterilizing food by exposing it to X-ray radiation.

Irrigation   Providing water to land where crops are grown.

Jet Engine   An engine that provides thrust from the burning of fuel and air and the rapid expansion of the burning gases.

Just-In-Time Manufacturing   A manufacturing process in which the raw materials and other required parts arrive at the factory just before they are needed in the assembly process.

Laser   Light Amplification by Stimulated Emission of Radiation. A laser is a source of very pure (single-color) light that is focused into a very narrow beam.

Lift   The upward force on an object in the air, resulting from the object's weight and volume (passive lift) or its shape and movement through the air (active lift).

Machines   With tools, one of the seven resources used in technological systems. Machines change the amount, speed, or direction of a force.

Magnetic   The property a material has that causes it to be attracted to a magnet.

Manufacturing The building of products in a workshop or factory.

Mass Production The manufacture of many goods of the same type at one time, frequently involving interchangeable parts and the use of an assembly line.

Material One of the seven resources used in technological systems. The physical substances (e.g., wood, iron, oil, water, sand, etc.) that are used in a process.

Microorganism A living organism that is so small it must be seen under a microscope.

Memory The place in a computer system where data are stored. Internal memory includes random access memory (RAM), and read only memory (ROM). External memory includes tapes, floppy disks, hard disks, and optical disks.

Modeling The testing of a problem solution or a system without building the solution or system itself. Modeling includes using small physical replicas of the solution (scale models) and intangible representations of the solution (mathematical models, computer models, etc.).

Monitor To observe the output of a system.

Mortgage A loan that is secured by a home, building, or other large physical asset.

Noise An imperfection in a communication channel or equipment. Noise makes the message more likely to be misunderstood.

Nuclear Energy Energy derived from the splitting (fission) or combining (fusion) of atoms.

Nuclear Magnetic Resonance (NMR) A medical tool that uses powerful magnets and radio waves to obtain a very accurate image of the part of the body being examined.

Numerical Control The control of manufacturing machines by punched tape.

Offset Printing A printing process in which the image to be printed is photographically placed on a metal plate, transferred to a cylinder covered by a rubber blanket, and then transferred (offset) to paper.

Open-Loop System A system that does not use information about the output(s) to affect the process.

Operating System Computer software that allows the user or other programs to access a computer's memories (disk, tape, semiconductor, and so on), printers, and other attached devices.

Optical Having to do with light or sight.

Optimization　In problem solving, the process of making an alternative work as well as it can.

Output　The actual result obtained from a system.

People　One of the seven resources in a technological system. People design systems, operate them, and benefit from them.

Piggyback　The use of railroad flat cars to carry truck trailers.

Pneumatic　Activated by air pressure.

Pollution　Outputs of technology that affect the environment negatively.

Prefabricate　In construction, to build a building or a portion of it at a location other than the construction site.

Problem Solving　A multi-step process used to reach a solution in response to a human need or want. It includes defining the problem, specifying results, gathering information, developing alternative ideas, selecting the best solution, implementing the solution, and evaluating the solution and making necessary changes.

Process　The part of a system that combines resources to produce an output in response to a command input.

Processor　In computers, the part that controls the flow, storage, and manipulation of data.

Production　The system people use to make products. Products may be made in a factory or workshop (manufacturing) or on site (construction).

Program　A list of instructions that directs a computer's activities.

Program Control　The changing of the command input to a system in a prearranged sequence.

Proportional Control　The control of a system by varying the process by an amount based on how much different the output is from the input (desired results).

Prototype　A model of a final product or structure that is built to help evaluate the soundness of a design and to discover unanticipated problems.

Quality Control　The act of testing for faults in a product and correcting their causes.

Radar　A device used to determine the location of an object by measuring the time it takes for a radio wave to travel to the object, reflect off it, and travel back.

Raw Materials   Materials that are found in nature and processed into basic industrial materials.

Receiver   The part of a communication system that accepts the message from the channel and presents the message to the destination.

Recycle   To reuse all or portions of a substance.

Research   A process carried out to discover information. Basic research involves investigating a subject to find out facts. Market research involves surveying people about their attitudes toward a product and surveying the prices and capabilities of competing products.

Resistance   Opposition to electrical flow.

Resources   The things needed to get a job done. In a technological system, seven types of resources are processed to produce outputs. The seven types of resources are people, information, materials, tools and machines, energy, capital, and time.

Robot   A multifunction, reprogrammable machine capable of movement.

Rocket Engine   An engine that provides thrust from the rapid expansion of burning fuel and oxygen (which the rocket must carry).

Satellite Communications   A form of communication that makes use of radio relay stations located high above the earth. These relay stations, called satellites, receive signals from transmitting stations on the earth (the uplink). The satellites then retransmit these signals at another frequency (the downlink) to satellite receivers. Because the satellites are very high, these signals can be picked up over a very large area of the earth.

Science   The study and description of natural phenomena.

Screen Printing   A printing process in which ink is pressed through holes in a master stencil (the screen) onto the paper or object to be imprinted.

Semiconductor   A material that is neither a good conductor nor a good insulator. Transistors, diodes, integrated circuits, and some other electronic components are made of semiconductor material.

Sensor   A device used to observe the output of a system.

Separating   A means of processing materials in which one portion of material is removed by cutting processes like sawing, shearing, drilling, grinding, shaping, or turning. Separation can also be accomplished by chemical means, magnetic means, and by filtering.

Shear   A pair of forces that act on an object in opposite directions along the same line or plane (for example, two blades on a pair of scissors).

Smart House   A home with computer control of many routine functions.

Solar   Coming from the sun.

Soldering   A method of joining two wires together by melting a metal called solder onto them.

Specification   A detailed statement of requirements. In problem solving, a goal is a broad or general statement of requirements; a specification is detailed and specific.

Speech Synthesis   Speech simulation by a computer or electronic circuit.

Steam Engine   An engine that delivers power generated from the expansion of steam created by boiling water.

Stone Age   The period during which people used stones to makes tools and weapons. In the Middle East, this period was before 3000 B.C.

Structure   Something built or constructed on a construction site. Structures can include buildings, bridges, dams, tunnels, roads, airports, canals, harbors, pipelines, and towers.

Subsystem   A system that, together with other subsystems, contributes to the functioning of a larger system.

Supercomputer   A computer vastly superior in size and speed to the mainframe computers in use at any given time. Because of the rapid progress in computer technology, supercomputer performance of yesterday is commonplace today.

Superstructure   In construction, that part of the structure above the foundation. The superstructure is usually the part of the structure that is visible above the ground.

Synthetic   Human-made; not occurring in nature.

System   (1) A combination of things that act together in a more powerful way than each would act individually. (2) A means of achieving a desired result through the processing of resources in response to a command input. A system may be open-loop (no feedback) or closed-loop (using feedback).

Technology   The use of accumulated knowledge to process resources to satisfy human needs and wants.

Telecommute   To work at home using computers, terminals, and data communications rather than physically traveling to work.

Teleconference   A meeting conducted by people located at different sites but linked together by communication technologies. The attendees can use voice only (telephone conference call), voice with still pictures, or full motion pictures and voice (videoconference).

Tension   A force that stretches an object.

Thermal   Having to do with heat or the transfer of heat.

Thrust   The force developed to move an airplane forward through the air.

Tools   With machines, one of the seven resources used in technological systems. Tools extend the natural capabilities of people and are used to process or maintain other resources in systems.

Torsion   A force that twists an object.

Toughness   The ability of a material to absorb shocks without breaking.

Trade-Off   An exchange of the benefits and disadvantages of one solution for those of another solution.

Transducer   A device that changes information in one form of energy to information in another form of energy.

Transistor   A three-terminal electronic component made of semiconductor material. A transistor enables a small amount of current to control a large amount of current.

Transmitter   The part of a communication system that accepts the message from the originator and places it on the channel.

Trial and Error   A method of solving problems in which many solutions are tried until one is found to be acceptable.

Ultrasound   A medical tool that generates ultrasonic waves (sound of higher pitch than people can hear) and displays their echoes from different body parts.

Union   An organization of working people joined together for common purposes, usually to maintain good working conditions.

Unity   Refers to the way in which all the parts of a design look as if they belong together and produce a single general effect.

Uplink   The part of a satellite communication link from a ground station to the satellite.

Vehicle   A container for transporting people, animals, or goods. Vehicles can operate on fixed routes (elevators, railroad trains, monorails, trolley cars) or random routes (bicycles, cars, trucks, airplanes, rockets).

Videoconference   A meeting between people at different locations which uses television signals and communication links to transmit voice and pictures from each meeting location to the others.

Voice Recognition   The technology of using the human voice as an input to a computer or a machine.

Voltage   The force necessary to make an electric current flow.

Wavelength   The distance covered by one cycle of an electromagnetic or other wave.

Weight   In aeronautics, the downward force on a plane. Weight opposes lift.

Word Processor   A machine that combines typing and computer technologies to help an operator create, store, retrieve, modify, and print text.

INDEX